Experimental Combustion
An Introduction

Experimental Combustion
An Introduction

D. P. MISHRA

CRC Press
Taylor & Francis Group
Boca Raton   London   New York

CRC Press is an imprint of the
Taylor & Francis Group, an **informa** business

CRC Press
Taylor & Francis Group
6000 Broken Sound Parkway NW, Suite 300
Boca Raton, FL 33487-2742

First issued in paperback 2017

© 2014 by Taylor & Francis Group, LLC
CRC Press is an imprint of Taylor & Francis Group, an Informa business

No claim to original U.S. Government works

Version Date: 20140213

ISBN 13: 978-1-4665-1735-6 (hbk)
ISBN 13: 978-1-138-07421-7 (pbk)

Library of Congress Cataloging-in-Publication Data

Mishra, D. P. (Aerospace engineer)
    Experimental combustion : an introduction / author, D.P. Mishra.
        pages cm
    Includes bibliographical references and index.
    ISBN 978-1-4665-1735-6 (hardback)
    1. Combustion--Experiments. 2. Combustion engineering--Research. I. Title.

    QD516.M685 2014
    541'.3610287--dc23   2014001308

Visit the Taylor & Francis Web site at
http://www.taylorandfrancis.com

and the CRC Press Web site at
http://www.crcpress.com

*Dedicated to my research students who help me
to dabble in experimental combustion*

Contents

7 Optical Combustion Diagnostics   247

List of Figures

List of Tables

Preface

I believe that discovery of fire by human beings in the prehistoric era and subsequent mastery over the combustion process has propelled the development of science and technology, the fruits of which we enjoy today. Of course, the potential of combustion devices was realized during the industrial revolution. We use combustion devices, directly or indirectly, in many activities. Whether cooking by gas burners, transport by automobiles, locomotive, airplane, or spacecraft, winning a war, sending a satellite into its orbit, processing materials, and so forth, combustion plays a major role in each of these activities. Hence, combustion engineering has been an important area influencing every facet of human life and will continue to be an important deterministic factor in shaping the socioeconomic structure of our society in the future.

Nevertheless, blatant use of fossil fuels in combustion devices has depleted their reservoirs. Environmental pollution created by combustion systems is one of the major social concerns and stringent rules have been promulgated to check its impact. In order to meet the increasingly stringent emission regulations in the future, there is an urgent need to design and develop new eco-friendly, energy-efficient combustion devices and to retrofit existing combustion devices with minimum modifications, both of which the demand immediate attention of researchers and engineers.

With this in mind, it is important to impart the basic skills in experimental combustion to students so that they can effectively carry out the onerous task of design and development of energy-efficient and low-emission combustion systems for the new generation of devices and processes in order to meet the current energy demand while mitigating the global ecological imbalance.

This book is intended as an introductory text on experimental aspects of combustion at the graduate and undergraduate levels. Various measurement systems are described systematically in simple and lucid language. Readers may refer to advanced books/monograms/manuals for more comprehensive and exhaustive treatment on each topic, many of which are listed at the end of each chapter. It has been assumed that the readers have prior knowledge on combustion apart from thermodynamics, fluid mechanics, and mathematics. However, Chapter 1 reviews the fundamental aspects of combustion and Chapter 2 introduces the general characteristics of instruments essential for grasping the foundations of measurement systems employed in laboratory course work in the field of combustion. Chapter 3 introduces data acquisition systems and covers data sampling, signal conditioning, and data transmission for storage and further analysis. Subsequently, readers are exposed to data analysis in Chapter 4, which covers requisite statistical tools and experimental uncertainty analysis. In Chapter 5, several aspects of pressure, sound pressure level, velocity, and temperature measurements are discussed. Chapter 6 is devoted to the measurement of gas composition of fundamental importance to understanding of combustion. In this chapter, sampling systems and various probe design concepts are covered briefly for measurement of exhaust gases in combustion system. Some of the important modern optical methods used in combustion systems, including laser doppler velocimetry (LDV), particle image velocimetry (PIV), planar laser induced fluorescence (PLIF), and Rayleigh scattering thermometry, are covered in Chapter 7. Basic principles of light scattering are expounded sufficiently to facilitate the discussion on optical measurement systems. Types of light sources including laser are also discussed briefly. Schlieren, shadowgraph, and interferometric optical techniques used for flame visualization in combustion systems are discussed in detail. Several examples, review questions, and problems are provided at the end of each chapter to foster better understanding of the concepts covered in the text. I hope and wish that this book will make the learning process of experimental combustion very enjoyable.

Several individuals have contributed directly or indirectly during the preparation of this manuscript. I appreciate the help rendered by many students at IIT Kanpur and other institutes. I am indebted to my graduate students Ezhil, Mahesh, and Manisha for suggesting corrections in the manuscript. I am also thankful to my students Pravendra, Satish, Vikas, and Rahul who have helped in correcting the solution manuals. I am indebted to Dr. Swarup Y. Jejurkar, Mr. Indramani Dhada, IIT Kanpur, and Dr. Ranjan, assistant professor IIT Patna, India, who have gone through some of the chapters and suggested corrections in spite of their busy schedules. The official support provided by Mr. Ankit Upadhyay especially in preparing diagrams is highly appreciated. I am also thankful to Dr. Gagandeep Singh and Ms. Laurie Schlags of Taylor and Francis Books Pvt Ltd. for providing their support during the publication of this book. Lastly, the unwavering support of my family for this time consuming project is highly appreciated.

D. P. Mishra

Author

Dr. D. P. Mishra is a professor in the Department of Aerospace Engineering at the Indian Institute of Technology (IIT) Kanpur, Kanpur, India, where he was instrumental in establishing a combustion laboratory. His areas of research interest include combustion, computational fluid dynamics, atomization, and so forth. He has taught several undergraduate and postgraduate courses such as applied combustion, engineering thermodynamics, air breathing missile propulsion, aircraft propulsion, and rocket engine design. He is the recipient of several awards, including the Young Scientist Award (1991), INSA-JSPS Fellowship (2002), Sir Rajendranath Mookerjee Memorial award, and Samanta Chadrasekhar award. Currently he serves as an assistant editor for the *International Journal of Hydrogen Energy*, Elsevier, serves as an editorial board member of *Journal of the Chinese Institute of Engineers*, Taylor & Francis, and *International Journal of Turbo and Jet Engines*. Dr. Mishra has six Indian patents and more than 178 research papers in refereed journals and in conference proceedings to his credit. He has authored two textbooks, *Fundamentals of Combustion*, published by Prentice Hall of India, and *Engineering Thermodynamics*, published by Cengage India Pvt Ltd., New Delhi.

1

Introduction to Combustion

Conscience is the celestial fire that propels us to the heaven of humanity.

D. P. Mishra

1.1 Introduction

Combustion is as old as human civilization, and I personally feel that the discovery of fire is one of the greatest findings of human civilization. Man's subsequent mastery over fire has made possible all the developments in science and technology that we enjoy today. It is believed that Indians were first to recognize the efficacy of the fire even in the ancient *Vedic* era. The description of fire goes back to the Rig Veda, one of the oldest scripture of human civilization.

In modern times, combustion continues to play a very important role in driving the human race toward the path of prosperity and progress, because 90% of our worldwide energy demand is met by the combustion of fuel. Hence, I believe that combustion will remain a very important subject of interest as long as human civilization exists.

In layman's terms, combustion can be thought of as the process of setting fire to a fuel, of course in a controlled manner. It is basically a chemical process in which fuel is burned in the presence of an oxidizer. The chemical reactions involved in the process of combustion must be exothermic in nature, which liberate enough heat to sustain the combustion process itself. Examples of combustion devices are candle flames, lighting of matchsticks, cigarette burning, wood burning, liquefied petroleum gas (LPG) burners for cooking, furnaces, piston engines, gas turbine engines, and rocket motors.

1.2 Definition of Fuel/Oxidizer

Since fuel and oxidizer are the main constituents for the combustion process to take place, we need to understand the scientific meaning of fuel and oxidizer. In other words, we must ask certain pertinent questions such as, What is a fuel? What is an oxidizer? Chemically, an oxidizer can be defined as an element that accepts an electron. In contrast, fuel can be defined as an element that provides an electron. This property of an element's ability to accept or donate an electron is the electronegativity that dictates whether an element can be used as a fuel or an oxidizer. The electronegativity of various elements is shown in Table 1.1.

Can oxygen act as a fuel at any time? The common notion will tell us that oxygen can never act as a fuel. But this is not true. For example, let us consider fluorine and oxygen, which are allowed to react together. In this situation, oxygen can act as a fuel because fluorine has the highest electronegativity (see Table 1.1) and hence has the largest capability for receiving electrons. Fluorine is the most powerful oxidizer. Fortunately, the earth has very little amounts of fluorine. The next oxidizer is oxygen. Although it is available abundantly in nature, it is mostly accompanied by nitrogen, which reduces the actual capability of oxygen being an inert gas. The electronegativity of chlorine, bromine, and iodine are in

Table 1.1 Electronegativity of Various Elements

Element	Electronegativity
F	4
O	3.5
N, Cl	3.0
Br	2.8
C, S, I	2.5
H, P	2.1
B	2.0
Be, Al	1.5
Mg	1.2
Li, Ca	1.0
Na, Ba	0.9
K	0.8

descending order. Fuels such as carbon, hydrogen, aluminum, and magnesium have lower electronegativity in comparison with oxygen. We know that oxygen is a very common oxidizer that is used frequently, while fuel elements such as carbon, hydrogen, aluminum, and boron are fuels. The most common fuels that we use in our day-to-day lives are based on hydrocarbon.

1.3 Kinds of Fuels and Oxidizers

Several types of the fuels and oxidizers listed in Table 1.2 are used in various combustion systems. On a physical basis, these fuels and oxidizers can be broadly divided into three categories: (i) gaseous, (ii) liquid, and (iii) solid forms. In medieval times, mostly wood and coal/charcoal were used as fuel and air was used as a common oxidizer. In recent times, gaseous fuels such as compressed natural gas (CNG) and LPG are preferred in domestic and transport sectors due to stringent emission regulations over kerosene, diesel, and petrol (see Table 1.3). However, the majority of transportation sectors still rely on liquid fuels such as kerosene, diesel, and petrol. Solid fuels and oxidizers used in day-to-day life are given in Table 1.4. The applications of combustion systems and their respective fuel-air systems are listed in Tables 1.2 through 1.4.

Table 1.2 Types of Fuels and Oxidizers

	Fuel	Oxidizer	Application
1	LPG	Air/O_2	Burner, furnace, engine
2	Natural gas	Air/O_2	Furnace, piston engine, gas turbine engine
3	Producer gas	Air/O_2	Furnace, piston engine, gas turbine engine
4	Ch_4, C_3H_8, H_2	Air/O_2	Furnace, piston engine, gas turbine engine
5	Biogas	Air/O_2	Engine, burner
6	Acetylene	Air/O_2	Gas welding, cutting

Note: LPG = liquefied petroleum gas, NG = natural gas.

Table 1.3 Types of Liquid Fuels/Oxidizer

1	Gasoline	Air	Piston engine, gas turbine engine
2	HSD	Air	CI engine
3	LDO	Air	Furnaces
4	Kerosene	Air	Aircraft, gas turbine, ramjet, domestic
5	Alcohols	Air	Piston engine

Note: CI = compression-ignition, HSD = high-speed diesel, LDO = light diesel oil.

Table 1.4 Types of Solid Fuel/Oxidizer

1	Biomass (wood, sawdust, etc.)	Air/O_2	Furnace, burner, gas turbine engine, piston engine with producer gas
2	Coal, coke	Air/O_2	Steam turbine engine, piston engine, furnace

1.4 Stoichiometry

Let us consider an example in which a certain amount of methane is burned in the presence of a certain amount of oxygen undergoing a chemical reaction, liberating a certain amount of heat energy. The chemical reaction that represents such a reaction is shown below:

$$H_2 + 1/2\, O_2 \rightarrow H_2O + \Delta H_R$$

$$(2\ g) + (16\ g) \rightarrow (18\ g)$$

(1.1)

It can be observed from this reaction that 1 mole of hydrogen is reacted with 1/2 mole of oxygen to produce 1 mole of water. This is a balanced equation in which the number of chemical elements are balanced. It is interesting to note that 2 g of hydrogen can react with 16 g of oxygen to produce 18 g of water. That means the mass in the left side of the reaction is the same as that of the right side, showing that mass is conserved. This is the underlying principle of a chemical reaction: that no mass is created or destroyed. In contrast, the number of moles need not be conserved. Such a reaction is known as a stoichiometric reaction. In this case, the quantity of an oxidizer just sufficient to burn a certain quantity of fuel completely in a chemical reaction is known as *stoichiometry*, which is the ratio of oxidizer to fuel that is sufficient to burn fuel, leading to the formation of complete products of combustion. In the above example, the stoichiometry, (m_{ox}/m_{fuel}), would be 8. In combustion, the oxidizer/fuel ratio may not always be in stoichiometric proportion. On several occasions, excess oxidizers are supplied to ensure complete combustion of fuel in actual devices. When more than a stoichiometric quantity of oxidizer is used, the mixture is known as fuel-lean or *lean mixture*. In contrast, if less than a stoichiometric quantity of oxidizer is present, then the mixture is known as fuel-rich or *rich mixture*. On several occasions, hydrocarbon fuels are burned in the presence of air. For a hydrocarbon fuel represented by C_xH_y, the stoichiometric relation is

$$C_xH_y + a\,(O_2 + 3.76\ N_2) \rightarrow x\ CO_2 + (y/2)\ H_2O + 3.76\ a\ N_2$$

(1.2)

In the above reaction, the air is assumed to consist of 21% O_2 and 79% N_2 by volume for simplicity, which will be used throughout this book. When the above reaction is balanced, a turns out to be $x + y/4$ $(a = x + y/4)$. Then the stoichiometric air-fuel ratio would be

$$(A/F)_{stoic} = (m_{ox}/m_{fuel}) = 4.76\ \mathbf{a}\ (MW_{air}/MW_{fuel})$$

(1.3)

where MW_{air} and MW_{fuel} are the molecular weights of air and fuel, respectively. The stoichiometric ratio with respect to air can be obtained using the above expression for any hydrocarbon fuels such as methane, propane, and butane. Note that stoichiometric air-fuel ratio for methane is 17.11 while for higher hydrocarbons such as propane and butane, it is around 15. Also, it must be appreciated that many more times of oxidizer than fuel has to be supplied for complete combustion of fuel even theoretically. But in practice, one has to supply the oxidizer in a larger proportion than the stoichiometric ratio to ensure complete combustion. When other than the stoichiometric air-fuel ratio is used, one of the useful quantities known as the equivalence ratio is used to describe the air-fuel mixture. The *equivalence ratio*, Φ, is defined as

$$\phi = \frac{(F/A)}{(F/A)_{stoic}}; \Rightarrow (F/A) = \phi(F/A)_{stoic} \qquad (1.4)$$

Note that equivalence ratio is a nondimensional number in which the fuel-air ratio is expressed in terms of mass. This ratio describes quantitatively whether a fuel-oxidizer mixture is rich, lean, or stoichiometric. From the above equation, it is very clear that the stoichiometric mixture has equivalence ratio of unity. For a fuel-rich mixture, the equivalence ratio is greater than unity (Φ > 1). A fuel-lean mixture has less than unity (Φ < 1). By lean and rich mixture, we mean basically the extent of fuel in the mixture. In several combustion applications, the equivalence ratio is one of the most important parameters that dictates the performance of the system. Another parameter that is used very often to describe relative stoichiometry is the *percent stoichiometric air*, which can be related to the equivalence ratio as

$$\% \text{ stoichiometric air} = \frac{100\%}{\phi} \qquad (1.5)$$

The other useful parameter to define relative stoichiometry of mixture is *percent excess air*

$$\% \text{ excess air} = \frac{(1-\phi)}{\phi}.100\% \qquad (1.6)$$

Example 1.1

Ethyl alcohol is burned with dry air. The volumetric analysis of products on dry basis is $CO_2 = 10.5\%$, $O_2 = 6.76\%$, $CO = 0.93\%$, and $N_2 = 61.53\%$. Determine (a) A/F ratio, (b) equivalence ratio, and (c) % stoichiometric air used.

Solution: $X\ C_2H_5OH + a\ (O_2 + 3.76\ N_2) \rightarrow 10.5\ CO_2 + 0.93\ CO + 6.76\ O_2 + 61.53\ N_2 + b\ H_2O$

By mass balance, we can have

N: $3.76\ a = 61.53$;   → a = 16.365
C: $2\ X = 10.5 + 0.93$;   → X = 5.72
H: $6\ x = 2b$;   → b = 3
O: $2a + X = 10.5 \times 2 + 0.93 + 2 \times 6.76 + b$
$38.45 = 38.45$ (checked)

Then the above balance equation becomes

$$5.72 \ C_2H_5OH + 16.365 \ (O_2 + 3.76 \ N_2) \rightarrow 10.5 \ CO_2 + 0.93 \ CO$$
$$+ 6.76 \ O_2 + 61.53 \ N_2 + 3 \ H_2O$$

Let us recast the above chemical reaction in terms of 1 mole of fuel as

$$C_2H_5OH + 2.861 \ (O_2 + 3.76 \ N_2) \rightarrow 1.836 \ CO_2$$
$$+ 0.162 \ CO + 1.18 \ O_2 + 10.757 N_2 + 0.524 \ H_2O$$

Then the fuel/air ratio by mass becomes

$$(F/A)_{actual} = m_{Fuel}/m_{Air} = (12 \times 2 + 6 + 16)/2.861 \ (32 + 3.76 \times 28) = 0.117$$

In order to find out the equivalence ratio, we will have to find out the stoichiometric fuel-air ratio by considering a balance equation as

$$C_2H_5OH + 3 \ (O_2 + 3.76 \ N_2) \rightarrow 2 \ CO_2 + 3 \ H_2O + 11.28 \ N_2$$

Then the stoichiometric fuel-air ratio becomes

$$\left(\frac{F}{A} \right)_{stoic} = \frac{m_F}{m_A} = \frac{(12 \times 2 + 6 + 16)}{3(32 + 3.76 \times 28)} = \frac{46}{411.84} = 0.1117$$

Note that the stoichiometric air fuel ratio is 15.05. Then the equivalence ratio becomes

$$\varphi = \frac{(A/F)_{stoic}}{(A/F)} = \frac{(F/A)}{(F/A)_{stoic}} = \frac{0.117}{0.11169} = 1.047$$

Hence, this mixture is a slightly rich one. Then, let us calculate the percent of stoichiometric air as

$$\% \ \text{stoichiometric air} = \frac{100\%}{\phi} = \frac{100}{1.047} = 95.51\%$$

1.5 Ideal Gas Mixture

We have learned that a fuel and an oxidizer can occur in three forms: solids, liquids, and gases. The actual measured properties of solid and liquid substances can be expressed in the form of the thermodynamic equation of state. This kind of equation of state may be quite restricted in nature, but the thermodynamic equation of state for the ideal gaseous fuel and oxidizer is quite broad and can be applied to a combustion system even though it deals with a mixture of gases. Note that the properties of a mixture can be found by assuming it to be an ideal gas. We know that an ideal gas obeys the equation of state that is given below:

$$PV = nR_uT \tag{1.7}$$

where P is the pressure, V is the volume of the gas, T is the temperature of the gas, n is the number of moles of gas, and R_u is the universal gas constant ($R_u = 8.314$ kJ/kmole K). But when we are dealing with a mixture of gases, we can find out the properties of the mixture from an individual gas by applying the Gibbs–Dalton law.

Let us consider a container C that contains a multicomponent mixture of gases composed of m_A grams of species A, m_B grams of species B, and m_i grams of ith species. Then total number of moles, m_{tot}, in the container would be given by

$$m_{tot} = m_A + m_B + \ldots\ldots + m_i \tag{1.8}$$

By dividing the above equation by the total number of mass, m_{tot}, we will get

$$1 = Y_A + Y_B + Y_C + \ldots + Y_i = \sum Y_i \tag{1.9}$$

where, $Y_A (= m_A/m_{tot})$ is the mass fraction of species, A. Note that the sum of all mass fractions of individual species in a mixture is equal to unity. Similarly, we can have a relationship for total number of moles

$$n_{tot} = n_A + n_B + \ldots\ldots + n_i \tag{1.10}$$

By dividing the above equation by total number of moles, n_{tot}, we will get

$$1 = X_A + X_B + X_C + \ldots + X_i = \sum X_i \tag{1.11}$$

where $X_A (= n_A/n_{tot})$ is the mole fraction of species A. Note that the sum of all mole fractions of individual species in a mixture is equal to unity. Similarly, the total mass of the mixture, m_{tot}, can be expressed as given below, and the mole fraction, X_i, and mass fraction, Y_i, can be related easily as:

$$Y_i = \frac{m_i}{m_{mix}} = \frac{n_i MW_i}{n\,MW_{mix}} = X_i\,\frac{MW_i}{MW_{mix}} \tag{1.12}$$

where MW_i is the molecular weight of the ith species. The mixture molecular weight can be easily estimated by knowing either the species mole or mass fractions as

$$MW_{mix} = \sum_i X_i MW_i \qquad (1.13)$$

$$MW_{mix} = \frac{1}{\sum_i Y_i / MW_i} \qquad (1.14)$$

We can express the pressure of a mixture of gases such as species A, B, C ..., as

$$P = \frac{n_A RT}{V} + \frac{n_B RT}{V} + \frac{n_C RT}{V} + ... + \frac{n_i RT}{V} = p_A + p_B + p_C ...... + p_i = \sum_i p_i$$

$$(1.15)$$

where p is the partial pressure of individual species by definition. In this case we have assumed that individual species are in the same temperature and volume of that of the mixture. This is known as Dalton's law of partial pressure, which states that

The pressure of a gaseous mixture is the sum of the pressure that each component would exert if it alone occupies the same volume of a mixture at the same temperature of the mixture.

It must be kept in mind that the partial pressure of a species can be determined by knowing the mole fraction of the corresponding species and total pressure of the mixture by the following relation:

$$p_i = X_i P \qquad (1.16)$$

The specific internal energy, u, of the mixture can be determined from knowledge of the respective values of the constituent species by invoking Gibb's theorem, which states that

The internal energy of a mixture of ideal gases is equal to the sum of internal energy of individual component of the mixture at same pressure and temperature of the mixture.

Then the specific internal energy, u, of the mixture can be obtained by either the mole fraction or mass fraction weighted sum of an individual component's specific internal energy as

$$\hat{u}_{mix} = \sum_i X_i \hat{u}_i \qquad (1.17)$$

$$u_{mix} = \sum_i Y_i u_i \qquad (1.18)$$

Similarly, the specific enthalpy, h, of the mixture can be obtained by similar relations as

$$h_{mix} = \sum_i X_i \hat{h}; \quad h_{mix} = \sum_i Y_i h_i \tag{1.19}$$

The enthalpy of a species can be expressed as

$$h^0_{i,T}(T) = h^0_{f,298.15} + \int_{298.15}^{T} C_{P,i}\, dT \tag{1.20}$$

Note that the enthalpy of any species at particular temperature is composed of two parts: (i) the heat of formation that represents the sum of enthalpy due to chemical energy associated with chemical bonds, and (ii) sensible enthalpy as it is associated with temperature. The other specific properties of the mixture such as entropy, s, Gibbs free energy, g, specific heat, C_P, and so forth can be obtained by a similar relation from the individual species.

1.6 Heat of Formation and Reaction

In the combustion process, several chemical reactions take place simultaneously. In some reactions, heat will be evolved and in others, heat will be absorbed. Hence, it is important to evaluate the heat liberated or absorbed in a chemical reaction. Let us take an example of a burner (see Figure 2.1) in which 1 mole of propane is reacting with 5 moles of oxygen to produce to 3 moles of carbon dioxide and 4 moles of water as per the following chemical reaction:

$$C_3H_8 + 5\,O_2 \rightarrow 3\,CO_2 + 4\,H_2O \tag{1.21}$$

We need to determine the amount of heat liberated during this reaction for which we must know the heat of formation of each participating species. In other words, the heat of formation of each species must be known to evaluate the heat of reaction. The *heat formation* of a particular species can be defined as the heat of the reaction per mole of product formed isothermally from elements in their standard states. It must be kept in mind that the heat of formation of elements in their standard states is assigned a value of zero as per international norms. For example, nitrogen gas at standard temperature and pressure is the most stable whose heat of formation at standard state is zero. In contrast, the *heat of reaction* can be defined as the difference between the enthalpy of the products and enthalpy of reactants at the specified states. Note that the heat of reaction is stated in the standard state. If one of the reactant happens to be fuel and the other is the oxidizer, then combustion takes place, liberating a certain amount of heat, and the heat of reaction is known as the *heat of combustion*. The heat of formation of

Table 1.5 Heat of Formation of Some Important Species

Chemical Formula	Species Name	State	Standard Heat of Formation kJ/mol
O_2	Oxygen	Gas	0.0
O	Element oxygen	Gas	247.4
H_2	Hydrogen	Gas	0.0
H	Element hydrogen	Gas	218.1
OH	Hydroxyl	Gas	42.3
H_2O	Water	Gas	−242.0
H_2O	Water	Liquid	−286.0
C	Graphite	Solid	0.0
CO	Carbon monoxide	Gas	−110.5
CO_2	Carbon dioxide	Gas	−394.0
CH_4	Methane	Gas	−74.5
C_3H_8	Propane	Gas	−103.8
C_4H_{10}	Butane(n)	Gas	−24.7
C_4H_{10}	Butane(iso)	Gas	−131.8
C_2H_2	Acetylene	Gas	226.9
N_2	Nitrogen	Gas	0
H_2O	Water	Gas	−242.0
H_2O	Water	Liquid	−272.0

some important pure substances is given in Table 1.5 at standard conditions. The main advantage of tabulating heat of formation is that one has to keep track of heat of formation for only a few numbers of species to calculate the heat of reaction of several reactions. In contrast, a large number of heat reactions has to be tabulated, which is possible for a large number of combinations of species.

In combustion calculation we need to deal with thermochemical systems, which can be handled easily by invoking Hess's law, also known as constant heat summation. Hess's law states that the resultant heat evolved or absorbed at constant pressure or constant volume for a given chemical reaction is the same whether it takes place in one or many steps. In other words, it does not depend on the intermediate paths that may occur between the reactants and product. Hence we can manage to add or subtract thermochemical equations algebraically for arriving at the final heat of reaction by using the Hess's law.

1.7 Adiabatic Flame Temperature

We need to know the theoretical/ideal flame temperature during the combustion process for a particular fuel-air ratio provided no liberated heat can be transferred from its system boundary. Let us consider the combustor in which a certain amount of fuel and air in a certain ratio is reacted, which leads to complete combustion (equilibrium state) at constant pressure under an adiabatic condition. The final temperature attained by the system is known as *adiabatic flame* *temperature, T_{ad}.* Note that T_{ad} depends on initial pressure P, initial unburned temperature T_u and composition of the reactants. When the final compositions of the products are known, the first law of thermodynamics is sufficient to determine the adiabatic flame temperature, T_{ad}. Let us consider the physical boundary of the combustor as the control volume. The process is considered to be adiabatic as it is insulated perfectly. Under this condition, the first law of thermodynamics turns out to be

$$H_P(T_{ad}, P) = H_R(T_u, P) \tag{1.22}$$

where H_p is the total enthalpy of products at adiabatic temperature T_{ad}, and pressure P and H_R is the total enthalpy of reactants at initial temperature T_u, ambient pressure. H_p and H_R are expressed in terms of the heat of formation and sensible enthalpy of a participating species as

$$H_R = \sum_{i=R} n_i \hat{h}_i = \sum_{i=R} n_i \left[\hat{h}_{f,i}^o + \int_{T_u}^{T} C_{Pi}(T) dT \right] \tag{1.23}$$

$$H_p = \sum_{i=P} n_i \hat{h}_i = \sum_{i=P} n_i \left[\hat{h}_{f,i}^o + \int_{T_u}^{T_{ad}} C_{Pi}(T) dT \right] \tag{1.24}$$

where $\hat{h}_{f,i}^o$ is the heat of formation of ith species that are available in Table 1.5, n_i is the number of moles of ith species, and C_{Pi} is the specific heat of ith species that are dependent on temperature. Note that the appropriate value of the specific heats of the products must be chosen judiciously to get the correct adiabatic temperature. In the above equation, the final composition of the product must be known. Otherwise, we cannot evaluate the adiabatic flame temperature T_{ad}. Generally the final equilibrium composition is used for evaluation of adiabatic temperature T_{ad}. Unfortunately the final equilibrium composition is dependent on the final temperature T_{ad} and hence, one has to resort to an iterative technique for determining both T_{ad} and equilibrium composition. Let us consider an example for determination of adiabatic temperature T_{ad} in which the final compositions are known.

The adiabatic flame temperature data for stoichiometric mixtures of a fuel-oxidizer system at their respective initial temperatures and pressures are shown in Table 1.6. Note that adiabatic flame temperature does not depend much on the nature of hydrocarbon fuel as long as air is used as an oxidizer. It can also be observed from Table 1.6 that the flame temperature of hydrocarbon-air is around 2200 + 100 K. However, if the air is replaced by oxygen, the flame temperature

Table 1.6 Adiabatic Flame Temperature of Typical Fuels at Stoichiometric Mixture

System at T_u (K)	P (MPa)	T_{ad} (K)
CH_4-air, 300 K	0.1	2200
CH_4-air, 300 K	2.0	2278
CH_4-air, 600 K	2.0	2500
CH_4-O_2, 300 K	0.1	3030
C_3H_8-air, 300 K	0.1	2278
C_4H_{10}-air	0.1	2240
H_2-air, 300 K	0.1	2400

increases by 500–600 K. Besides this, an increase in initial temperature enhances the flame temperature by almost the same amount.

Example 1.2

Determine the adiabatic flame temperature of the stoichiometric isobutane–air mixture at 298 K, 0.1 MPA, assuming no dissociation of the products for the following two cases: (i) evaluating C_P of each species at 298 K and (ii) evaluating C_P of each species at 2000 K.

Given: Stoichiometric butane–air mixture $T_u = 298$ K, $P = 0.1$ MPa, etc.
To Find: Adiabatic flame temperature T_{ad}.
Solution:

$$C_4H_{10} + \frac{13}{2}(O_2 + 3.76N_2) \rightarrow 4CO_2 + 5H_2O + 24.44N_2 \qquad (E1.1)$$

Applying the first law of thermodynamics for control volume (CV), we have from energy balance:

$$\int_{298}^{T_{ad}} \sum n_i C_{Pi} dT = \left[\sum_{i=R} n_i \bar{h}_{fi,298} - \sum_{i=P} n_i \bar{h}_{fi,298} \right]$$

$$= \bar{h}_{f,C_4H_{10}} + 6.5\bar{h}_{f,O_2} + 24.44\bar{h}_{f,N_2} - 4\bar{h}f, w_2 - 5\bar{h}_{f,H_2O} - 24.44\bar{h}_{f,N_2}$$

$$= -131.8 - (4\times(-394.0) + 5\times(-247))$$

$$= 2654.2 \text{ kJ}$$

$$(E1.2)$$

For standard heat of formation see [1].

1. We can consider C_p at 298 K:

$$C_{P,CO_2} = 37.129 \text{ J/kmol K}$$

$$C_{P,H_2O} = 35.59 \text{ J/kmol K}$$

$$C_{P,N_2} = 29.124 \text{ J/kmol K}$$

Substituting these values in Equation E1.2, we can have

$$n_{CO_2}C_{P,CO_2} + n_{H_2O}C_{P,H_2O} + n_{N_2}C_{P,N_2}(T_{ad} - 298) = (4 \times 37.129 + 5 \times 35.59 + 24.44 \times 29.12)(T_{ad} - 288) = 2654.2 \times 10^3$$

By solving the above equation, we can determine the adiabatic flame temperature, T_{ad} as

$$T_{ad} = 2844.64 \text{ K}$$

2. By considering the value of species C_p at 2000 K, we have

$$C_{P,CO_2} = 60.35 \text{ J/kmol K}$$

$$C_{P,H_2O} = 51.18 \text{ J/kmol K}$$

$$C_{P,N_2} = 35.97 \text{ J/kmol K}$$

Substituting these values in Equation E1.2, we can determine the adiabatic flame temperature T_{ad} as

$$(4 \times 60.350 + 5 \times 51.180 + 24.44 \times 35.971)(T_{ad} - 298) = 2654.2 \times 10^3$$

$$T_{ad} = 2226.41 \text{ K}$$

Note that this estimated adiabatic temperature of 2226.41 is much closer to the adiabatic temperature obtained from equilibrium calculation (see Table 1.6) as compared to 2844.6 K obtained for case 1. Hence, it is important to choose specific heat values at the appropriate temperature.

1.8 Chemical Kinetics

During thermodynamic calculation, we need not bother about the actual paths taken by the chemical reactions but rather will focus on reactants and product composition. However, to characterize the combustion process we need to know the path of reactions and the rate at which chemical reactions proceed from their initial state to their final state. For this purpose we need to understand the chemical kinetics, which is discussed briefly in this section. Interested readers may refer to [1,3,4] for more details. We know that reactions can occur only when they collide, which may be true in our present discussion on chemical kinetics.

In other words, new bonds are formed and existing bonds are broken during the chemical reactions, which is possible due to collision of molecules when they are in random motion. The reaction can proceed successfully only when suitable molecules can collide with the proper orientation such that bonds can be broken. As well, the molecules must possess a higher energy level than the mean energy level so that collision can precede, which will lead to the breaking of bonds. As just mentioned, the energy of the colliding molecules must exceed the threshold energy, E, for the reaction to be successful. All molecules may not have higher energy levels. One may ask what is the fraction of molecules that have energy greater than or equal to the threshold energy, E, often known as the activation energy of a particular reaction. Note that the probability of a molecule possessing energy, E, is proportional to $\exp(-E/RT)$ as per Boltzman's energy distribution law. In summary we can state that for a successful chemical reaction, following conditions must be satisfied:

    i. Suitable molecules must collide each other
   ii. They must collide with proper orientation
  iii. Colliding molecules must have energy greater than activation energy, E

Let us consider an arbitrary bimolecular chemical reaction:

$$mA + nB \rightarrow xC + yD \tag{1.25}$$

The rate of change of concentration of participating species can be related as

$$-\frac{1}{m}\frac{dC_A}{dt} = -\frac{1}{n}\frac{dC_B}{dt} = \frac{1}{x}\frac{dC_C}{dt} = \frac{1}{y}\frac{dC_D}{dt} \tag{1.26}$$

where C is the concentration of ith species. Let us consider a real example of bimolecular reaction, which would be the reaction between a hydrogen atom and an oxygen atom to produce an OH species:

$$H + O \rightarrow OH \tag{1.27}$$

As discussed above, the chemical reaction takes place due to collision of participating species, so we can intuitively say that the rate of reaction is proportional to the concentration of each participating species. In the above example, the rate of reaction, RR of the reactant, H is proportional to C_H and C_O, which is given by

$$RR = \frac{dC_H}{dt} \propto C_H C_O \tag{1.28}$$

This is known as the *law of mass action,* which states that *the rate of reaction, RR, of a chemical species is proportional to the product of the concentrations of the participating chemical species, where each concentration is being raised to the power equal to the corresponding stoichiometric coefficient in the chemical reaction.* This has been confirmed by several experimental observations. By applying the law of mass action, we can link the rate of a reaction to the concentration of each reactant in the arbitrary bimolecular chemical reaction (Equation 1.25) as given below:

$$RR = \frac{dC_C}{dt} = kC_A^m C_B^n \tag{1.29}$$

where k is the *reaction rate constant.* Note that it is not really a constant that is dependent on temperature, collision orientation, activation energy, and so forth, as discussed above in this section. In order to get an expression for *reaction rate constant, k,* Svante Arrhenius conducted a series of experiments around 1889 and was the first to come up with this Boltzman factor, $\exp(-E/RT)$, to calculate the chemical reaction rate constant, as

$$k = BT^{0.5} e^{(-E/RT)} = Ae^{(-E/RT)} \tag{1.30}$$

It can be observed from the above equation that the specific reaction rate constant/coefficient, k, does depend on the temperature and is independent of concentration. The factor A ($= BT^{0.5}$) in Equation 1.30 is called the kinetic pre-exponential factor, which has a very weak temperature dependence in comparison to the exponential term. Hence, it can be treated as a constant, particularly when evaluating its value from the experimental data. Note that this form of equation, known as the Arrhenius law, holds true for many combustion reactions.

1.9 Flame

Flame is the genesis of any combustion process, so it is important to understand the flame and its structure. Then question arises: What is a flame? It can be represented as a spatial domain in which rapid exothermic chemical reactions take place, often emitting light. There are several key words in this definition of flame. First, it is required that the flame should be located in spatial domain. It means that the flame occupies only in spatial domain. It means that the flame occupies only small portion of the combustible mixture. This is in contrast to the various homogeneous reactors, where reaction is assumed to occur uniformly throughout the reaction vessel. That means the flame is basically a self-sustainable localized combustion zone that moves at a certain velocity into the fuel-air mixture. This is also known as a combustion wave since it moves and there is sudden change in properties like temperature, T, pressure, P, and mass fraction, Y_i, of constituent species. Generally the flame can be broadly divided into two categories: (i) diffusion flame and (ii) premixed flame.

1.10 Premixed Flame

In the case of a premixed flame, the fuel and the oxidizer are mixed well at the molecular level before combustion takes place. Examples of premixed flames that we encounter in our day-to-day lives are the flames in laboratory Bunsen burners, domestic burners, heating appliances, and jet engine afterburners. These flames do not occur very commonly in nature. The first laboratory premixed flame burner was invented by Bunsen in 1855, and premixed flames are still preferred today in many residential, commercial, and industrial devices to meet the demand of current stringent environmental regulation. Let us consider a line diagram of a typical Bunsen burner as shown in Figure 1.1 depicting its various parts. Note that fuel enters into the burner at its base as a jet through an orifice. As a result, it entrains ambient air through primary air ports and thus forms a homogeneous mixture of fuel and air. On ignition, it forms a conical flame at the mouth of the Bunsen burner.

Let us now consider a glass tube filled with fuel and oxidizer mixtures. A spark is created by using a high-voltage source. Then the flame is established, which will move into unburned fuel/air mixtures at a certain velocity. Thus, the flame can be considered as a combustion wave. If the flame moves at subsonic speed, it is termed a *deflagration*. When the flame propagates at supersonic speed, it is known as *detonation*. We will restrict our discussion to deflagration only. If we want to analyze the flame, we must establish a reference frame for our coordinate system. The

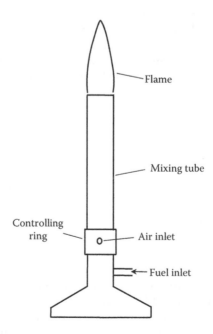

Figure 1.1

Schematic of a typical Bunsen burner premixed burner.

1. Introduction to Combustion

appropriate coordinate system would be fixing to the propagating flame. Then we can treat the flame as standing with respect to the reference system. In other words, if we can feed the fuel/air unburned mixture at the same speed with which the laminar flame is propagating, then the flame will be a stationary one. Thus, we can define *laminar flame speed S_L as the velocity of a laminar flame front in the direction normal to itself with respect to the unburned mixture.* In this case, it has been assumed as a one-dimensional (1-D) steady flame. In laboratory, one can establish a 1-D steady flame for measuring flame speed.

1.10.1 Thermal Analysis of a Planar Laminar Flame

In this analysis we are considering the flame to be a steady, laminar, and planar subsonic combustion wave. Let us consider the temperature profile of a typical laminar premixed flame as shown in Figure 1.2. The temperature at the upstream T_u corresponds to an unburned fuel and air mixture. The temperature increases monotonically with a certain gradient along the x-direction and reaches peak temperature known as flame temperature, T_f, indicating the completion of combustion in the ideal sense. Note that most chemical reactions in the case of limit mixture hydrocarbon-air mixture can occur at a high temperature zone due to high activation energy reactions. Hence the flame can be conveniently divided into two zones (see Figure 1.2): (i) the preheat zone, δ_f, where thermal conduction will be predominant, and (ii) the reaction zone, where heat release due to chemical reaction occurs. These two zones can be divided based on the ignition temperature, as shown in Figure 1.2. Note that the reaction/flame zone, δ_f, is quite thin as compared to the preheat zone, δ_c. This ignition temperature is considered to be the property of reactants in the fuel-air mixture. Let us now carry out this thermal analysis of a laminar premixed flame with the following assumption in order to arrive at an expression for laminar flame speed.

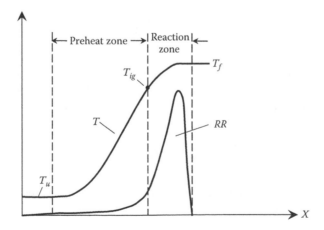

Figure 1.2

Schematic of the temperature profile of a 1-D steady flame.

Assumption:

i. One dimensional (1-D) steady laminar flame
ii. No radiation/convection heat loss
iii. Constant thermodynamics properties
iv. Change in kinetic energy is quite negligible

As discussed earlier, the heat flux at the interface of the preheat zone is generated due to the occurrence of an exothermic chemical reaction in the reaction zone. As a result there will be an increase of temperature from unburned temperature, T_u, to ignition temperature, T_i. Let us carry out an energy balance in the preheat zone, as given below:

$$\dot{m}'' C_P (T_i - T_u) = k \frac{T_f - T_i}{\delta_f} \tag{1.31}$$

where $\dot{m}'' = \rho_u S_L$ is the rate of mixture consumption per unit area, δ_f is flame thickness, and k is the thermal conductivity of the fuel-air mixture. Then the above equation can be expressed as

$$S_L = \frac{k}{\rho_u C_P} \frac{T_f - T_i}{(T_i - T_u)\delta_f} \tag{1.32}$$

We can determine other parameters in the above expression except the flame thickness δ_f. As the activation energy is small, most chemical reactions will be occurring in the reaction zone. We can express flame thickness, δ_f approximately as

$$\delta_f \approx V_b t_r = \frac{\rho_u}{\rho_b} S_L t_r \tag{1.33}$$

where V_b is the burning velocity, ρ_u is the density of mixture at T_u, ρ_b is the density of mixture at T_f, and t_r is the reaction time. By considering a single-step model ($F + Ox \rightarrow Pr$), the average reaction rate, RR_{av} can be expressed as

$$RR_{av} = \frac{dC_{Pr}}{dt} = A C_F^m C_{Ox}^n e^{(-E/RT)} \tag{1.34}$$

By integrating Equation 1.34, over time t_r, we can have an expression for product concentration, C_{Pr} as

$$C_{Pr} = RR_{av} t_r = t_r A C_F^m C_{Ox}^n e^{(-E/RT)} \tag{1.35}$$

Substituting the reaction time, t_r, from Equation 1.35 in Equation 1.33, we can have

$$\delta_f \approx V_b t_r = \frac{\rho_u}{\rho_b} S_L \frac{C_{Pr}}{RR_{av}} = \frac{\rho_u S_L}{MW_b RR_{av}} \tag{1.36}$$

where MW_b is the molecular weight of the products. Now combining Equations 1.36 and 1.32, we can have an expression for laminar burning velocity:

$$S_L = \sqrt{\frac{k}{\rho_u C_P} \frac{(T_f - T_i)}{(T_i - T_u)} \frac{MW_b RR_{av}}{\rho_u}} = \sqrt{\frac{k}{\rho_u C_P} \frac{(T_f - T_i)}{(T_i - T_u)} \frac{MW_b A C_F^m C_{Ox}^n e^{(-E/RT)}}{\rho_u}} \tag{1.37}$$

We can observe from the above equation that the flame speed depends on the type of fuel-oxidizer mixture, the equivalence ratio, initial pressure, and temperature. The experimental data for some fuel-air mixtures is shown in Figure 1.3. Note that the peak flame speed for CH_4-air mixture is around 40 cm/s, which occurs around the stoichiometric mixture. But for CO-air and H_2-air, the peak flame speed occurs at rich mixture. The value of a hydrogen-air system is one order of magnitude higher in comparison to hydrocarbon flames. The laminar flame speed is dependent on the pressure $\left(S_L \cong P^{\frac{q}{2}-1}\right)$. For a hydrocarbon-air

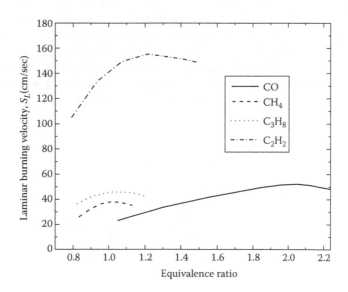

Figure 1.3

Variation of flame speed, S_L, with equivalence ratio. (From Strahle, W. C., *An Introduction to Combustion*, First Edition, Gordon and Breach Publishers, 1993.)

flame, $q = m + n$ will vary from 1.8 to 2.2 and therefore the pressure index, n, varies from ±0.1. Hence it can be concluded that laminar flame speed is almost independent of pressure. This trend is commensurate with experimental data, particularly for hydrocarbon-air flames. But the dependence of initial temperature on flame speed is caused due to change in flame temperature.

1.10.2 Methods of Measuring Flame Speed

As discussed above, flame speed is an important parameter that characterizes the premixed flame as it is considered to be the physiochemical property of the specific combustible fuel-air mixture. Recall that the flame speed can be defined as the relative velocity normal to the flame front with which the unburned gas moves into this flame front. In other words, the flame speed can be obtained experimentally from the rate of propagation of combustion wave while passing through the quiescent gas. Several measurement methods have been developed for measuring flame speed over the years. These methods can be broadly classified into two categories: (i) propagation method and (ii) stationary method. In the case of the stationary method, the flame front would not propagate but rather remains stationary in space. In contrast, in the propagation method, the flame front propagates with respect to some fixed coordinate system in space. Some of these methods are discussed briefly. Interested readers may consult other books on combustion [1,3,4–7].

i. Tube Method

In this tube method, the combustible mixture is filled in a tube, as shown in Figure 1.4. Keep in mind that the inner diameter of the tube must be greater than critical diameter below in which the flame is quenched. When the mixture is ignited at one end of the tube, a flame is initiated that propagates through the tube (see Figure 1.4). It can be readily observed from this figure that the flame becomes a parabolic shape from the initial planar kernel as it moves further downstream toward the unburned mixture for two reasons:

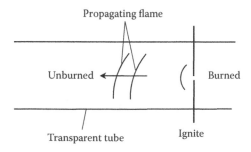

Figure 1.4

Schematic of the tube method.

i. Natural convection distorts the planar flame front due to the difference in densities of burned and unburned mixtures

ii. Due to friction at the tube wall, the flame front propagates at a higher velocity along the tube axis, which causes the flame front to be parabolic in shape

Generally a small opening is used at one end of the tube to reduce reflected pressure waves that ensure uniform laminar flame front velocity measurement over the large portion of the tube. This small opening is connected to a soap bubble meter to measure unburned gas velocity caused by movement of the flame front and it also helps in maintaining a constant shape of the flame front. From the measurement of flame photographs over time, the laminar flame speed, S_L, can be determined as

$$S_L = (V_S - V_g) A_t/A_f \qquad (1.38)$$

where V_S is the flame front velocity, V_g is the velocity of unburned gas ahead of the flame, A_t is the cross-sectional area of the tube, and A_f is the flame surface area. Keep in mind that this is not a very accurate method due to wall effects and distortion of flame surface as a result of buoyancy. The buoyancy effect can be avoided by using a vertical tube, which is also employed for determining the flammability limits of fuel-air mixture.

ii. Combustion Bomb Method

In the combustion bomb method, the combustible mixture is ignited at the center of a rigid spherical vessel around 25 cm in diameter, as shown in Figure 1.5. When flame travels toward unburned gas in the vessel, the expansion of gas due to high temperature causes its pressure and temperature to increase due to adiabatic compression. Initially, the rise in pressure and temperature is quite insignificant, particularly when the ratio of flame size to vessel volume is small. But later, there will be a significant temperature rise that causes the flame speed to increase

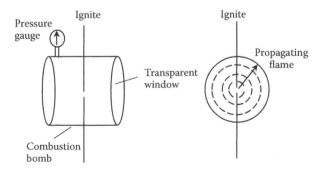

Figure 1.5

Schematic of a combustion bomb apparatus.

continuously from the center to the wall of the vessel. In this case the flame from velocity and pressure change with time is being measured to determine the flame speed of a particular fuel-air mixture.

For combustion bomb apparatus, the flame speed can be expressed as [1]

$$S_L = \frac{dR_F}{dt} - \frac{R^3 - R_f^3}{3P\gamma_u R_f^2}\frac{dP}{dt} \tag{1.39}$$

where $\gamma = C_p/C_v$ = specific heat ratio with respect to unburned gas, P is the instantaneous pressure, R is the radius of sphere, and R_f is the instantaneous radius of the spherical flame. The above expression is based on the following assumptions:

i. The effect of flame front thickness and curvature are negligibly small, which means that the flame front remains smooth, spherical, and centered on the point of ignition
ii. The pressure at any instant is uniform throughout the vessel
iii. There is no heat loss, including radiation
iv. Chemical equilibrium is achieved behind the flame front

iii. Soap Bubble Method

In the soap bubble method, the homogeneous fuel-oxidizer mixture confined in a soap bubble, as shown in Figure 1.6, is ignited at the center by a spark. A spherical flame starts propagating along the radial direction through the mixture. Due to high temperature, the burned gas gets expanded in the flexible soap bubble as shown in Figure 1.6. As a result the pressure of burned gas remains almost constant when the flame propagates outwardly. This flame front can be visualized experimentally by any of the three commonly used methods namely direct, shadow, and schlieren photography. Assuming the flame to be spherical and the pressure remaining constant, a mass balance through the flame results in

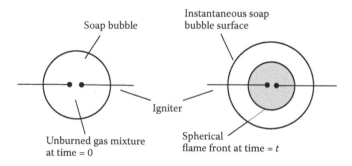

Figure 1.6

Schematic of the soap bubble method.

$$S_L A \rho_u = V_f A \rho_b \Rightarrow S_L = V_f \left(\frac{\rho_b}{\rho_u} \right) = V_f \left(\frac{T_u}{T_b} \right) \qquad (1.40)$$

where V_f is the flame front velocity that can be obtained by photography, ρ_b is the density of gas mixture at burned state, and ρ_u is the density of unburned mixture.

The density ratio, ρ_b/ρ_u, can be estimated from the temperature measurement. Besides this, one should estimate the initial and final size of the bubble accurately. But the final size is difficult to measure as it bursts instantly. Apart from these two problems, the soap bubble method has several other disadvantages:

   i. It cannot be used for measurement of the flame speed of a dry mixture
   ii. The flame front may not retain its spherical shape during its propagation
   iii. The flame front would not be smooth for fast flames
   iv. Heat loss to the electrodes and ambient incurs error in measurement

iv. Stationary Flame Method

In the stationary flame method, the flame will remain stationary relative to the laboratory coordinates. Based on this basic principle, several burners have been designed and developed for measurement of the flame speed. We will restrict our discussion only to two widely used burners: (i) the *Bunsen burner* and (ii) the *flat flame burner*.

Bunsen burner: We have already briefly discussed the Bunsen flame in the previous section. Keep in mind that the flame is being stabilized on the burner rim, and that the local flame speed must be equal to the local flow velocity of the unburned gas. The shape of the flame is influenced by the velocity profile at the exit of the burner tube and heat losses to the tube wall. For the stationary flame, a mass balance across the flame can provide an expression for flame speed, S_L, as

$$S_L = V_t \frac{A_t}{A_f} \qquad (1.41)$$

where V_t is the average flow velocity in the tube, A_t is the tube cross-sectional area, and A_f is the conical surface area of the flame. This method is known as the *area method*, which will be discussed in detail in Chapter 8.

Besides the above method, high-contraction contoured nozzles can be used in which the uniform flow can be established at exit. As a result, a flame cone with straight edges is established as shown in Figure 1.7. Note that the shape of the flame is affected by the velocity profile at its mouth and heat losses to the nozzle wall. The angle of the cone slant with respect to the burner axis is determined as shown in Figure 1.7 at the central portion of the cone from the flame photograph. Hence, the expression for the laminar flame speed can be obtained as

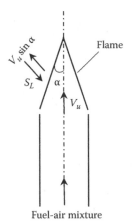

Figure 1.7

Schematic of Bunsen burner flame with velocity vector.

$$S_L = V_u \sin \alpha \qquad (1.42)$$

where V_u is the velocity of the unburned mixture and α is the flame cone half angle.

The main advantage of the Bunsen burner is that it is simple and easy to fabricate. Apart from this, it can be used easily for the measurement of flame stability, flammability limits, and so forth by varying temperature and pressure. However, it has several disadvantages, which are as follows [1]:

i. Heat loss to the wall cannot be avoided completely
ii. The flame speed does not remain constant along its surface
iii. The diffusion of secondary atmospheric air alters the mixture ratio, which affects flame speed measurement
iv. Flame stabilization is difficult for larger diameter due to a flashback problem

The stationery flame method as applied to a Bunsen flame is discussed further in Chapter 8.

Flat flame burner: This flat flame burner is often used to establish a 1-D steady flame in the laboratory to measure burning velocity. It is one of the simplest and most accurate methods that we have discussed so far. The schematic of a typical flat flame burner is shown in Figure 1.8. It is very essential to have a 1-D velocity profile at the mouth of the burner, which can be achieved by using either a honeycomb or a porous plug. In order to prevent diffusion of atmospheric oxidizer, N_2 gas is used around the flame. The burning velocity can be easily determined by dividing flame area with volumetric flow rate of the unburned gas mixture.

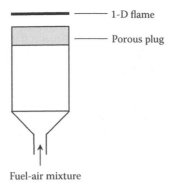

Figure 1.8

Schematic of a flat flame burner.

The area of the flat flame can be measured by using schlieren/shadow/luminous photography. It is interesting to note that the flame area obtained by visible or shadow or schlieren photograph is the same. As a result, uncertainty in obtaining the flame area is eliminated in this burner. This burner is appropriate for mixtures having a low burning velocity, particularly in the range of 15–20 cm/s or less. However, it can also be used for higher burning velocity measurements by employing a *water cooled porous plug burner.*

1.11 Diffusion Flames

In the last section, we discussed premixed flames. In this section, we will learn about various kinds of diffusion flames. We know that in the case of diffusion flames, the fuel and oxidizer remain unmixed up to the flame surface. The fuel and oxidizer diffuse to the reaction zones due to the molecular and turbulent diffusion and mix at the flame surface itself. As the burning rate is influenced by the rate at which the fuel and oxidizer can reach together at the flame surface in proper proportions, diffusion flames can be considered as *diffusion-limited.* On the other hand, the premixed flames are *kinetically controlled,* because the reaction rates are most important in deciding the flame structure. Note that diffusion flames are likely to take place more often in nature. For example, forest fires occur more commonly in tropical forests. Some examples of diffusion flames are

- Candle flame
- Industrial furnace
- Gas turbine combustor
- Liquid and solid propellant rocket motors
- Burning of solid fuels such as coal, biomass, etc.

1.12 Gaseous Jet Flame

As discussed in the previous section, a jet diffusion flame can be established easily in a Bunsen burner when the air vent is closed. Besides this, a diffusion flame can occur in various residential appliances such as gas-fired furnaces and cooking ovens. Jet diffusion flames can also be found in petrochemical complexes, and unwanted gas-duct explosion, etc. The effectiveness of jet diffusion flame-based combustion systems is influenced by the shape and size of the jet diffusion flames, their emission levels, and the temperatures across them. Hence we need to understand the structure of jet diffusion flames, and therefore laminar jet diffusion flames are discussed in subsequent sections.

1.12.1 Physical Description of a Jet Flame

Let us consider a gaseous fuel jet, which is issued from a jet diameter d_j into a quiescent oxidizing atmosphere, as shown in Figure 1.9. For the purpose of discussion, let us consider that hydrogen gas as a fuel is issued from a jet to an atmosphere containing oxygen as oxidizer. In this case, the fuel diffuses radially outward while the oxidizer diffuses radially inward toward its center. As a result, there will be interdiffusion in the mixing layer that grows along the vertical direction. But near the jet exit there is a region known as the *potential core* (see Figure 1.9), which is unaffected by the mixing. However, H_2 gaseous fuel diffuses into the oxidizer and vice versa at a downstream region and they get mixed at a

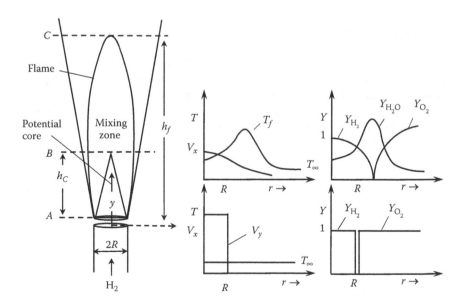

Figure 1.9

The flame structures of laminar gaseous jet diffusion flames.

1. Introduction to Combustion

certain location where the flame is being established on ignition. Hence the flame is established at the interface of the mixing of the fuel and oxidizer. Note that the *flame surface corresponds to the location of points where the equivalence ratio, (φ) is equal to 1 (stoichiometry)*. As the chemical reactions occur in the flame in a thin region, the *flame can be considered to be a thin surface*. It is interesting to note that the reaction zone occurs in an annular region inside the flame surface. Let us understand the structure of diffusion flames as depicted in Figure 1.9. The variation of axial velocity, V_x, mass fractions of fuel, Y_F, and oxidizers, Y_{Ox}, and temperature, T, along with radial distances from the center are depicted at three locations along the y-direction ($y = 0$, $y = h_C$, $y = h_f$). We can observe that the peak temperature occurs at a particular radial location between the flame's tip and inlet (e.g., $y = h_C$). Note that both the fuel mass fraction, Y_F, and oxidizer mass fractions, Y_{Ox}, become almost zero value at the flame surface and product mass fraction, Y_P, attains its maximum values. However at the flame tip, the temperature drops monotonously from its peak value at its center along the radial direction. The mass fraction of fuel Y_F reduces from the peak value at its center asymptotically to zero at infinite radial distance while the oxidizer mass fraction increases from zero at the center to unity toward the infinite radial distance.

1.12.2 Phenomenological Analysis

We will be learning about a simple phenomenological analysis based on certain logic and assumptions that are useful in providing a viable but simple tool for analyzing complex overventilated diffusion flames. Note that the upper region of a vertical flame contains a sufficient quantity of hot gases. As a result, the buoyancy force accelerates the flow, which results in a conical diffusion flame shape, as shown in Figure 1.9. With increase in the velocity the streamlines become closer together to satisfy the mass conservation principle and thus there will be an increase in the fuel concentration gradient (dY_F/dr), which enhances the diffusion of fuel. It is interesting to note that the effects of these two phenomena on the length of diffusion flame, h_f, issuing from a circular port cancel each other. Hence, the buoyancy effect can be neglected, particularly for circular and two-dimensional diffusion flames. Besides this, we can assume that the burning process in a diffusion flame is not being affected by the mixing rate between the fuel jet and the oxidizer. In other words, whenever the fuel and oxidizer come in contact with each other, they react immediately.

For a cylindrical jet diffusion flame, the flame height can be defined as the point along the axis on which interdiffusion of the fuel and oxidizer is reached for the first time. That means the end of the flame is reached by a fluid element on the axis in a time magnitude (h_F/V) where h_f is the flame height and V is the inlet velocity of fuel jet at its inlet. But the average square displacement ($\bar{x}^2$) due to diffusion as per the *Einstein diffusion equation* is given by [1]

$$\bar{x}^2 = 2Dt \tag{1.43}$$

where D is the fuel-air diffusion coefficient. We know that the height of the flame is the point where the average depths of interdiffusion is equal to tube radius, R. Note that $\bar{x}^2$ can be approximated as R^2 and time is also given by flame height, h_f, divided by gas velocity V. Then

$$h_f = \frac{VR^2}{2D} \tag{1.44}$$

Note that this relation is indeed linear in terms of axial flow velocity. We can express the above relation in terms of volumetric flow rate, Q, as

$$h_f = \frac{V\pi R^2}{2\pi D} = \frac{Q}{2\pi D} \tag{1.45}$$

Note that the experimental data is often reported in terms of volumetric flow rate. Now if we multiply the numerator and denominator of the equation by ρ, it becomes

$$h_f = \frac{\rho V\pi R^2}{2\pi\rho D} = \frac{\rho Q}{2\pi k/C_p} = \frac{\dot{m}}{2\pi k/C_p} \tag{1.46}$$

Keep in mind that we have assumed in the present analysis that the Lewis number is equal to 1 ($\rho D = k/C_p$). It is interesting to note that the flame height, h_f, is independent of the burner diameter for a particular volume flow rate. We know that (k/C_p) is highly insensitive to the thermodynamic conditions. Hence, it can be concluded that the flame height, h_f, is only dependent on the mass flow rate of fuel. The flame height is also independent of pressure at a constant mass flow rate because the diffusion coefficient is inversely proportional to pressure.

Although this analysis is very crude in nature, it provides a qualitative prediction of flame height, h_f, and its dependence on mass flow rate and thermal properties, which is supported by experimental evidence. We will discuss a method for measuring flame height and verify the predictions of phenomenological analysis in Section 8.4.

For the turbulent diffusion flame, we can extend this simple phenomenological analysis and arrive at a similar relationship as given above, in which the molecular diffusivity, D, is replaced by eddy diffusively, v_t, which is the product of a turbulent scale and root mean square (rms) velocity of fluctuating component, V'_{rms}. Note that the turbulent scale is proportional to the tube radius, R and V'_{rms}, which can be approximated to axial mean flow velocity, $\bar{V}$. Then the eddy diffusivity, v_t, becomes

$$v_t \propto R\bar{V} \tag{1.47}$$

Then the flame height, h_f, becomes

$$h_f \propto \frac{\overline{V}R^2}{\upsilon_t} \propto \frac{\overline{V}R^2}{R\overline{V}} \propto R \qquad (1.48)$$

The above expression for flame height for turbulent diffusion flame indicates that it is proportional to the tube radius only. Note that this simple analysis can provide the qualitative prediction of flame height for a jet diffusion flame.

The variation of experimental diffusion flame height, h_f, with jet velocity, $\overline{V}$, is plotted in Figure 1.10. It can be observed from this figure that the laminar flame height increases linearly with an increase in jet velocity until the turbulent mixing occurs. Such a phenomenon appears first at the flame tip and moves down progressively as velocity increases further. Note that the flame height falls in this region while the visible flame boundary spreads outwardly. This position of transition from laminar to turbulent is called the *break point* (see Figure 1.10). The transition to fully developed turbulent flame can be characterized by a transition Reynolds number. The transition Reynolds numbers for typical fuels are given in Table 1.7. It can be observed from these that the transition Reynolds number is different for different fuels indicating the importance of both chemical kinetics and fluid mechanics. Note that the turbulent flame produces more noise in comparison to a laminar flame and less soot. As a result, there is a decrease in the yellow luminosity in a turbulent flame as compared to a laminar diffusion flame. This break point remains constant beyond certain velocity. However, the flame gets lifted off from the burner rims beyond a certain critical jet velocity. As the jet velocity increases further, the liftoff height of the flame

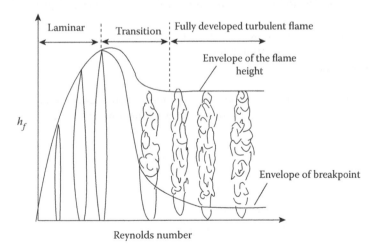

Figure 1.10

Variation of flame height with Reynolds number in the diffusion flame. (From Hawthorne, W. R., Weddell, D. S., and Hottel, H. C., Mixing and combustion in turbulent gas jets, *Proceedings of Combustion Flame and Explosions Phenomena*, 3, pp. 266–288, 1949.)

Table 1.7 Transition Reynolds Number in Gaseous Jet Diffusion Flames for Typical Fuels

Fuel Into air	Transition Reynolds Number
Acetylene (C_2H_2)	9500
Propane (C_3H_8)	9500
City gas	3500
Carbon monoxide (CO)	4800
Hydrogen (H_2)	2000

from the burner rims increases while the flame's height decreases. It becomes smaller and smaller while in oscillating mode and the flame blows off with an increase in jet velocity. The threshold velocity beyond which the flame is completely blown off is known as the blowoff limit, which is dependent on the type of fuel-oxidizer systems involved. Interested readers can consult books on combustion [1–7].

Example 1.3

Propane gas is issued from a tube of 2 mm diameter at 298 K and 0.1 MPa. The flow rate of methane gas is 1.5 LPM. Estimate the flame height by phenomenological analysis assuming the Lewis number (Le) is equal to one.

Given: $Q = 1.5$ LPM, $P = 0.1$ MPa, $T = 298$ K, d = 2 mm.

To Find: Flame height, h_F.

Solution: The following assumptions are made to solve this problem:

    i. Laminar steady flow exits at the tube's end
   ii. The ideal gas equation holds up
  iii. Constant thermodynamic properties

Let us consider the thermodynamic properties such as thermal conductivity, k_g, and specific heat corresponding to the ambient temperature, as given below:

$$k_g = \text{thermal conductivity} = 0.016 \text{ J/(m K s)}$$

$$C_P = \text{Specific heat} = 1.67 \text{ J/kg K}$$

Since Le = 1; $\rho D = k_g/C_P = 0.016/1.67 = 0.0096$
By the equation of state, we can evaluate the density of gas as

$$\rho = P\,MW/RT = 10^5 \times 44/(8.314 \times 298 \times 1000) = 1.77 \text{ kg/m}^3$$

The flame height, h_f, can be estimated as

$$h_f = \rho Q/2\pi \ (k_g/C_P) = (1.77 \times 1.5/60)/(2 \times 3.14 \times 0.0096) = 0.73 \text{ m}$$

Note that the diffusion flame height by this phenomenological analysis indicates the dependence of the flame height on a variable such as flow rate; however, it may not provide an accurate prediction of experimental data.

1.13 Liquid Droplet Combustion

Liquid fuels have a high-energy density and are therefore used extensively in several combustion systems (e.g., piston engines, gas turbines engines, rocket motors, furnaces, and boilers) as they can be easily stored and transported. But the combustion of liquid fuel cannot be accomplished effectively unless it is converted into arrays of small droplets (fuel spheres) known as spray. Generally atomizers are designed and developed to facilitate the disintegration of liquid fuel into a spray of droplets in order to increase the fuel surface area and thus enhance the combustion rate due to increase in vaporization rate.

But the process involved in combustion of fuel spray is very complex as it undergoes simultaneous heat, mass, and momentum transfer and also complex chemical reactions [1]. Some important factors that govern the combustion of liquid fuel spray are (i) *drop size,* (ii) *composition of the fuel,* (iii) *ambient gas temperature, pressure, and its composition,* and (iv) *relative velocity between the droplet and the surrounding gas.* As the processes involved in the combustion of liquid spray are quite complex, we will be discussing single isolated droplet combustion to understand the physical processes involved during the burning of a fuel droplet. When a droplet is suspended in a hot environment, it will first undergo evaporation at its surface and the vapor subsequently diffuses to the flame front. At the same time, oxygen also has to diffuse toward the flame front. As a result, the flame is established at a certain distance away from the droplet surface (see Figure 1.11). Moreover, the heat liberated in the flame front is transferred back to the liquid fuel surface such that its surface temperature can be maintained at the boiling point of the fuel. The shape of the flame front is dependent on the condition under which combustion takes place. The shape of the flame front is almost spherical under zero-gravity conditions (see Figure 1.11a) due to the absence of any buoyancy, but under normal gravity conditions, the flame shape gets elongated along the gravitational force direction (see Figure 1.11b), which is caused by natural convection. In practical combustion devices the fuel droplet moves along with gas flow and the flame is aligned with the flow, resulting in a flame front as shown in Figure 1.11c. We will be limiting our discussion here to the burning of a single droplet under zero-gravity conditions.

Let us discuss the theoretical aspects of a single droplet surrounded by quiescent air, as shown in Figure 1.11a. For this analysis, we can also assume that liquid droplet's surface is likely to be at its normal boiling temperature. The uniform temperature can be assumed within the droplet. Keeping these points in mind, we will carry out a simplified analysis for the burning of a droplet in a quiescent atmosphere, for which the following assumptions are made:

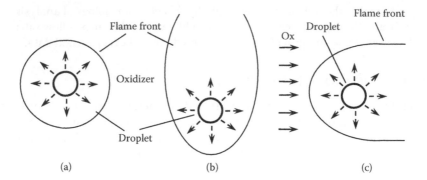

Figure 1.11

Schematic of liquid fuel droplet combustion: (a) zero gravity, (b) normal gravity, and (c) forced convection.

   i. A single droplet is burned in a quiescent, infinite medium surrounded by a spherically symmetric flame (see Figure 1.12).

  ii. The droplet temperature is uniform and is the same as that of the boiling point temperature of liquid fuel.

 iii. The density of liquid fuel is much higher than gas phase density. That means that the regression rate (the rate of radius change of droplet, $[dr_d/dt]$) will be quite small as compared to gas phase diffusion velocities around the droplet. Hence, the entire flow outside the droplet can

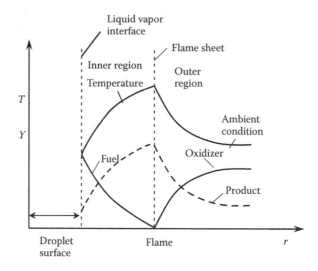

Figure 1.12

The variation of properties Y_F, Y_{Ox}, Y_P, and T along radial direction for a single droplet-burning model.

                                                           1. Introduction to Combustion

be treated as steady flow provided that the evaporation rate remains constant.

iv. The combustion reaction can be modeled as a single-step irreversible reaction with stoichiometry $(F + \upsilon\,Ox \rightarrow (\upsilon + 1)\mathrm{Pr})$, which is infinitely rapid at the flame surface. Hence, the gas phase region can be divided into two regions—inner $(r_S < r < r_F)$ and outer $(r_F < r < \infty)$ (see Figure 1.12)—where r_F is the flame radius and r_S is the radius of the droplet surface. Three temperatures—droplet surface temperature, T_S, flame temperature, T_f, and ambient temperature, T_∞—can be observed in Figure 1.12.

v. Liquid vapor density and heat of vaporization is assumed to be constant as the droplet is surrounded by quiescent air at constant pressure.

1.13.1 Droplet Burning Time

When designing any combustion system, we need to determine the droplet burning time because the residence time must be greater than the lifetime of the largest droplet in the spray so that complete combustion can be accomplished. We are interested here in determining the lifetime of a droplet under quiescent atmosphere for which we can invoke an energy balance at the surface of a spherical droplet as

$$\frac{d(m_d h_{fg})}{dt} = -\dot{q}_s''\pi D^2 \tag{1.49}$$

where $\dot{q}_s''$ is the heat flux from the flame at the droplet surface, h_{fg} is the heat of vaporization, and m_d is droplet mass, which can be evaluated as

$$m_d = \rho_l\,V = \rho_l \pi D^3/6 \tag{1.50}$$

where D is the droplet diameter at any instant. The heat flux from the flame toward the droplet surface can be determined as

$$\dot{q}_s'' = k_g\frac{dT}{dr} \approx k_g\frac{T_f - T_s}{\delta_r} \tag{1.51}$$

where the thermal conductivity of gas δ_r is the thickness of the thermal layer surrounding the droplet.

Note that the value of δ_r is dependent on the physical processes and can be obtained by solving the governing partial differential equation. Interested readers may refer to advanced literature on combustion [1,3,4,6,7]. However, we will simplify it by assuming that the value of δ_r is proportional to the diameter (characteristic length scale). Then the value of δ_r is approximately equal to $C_c * D$, where

---

C_c is the constant that can assumed to be 1/2 for simplicity. Substituting $\delta_r = C_c D = 1/2D$, Equation 1.51 and by combining Equations 1.49 through 1.51, we can have

$$2D\frac{dD}{dt} = \frac{dD^2}{dt} = \frac{-4k_g(T_f - T_s)}{\rho_l h_{fg} C_c} = -K_C \tag{1.52}$$

The above expression indicates that the time derivative of droplet diameter square D^2 is a constant and D^2 varies linearly with time, as shown in Figure 1.12. By integrating Equation 1.52 with time, we get

$$D^2(t) = D_o^2 - K_C t \tag{1.53}$$

where D_o is the initial droplet diameter and K_C is the burning rate constant as expressed in Equation 1.52. Typical values of burning constant, K_C, calculated from Equation 1.52 for certain hydrocarbon fuels are provided in Table 1.8 along with the respective experimental data.

We can see from this table that the prediction is in good agreement with the experimental data. It can also be observed that the wide variations in the chemical structure of fuels do not have much influence on the value of burning constant K. Besides this, K_C for hydrocarbons burning in air is around 10^{-2} cm²/s. Note that Equation 1.53 is known as the D^2 law. The experimental measurements of various researchers indicate that the D^2 law is a good predictor of the experimental data,

Table 1.8 Burning Constant for Various Hydrocarbons with Air at 20°C and 1 atm

Fuel	$\lambda_{calc.}$ 10^{-3}cm²/ sec	λ_{means}	Researcher
Ethyl alcohol	9.3	8.1	Godsave
Ethyl alcohol	9.3	8.6	Goldsmith
Ethyl alcohol	9.3	8.5	Wise
n-Heptane	14.2	9.7	Godsave
n-Heptane	14.2	8.4	Goldsmith
Isooctane	14.4	9.5	Godsave
Isooctane	14.4	11.4	Graves
Decane	11.6	10.1	Hall
Petroleum ether ($T = 100°C–120°C$)	—	9.9	Godsave
Kerosene ($\rho = 0.805$)	9.7	9.6	Godsave
Diesel ($\rho = 0.850$)	8.5	7.9	Godsave

Source: Godsave, G. A. E., Burning of fuel droplets, *Proceedings of the Fourth Symposium on Combustion*, p. 818, William and Wilkins, Baltimore, 1953.

1. Introduction to Combustion

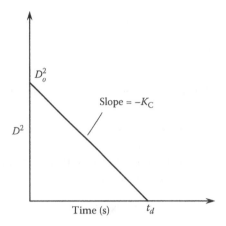

Figure 1.13

The square law of droplet combustion and definition of combustion constant.

particularly beyond the initial transient period. The lifetime of a droplet can be obtained by letting $D^2(t_d) = 0$ as follows:

$$t_d = D_o^2/K_C \qquad (1.54)$$

The lifetime of a droplet, t_d, can be obtained from the experimental data by plotting data shown in Figure 1.13. Note that the life time of droplet can be useful in designing liquid fuel combustion.

Example 1.4

An n-octane droplet with an initial droplet size of 300 μm is burned in a combustor at $P = 0.1$ MPa, 300 K with a stoichiometric flame around it. Determine the lifetime of this droplet.

Given: $D_0 = 300$ μm at 0.1 MPa and 300 K.

To Find: t_d = lifetime of the droplet.

Solution: Let us assume quiescent atmosphere assuming the velocity in the combustor is negligibly small.

We know that the lifetime of droplet t_d can be evaluated as

$$t_d = \frac{D_o^2}{K_C}$$

where the burning constant, K_C, is equal to $K_C = \dfrac{4k_g(T_f - T_s)}{\rho_l h_{fg} C_c}$.

The thermodynamic properties of a gaseous mixture are to be evaluated at average temperature, $\bar{T}$, considering the boiling point temperature of hexane to be 398.7 K

$$\bar{T} = 0.5(T_f + T_B) = 0.5(2200 + 398.7) = 1299.4 \text{ K}$$

Note that the flame temperature T_f is taken as adiabatic temperature T_{ad} (2200 K) of the octane air system.

The thermal conductivity of hexane and air at average temperature $\bar{T}$ can be obtained from Appendix F of [1], as

$$k_f = 0.178 \text{ W/mK}$$

$$k_{Ox} = 0.08 \text{ W/mK}$$

Then the gas thermal conductivity at average temperature $\bar{T}$

$$k_g = 0.4 \, k_f + 0.6 k_{Ox} = 0.4 \times 0.178 + 0.6 \times 0.08 = 0.119 \text{ W/m} - \text{K}$$

Then specific heat of the gas at average temperature $\bar{T}$

$$C_{p,F} = 1.21 \text{ kJ/kg·K}$$

The heat of vaporization of fuel is

$$h_{fg} = 300 \text{ kJ/kg}$$

Then we can evaluate the burning constant, K_C as

$$K_C = \frac{4k_g(T_f - T_s)}{\rho_l h_{fg} C_c} = \frac{4 \times 0.119(2200 - 398.7)}{703 \times 300 \times 1000 \times 0.5} = 0.813 \times 10^{-5}$$

The lifetime of droplet t_d can be evaluated as

$$t_d = \frac{D_o^2}{K} = \frac{(300 \times 10^{-6})^2}{0.813 \times 10^{-5}} = 0.011 \text{ s}$$

The lifetime of a droplet of 300 μm is 0.011 s, which is quite small.

1.13.2 Droplet Combustion in Convective Flow

So far we have discussed the burning of a fuel droplet under quiescent atmosphere without any convection. However, in practical application, a droplet is burned under the influence of both natural and forced convection. A flame is being formed around the droplet, as shown in Figure 1.11b and c for natural and forced convection respectively. Particularly in a forced convection case, when the droplet's Reynolds number, Re_d, goes beyond 20, a boundary layer prevails around the front portion of the droplet followed by a wake region behind the droplet (see Figure 1.11c). The heat flux at the surface of the droplet due to forced convective heat transfer can be expressed as

$$\dot{q}''_s = \bar{h}_C \Delta T \tag{1.55}$$

where $\dot{q}''_s$ is the heat flux from the flame at the droplet surface, $\bar{h}_c$ is the convective heat transfer coefficient, and ΔT is the temperature difference between the flame and droplet surface temperature that can be assumed to be approximately equal to the boiling temperature of fuel. For a spherical droplet, the convective heat transfer coefficient, $\bar{h}_c$, can be evaluated using an empirical relationship for Nusselt number, Nu, as

$$Nu = \frac{\bar{h}_c D}{k} = 2 + 0.4 \, Re_D^{1/2} \, Pr^{1/3} \tag{1.56}$$

where D is the droplet diameter at any instant, $Re = \rho V_D / \mu$ is the Reynolds number, ρ is the density of gas, V_D is the relative velocity of droplet, μ is the dynamic viscosity of gas, and $Pr = \nu / \alpha$ is the Prandtl number, which is the ratio of viscous diffusivity, ν, and thermal diffusivity, α and $\alpha = k_g / \rho C_P$. By following a similar procedure as in the case of droplet combustion under quiescent atmosphere, we can have

$$\frac{d(\rho_l [\pi D^3 / 6] h_{fg})}{dt} = -\dot{q}''_s \pi D^2 = \pi D^2 \bar{h}_c (T_f - T_s) \tag{1.57}$$

where D is the droplet diameter at any instant. By substituting the expression (Equation 1.56) of convective heat transfer coefficient $\bar{h}_c$ in Equation 1.55, we have

$$\frac{dD^2}{dt} = \frac{-4k_g (T_f - T_s) \left[2 + 0.4 \, Re_D^{1/2} \, Pr^{1/3} \right]}{\rho_l h_{fg}} = -K_{FC} \tag{1.58}$$

By integrating Equation 1.58 with time, we get

$$D^2(t) = D_o^2 - K_{FC} t \tag{1.59}$$

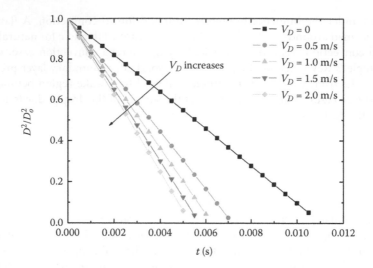

Figure 1.14

Effect of relative velocity on the droplet evolution during combustion at 1 MPa and 1300 K.

where D_o is the initial droplet diameter and K_{FC} is the burning rate constant under a forced convection condition. Note that as the diameter decreases with time, the Reynolds number will decrease; hence the average Reynolds number based on initial droplet diameter, D_o, is used for the evaluation of K_{FC}. The relative velocity V_D is considered for evaluation of the Reynolds number. The variation of n-octane droplet diameter ($D_o = 300$ μm) is plotted in Figure 1.14 for certain range of V_D. It can be observed that the droplet burns at a faster rate with an increase in the relative velocity V_D, indicating the effects of convective heat transfer.

Review Questions

1. What is equivalence ratio? How is it different from fuel/air ratio? What is its physical interpretation?

2. What is adiabatic flame? Can it be determined without knowing the composition of products?

3. Suppose combustion is taking place in an internal combustion (IC) engine. You need to determine the adiabatic temperature. Which is the assumption you need to consider: (i) constant pressure or (ii) constant volume? Justify your answer.

4. What is flame? How many ways is flame classified?

5. How is diffusion flame different from premixed flame?

1. Introduction to Combustion

6. Does the domestic LPG burner work based on a premixed/diffusion flame? Justify your answer.

7. What is the structure of a laminar premixed flame? Explain.

8. What is flame speed? Derive an expression for flame speed.

9. Why is flame height important for a jet diffusion flame? Derive an expression for flame height.

10. Why is liquid fuel converted from bulk to spray form for large-scale combustion systems?

11. Describe the combustion processes involved in droplet combustion in a quiescent atmosphere.

12. What is the lifetime of a droplet under a quiescent atmosphere? Derive an expression for it.

13. Suppose a fuel droplet is evaporated under a quiescent atmosphere at 600 K and 0.1 MPa. If a flame is surrounding it, then what will happen to the droplet lifetime as compared to evaporation only? Justify your answer.

14. What are the factors that influence the mass burning rate of a droplet under convection conditions as compared to a quiescent atmosphere condition?

Problems

1. Propane (C_3H_8) is burned with dry air. The volumetric analysis of products on a dry basis is $CO_2 = 11.23\%$, $O_2 = 2.215\%$, $CO = 0.013\%$, and $N_2 = 78.8\%$. Determine (a) A/F ratio, (b) equivalence ratio, and (c) percent of stoichiometric air used.

2. Gasoline (octane = C_8H_{18}) is burned with dry air. The volumetric analysis of products on a dry basis is $CO_2 = 10.02\%$, $O_2 = 5.62\%$, $CO = 0.88\%$, and $N_2 = 83.48\%$. Determine (a) A/F ratio, (b) equivalence ratio, and (c) percent of stoichiometric air used.

3. One mole of methane reacts with oxygen in a stoichiometric ratio. Consider that the reactants are at a temperature of 298.15 K and a pressure of 101,325 Pa. Determine the heat of combustion.

4. One mole of propane reacts with oxygen in a stoichiometric ratio. Consider that the reactants are at a temperature of 298.15 K and a pressure of 101,325 Pa. Determine the heat of combustion.

5. The following heats of reaction at 298.16 K are given below:

$$CO_2 \rightarrow C + O_2 \quad \Delta H_{R,298}^0 = 393.51 \text{ kJ} \qquad (P1.1)$$

$$C + \frac{1}{2}O_2 \rightarrow CO \quad \Delta H_{R,298}^0 = -110.59 \text{ kJ} \qquad (P1.2)$$

$$H_2 + \frac{1}{2}O_2 \rightarrow H_2O \quad \Delta H_{298}^0 = -241.82 \text{ kJ} \qquad (P1.3)$$

Determine the standard heat of reaction at 298 K for the famous water-shift reaction as given below:

$$CO_2(g) + H_2(g) \rightarrow CO(g) + H_2O(g) \qquad (P1.4)$$

6. In a reactor, LPG (30% (C_3H_8) and 70% (C_4H_{10})) and air at $T = 298$ K and $P = 0.1$ MPa undergo the combustion process at a stoichiometric fuel/air ratio to produce carbon dioxide, water, and nitrogen as a combustion product. Estimate the adiabatic temperature, T_{ad}.

7. Determine the laminar burning velocity of a propane-air stoichiometric mixture at $T = 298$ K and $P = 0.1$ MPa. The average reaction rate considering a single step chemistry model as given below can be used for this.

$$\dot{N}_F''' = AC_F^{-0.1}C_{OX}^{1.0}e^{(-E/RT)}$$

where A = activation energy = 1.7×10^8 and $E/R_u = 15,098$ K.
The following data can be used for solving these problems: $T_i = 743$ K, $T_u = 298$ K, $MW_{C_3H_8} = 44$, $C_P = 1.63$ kJ/kg·K.

8. Propane gas is issued from a tube of 4 mm diameter at 298 K and 0.1 MPa. The flow rate of methane gas is 15 LPM. Estimate the flame height by phenomenological analysis assuming the Lewis number (Le) is equal to 1.

9. The methane gas is issued from a tube of 0.5 mm diameter at 298 K and 0.1 MPa. The flow rate of methane gas is 10 LPM. Estimate the flame height by phenomenological analysis assuming the Lewis number (Le) is equal to 1.

10. An n-octane droplet with an initial droplet size of 150 μm is burned under a quiescent atmosphere at $P = 0.1$ MPa, with a stoichiometric flame around it. Determine the lifetime of this droplet.

11. An n-octane droplet with an initial droplet size of 150 μm is burned in a combustor at $P = 0.1$ MPa, 300 K, in which the average velocity is around 3 m/s with a stoichiometric flame around it. Determine the lifetime of this droplet.

References

1. Mishra, D. P., *Fundamentals of Combustion,* PHI Ltd, New Delhi, 2010.
2. Mukunda, H. S., *Understanding Combustion,* University Press, Hyderabad, India, 2009.
3. Turns, S. R., *An Introduction to Combustion: Concepts and Applications,* Third Edition, McGraw-Hill, New York, 2011.
4. Kuo, K. K., *Principles of Combustion,* John Wiley & Sons, New York, 1986.
5. Gaydon, A. G. and Wolfhard, H. G., *Flames, Their Structure, Radiation, and Temperature,* Fourth Edition, Chapman & Hall, 1979.
6. Williams, F. A., *Combustion Theory,* Second Edition, Addison-Wesley Publishing Co., Redwood City, CA, 1985.
7. Glassman, I., *Combustion,* Third Edition, Academic Press, San Diego, 1996.
8. Strahle W. C., *An Introduction to Combustion,* First Edition, Gordon and Breach Publishers, 1993.
9. Hawthorne, W. R., Weddell, D. S., and Hottel, H. C., Mixing and combustion in turbulent gas jets, *Proceedings of Combustion Flame and Explosions Phenomena,* 3, pp. 266–288, 1949.
10. Godsave, G. A. E., Burning of fuel droplets, *Proceedings of the Fourth Symposium on Combustion,* p. 818, William and Wilkins, Baltimore, 1953.

2

Elements of Measurement Systems

If you want to see the true measure of a man, watch how he behaves when he assumes power.

D. P. Mishra

2.1 Introduction

The measurement of physical quantities plays a vital role in all branches of science and technology, and combustion is no exception. Lord Kelvin stated in 1891, "when you measure what you are speaking about and express them in numbers, you know something about them and when you cannot measure them or when you cannot express in numbers, your knowledge is of meager and unsatisfactory kind. It may be the beginning of knowledge, but you have scarcely in your thought advanced to the stage of science." Hence, it is important to learn and master the measurement techniques for the advancement of combustion science and engineering. In this chapter, we will learn about the basic concepts of a generic measurement system that can be applied as well to the measurements in a combustion system.

2.2 What Is Measurement?

Measurement is a method of determining a physical quantity of a system or device during a process. The word *measurement* stems from the Latin word *mēnsūra* and the verb *metiri*, its subsequent transformation through the French language. It is basically a process of obtaining experimentally the magnitude of a quantity that can be attributed to a physical property. This means that measurement presumes a description of the quantity, which can be achieved by adopting a comparison of one quantity with a certain standard state under the same measurement condition. For example, the temperature of a flame can be measured using a fine-wire thermocouple that produces voltage due to the temperature difference between two junctions and assigning a certain value to it through calibration.

The measurement itself may change the phenomenon or substance such that the quantity being measured may be different from the actual measurand. Before proceeding further, let us understand the term *measurand*. It is basically the quantity that is intended to be measured by the instrument. Of course, a measurand can be specified with the knowledge of the kind of quantity, the state of the phenomenon, constituent of substance, and so forth. It should be noted that the actual difference between the quantity being measured and the actual measurand may be caused by the measuring system, the condition under which the measurement is being carried out, or a combination of both. Hence, care must be taken so that the quantity does not deviate greatly from the actual measurand. However, we will need to understand the basic phenomenon of a measurand. For measuring a specific measurand, the basic principle that governs the phenomenon and causes a change in sensing this phenomenon must be used as the basis of the measurement. For example, a change in electromotive force (EMF) across dissimilar metals in a thermocouple caused by the Peltier, Seebeck, and Thomson effects is used as the basic principle for the measurement of temperature. We will further discuss measurement principles in this chapter as we go along and also in subsequent chapters.

For the successful measurement of any quantity, we need to adopt a detailed measurement procedure based on one or more measurement principles and a given measurement method. Of course, it must be based on the measurement model that includes the appropriate calculation procedure to determine the measurement result. This procedure must be short but sufficient to enable the experimenter to perform a measurement. It is important to include a statement about the level of uncertainty in measurement that is expected so that the experimenter can develop a judgment about his or her measured value.

2.3 Fundamentals of Measurement Methods

The process of measurement involves the detection, acquisition, and control analysis of data of the relevant properties. Any measurand to be detected and measured must have a magnitude that can be expressed in terms of a certain unit. For example, if we want to measure the temperature of a flame, we can measure it using the appropriate temperature measuring instrument and express its magnitude in certain units of measurement (flame temperature, $T_f = 2000$ K). For arriving at the magnitude of the measured temperature and its unit, we need to use certain measurement methods that can be demonstrated by standard instruments and procedures so that the scientific facts can be replicated without any ambiguity and personal bias. Hence, it is important to follow the measurement method that will describe the logical organization of operations used in the measurement. Measurement methods developed over years can be fundamentally divided into (i) direct comparison and (ii) indirect comparison methods [1,2].

i. *Direct comparison method.* In this method, measurement of a physical quantity can be carried out by direct comparison with a specific standard quantity accepted by a group of people or an agency. For example, we can measure the length of a cloth by comparing the cloth with a steel bar. For this, we need to ascertain its accuracy by visual judgment. This steel bar is considered as the standard indicating the length in meters/centimeters. However, this standard must be repeated and accepted by the community. Hence, this measurement of cloth or any other item can be carried out using the direct comparison method. This direct method is used frequently in our day-to-day life for measuring various quantities such as length, mass, volume, and so forth. Note that for the measurement of any measurand by this direct comparison, one has to depend on the sensory organs of human beings, and so this method may not be adequate to measure all the quantities in science and technology, particularly when greater accuracy is essential. For example, if we need to measure the length of an item with an accuracy of 0.1 mm, the standard meter scale used for measurement of length would not be adequate. Therefore, other measurement methods have been devised that are discussed below.

ii. *Indirect comparison method.* There are several occasions in which requisite measured quantities cannot be detected by our sensing organs and thus cannot be measured by the direct comparison method as discussed above. For example, if we need to measure pressure in a combustion chamber, we cannot measure using the direct comparison method in which the measurand is sensed indirectly and converted into a certain signal that can be transferred to a display or storage unit. The readings obtained from the sensor must be calibrated using the relations between the input and the output of the instrument. Hence, this conversion is essential for making this measured value understandable to the users. Note that most modern instruments are based on this indirect comparison method. The details of modern measuring systems will be discussed in the next section.

2.4 Elements of Generic Measuring System

Before getting into specific measuring systems, let us look at a generic measuring system, which is meant to measure a physical variable during a process. It assigns a certain specific value to the physical variable commonly known as the measured variable. In other words, the measuring system is a tool that is used to quantify a physical variable during a process either by direct comparison or calibration method. In order to quantify a physical variable, the measuring system must have the capability of sensing the change of the physical variable during a process. It must also generate certain data that can be deciphered by humans, mostly through sensory perceptions. A generic measuring system must have four functional units: (i) sensor–transducer, (ii) signal–conditioning, (iii) output, and (iv) feedback control, as shown in Figure 2.1. This measuring system may not contain all four functional units, but it must have a sensor–transducer and an output unit. Note that these units in the measuring system act as facilitating links between the input signal and output to be deciphered by the human observer. In order to establish a relationship between input and output signals, one has to undertake calibration of the measuring system, as discussed briefly in this chapter (see Section 2.7). In the following we discuss further each of the four functional units of a generic measuring system:

i. *Sensor–transducer unit:* The main function of this unit is to detect the change in measurand (measured quantity/property/condition) during certain natural phenomena or processes. That means this unit must be sensitive to a certain physical quantity only. In other words, it must be insensitive to any change in all other physical quantities except the measured variable. For example, an instrument meant for pressure measurement must only be sensitive to pressure change and insensitive to temperature change during a process. However, it is not possible in a practical situation to have such an ideal sensor that is completely selective to one measuring variable. Hence, we need to choose a sensor that is more sensitive to a specific physical quantity compared to any other quantity. The unwanted sensitivity to other variables can cause an error in measurement, commonly known as *noise*, which must be minimized

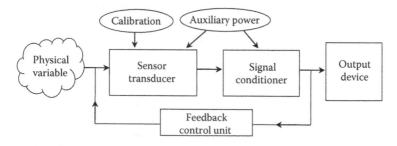

Figure 2.1

Block diagram of a generic measuring system.

2. Elements of Measurement Systems

during the design of the instrument. Therefore, it is important to select the proper sensor for a particular measuring system. The placement and installation of the sensor also plays an important role in the performance of the measuring system. We need to convert the information gathered by the sensor into a detectable signal that can be transmitted easily to the output device. For this purpose, we employ a transducer, which is a device capable of converting the information detected by a sensor into a signal form. Generally, a transducer converts a mechanical/optical measurand into electrical quantities so that any requisite operation on the sensed data can be understood by humans.

ii. *Signal–conditioning unit:* The main function of this unit is to transform the transducer signal into the desired form that can be deciphered by the sense organs of humans. Of course, this is an optional unit for a measuring system undertaking certain basic operations such as amplification of signal, selective filtering to eliminate noise, integration, differentiation, providing optical/mechanical/electrical linkage between the transducer and output devices, and so forth. Generally, the most common function of this unit is to amplify a signal by increasing its amplitude, power level, or both to the requisite level that can be displayed in the output device. For example, in case of a simple liquid-in-glass thermometer, a small-diameter capillary tube is used to amplify the change in volume of liquid due to a change in temperature so that the user can read it easily. The signal conditioning can be accomplished by a simple capillary tube or simple resistance network to complex multistage amplifiers with electrical/optical filters, demodulators, and so forth. The output signal from the signal conditioning can be either analog or digital.

iii. *Output device:* This unit is provided to help in deciphering the acquired data in the signal form by the senses of humans. As mentioned above, the signal can be either analog or digital. If the measured data are in the analog form, they can be displayed on a dial or panel meter with a pointer on a calibrated scale. For example, a pointer on a calibrated scale indicates the measured pressure in a pressure gauge. Note that measured analog output displayed visually has to be recorded by humans. The accuracy of these data will be influenced not only by the errors involved in the instrument but also by the person who records the data. In order to avoid human error, several devices, such as a self-balancing type potentiometric strip chart recorder, graphic pen type galvanometer recorder, and oscillograph, have been devised for recording analog data. However, in recent times, by using an analog-to-digital converter, analog signals are mostly converted into digital form that can be easily stored and transmitted for further analysis. The digital form of data can be displayed using a digital panel made of either a light-emitting diode (LED) or liquid crystal display (LCD). Digital data can also be printed out using a printer or chart recorder or stored in a computer, where it will be readily available for further analysis. Digital devices are preferred for data storage and display over the analog ones due to their higher accuracy, higher speed, and lower operating errors.

iv. *Feedback control unit:* This unit was not common in most of instruments, particularly in earlier days. In recent times, feedback control units are being included in specialized instruments where control of the process variable is essential, particularly in an industrial environment. A typical feedback control unit contains a controller to interpret the measured signal and make an appropriate decision for controlling the process so the desired output signal can be generated. In order to make a proper decision, several control strategies and optimization techniques have evolved and been implemented. A simple way of controlling a process parameter is to set a certain limit for its magnitude. If the measured process variable goes outside of the set limit, the controller will take an appropriate action to bring the process back to the desired operating condition. For example, in the case of a refrigerator, a simple thermostat is used as a controller. If the measured local temperature falls below a preset limit, the compressor will be switched off. When the local temperature exceeds a certain limit, the compressor will be switched on again. Several complex controller systems based on artificial intelligence, neutral networks, and so forth exist in practice but their discussion is beyond the scope of this book.

2.5 Types of Transducers

We have already explained that a transducer is a device that can be actuated by an energizing input (measurand), by another external energy source, or both, so that the sensed signal can be transmitted and displayed as the measured quantity for proper interpretation. Various forms of signals (mechanical, optical, electrical, chemical, etc.) can be detected by sensors [3,4]. The sensed signal has to be transduced into another convenient form that can be deciphered by humans. Hence, transducers can be classified into various types depending on the types of signal they can handle. Today, electrical transducers are preferred because other forms of signals such as mechanical, optical, and chemical can be easily converted into electrical form. An electrical signal is also easier to amplify, transmit, filter, and record, without incurring significant error during signal conditioning and its transmission to the output device. Electrical transducers can be divided into three categories: passive, active, and digital, as indicated in Table 2.1. In the case of an active transducer, there is no need to give any external form of energy, as it is a self-operating device based on the principle of energy conversion. Active transducers generate an equivalent electrical signal from another form of input. For example, a thermocouple is an active transducer in which electrical current flows due to the temperature gradient between its cold and hot junctions. In the case of a pressure transducer, a pressure signal can be converted into an electric charge, leading to an electrical signal. Various kinds of active transducers are listed in Table 2.1. On the other hand, in the case of a passive transducer, an external form

Table 2.1 Classifications of Transducers

Passive	Active	Digital
Resistive (strain gauge, thermometer, potentiometer)	Electromagnetive (tachometer)	Event counter (nucleonic)
	Photovoltaic (photodiode)	Frequency output
Inductive (LVDT)	Piezoelectric (pressure sensor)	(vibrating tube)
Capacitive (microphone)	Thermoelectric (thermocouple)	Encoder
Piezoresistive (pressure sensor)	Galvanic (galvanometer)	
Thermoresistive (thermometer)	Pyroelectric	
Elastoresistive (pressure sensor)		

Note: LVDT = linear variable differential transformer.

of energy is required to actuate the transducer. In other words, a passive transducer works on the energy-controlling principle. A passive transducer can operate only when excited by a secondary electrical energy from an external source that can effect changes in the signal of the measured variable. Capacitance is one example of a passive transducer for which an external form of energy needs to be supplied. Several kinds of possible transducers are provided in Table 2.1, some of which will be invoked during the discussion about various measurement systems in this book. Table 2.1 also lists digital transducers used in instruments.

2.6 Basic Requirements of a Transducer

A transducer is meant to sense a specific measuring quantity, and its functioning should not be affected by other variables. The information about the change of a variable during a process/event must be converted into a suitable form that can be transmitted, filtered, conditioned, and displayed in a suitable form. In order to choose a particular transducer for an application, we need to have sound knowledge about its characteristics. This will be helpful in the design of an experimental setup for studying combustion. Some of the fundamental requirements of a generic transducer are [1–4]:

    i. A linear relationship between input and output is desirable.
    ii. It must reproduce the output signal exactly under the same measuring conditions. In other words, it must have a higher repeatability.
    iii. Higher signal-to-noise ratio.
    iv. It must have good dynamic response.
    v. It must have higher stability and reliability.
    vi. It must have capability to withstand overloads.
    vii. It must have excellent performance in static, quasi-static, and dynamic states.

2.7 Performance Characteristics of a Measuring Device

We have already learned that the function of a measuring device is to sense any change in physical quantity such as pressure, temperature, and velocity, and quantify this change in the measuring parameter by assigning a certain value to it. In other words, it can quantify the change in a physical quantity during a process and express it in a certain standard manner so that it can be used for developing a better understanding of the process. Any change in input to a measuring instrument due to change in a physical quantity during a process must have a definite relationship with its output. The relationship between input and output of a measuring instrument that can be established using a reference standard is known as calibration. During calibration, a known value of a physical quantity is applied to an instrument that can have a specific output. Hence, it can be considered as an act of applying a known input to an instrument for creating a relationship with a desired output as per a certain standard. Note that the measured quantity can remain constant with time or vary rapidly with time, so calibration of an instrument can be divided into two categories: (i) static calibration and (ii) dynamic calibration. In the case of static calibration, the input to the measuring instrument remains either almost constant or varies very slowly with time. The procedure of establishing a functional relationship between static input quantity (q_i) and output quantity (q_o) is known as static calibration. This relationship between input and output can be expressed analytically, graphically, or in tabular form. It is desirable to have an analytical relationship so that it can be used easily. Generally, during calibration, input to an instrument must be an independent variable while the output variable is a dependent variable. The variation of output with input forms a typical calibration curve as shown in Figure 2.2, which can be curve-fitted to obtain a relationship between the input and output of a measuring instrument. The calibration curve is the basis by which

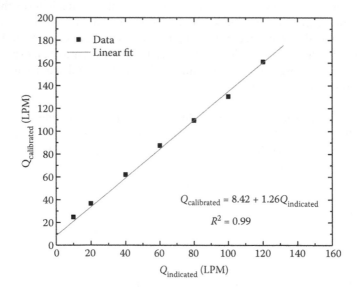

Figure 2.2

A typical static calibration curve.

2. Elements of Measurement Systems

an output display scale is devised in a measuring instrument. Static calibration is the widely used procedure for developing a relationship between the input and output of a measuring instrument. The input applied to a measuring instrument need not be static; rather, it varies with time. The input can be time-dependent in both frequency and amplitude content. If the input to a measuring instrument varies with time, then the output is expected to be dynamic in nature. Therefore, it is important to carry out a dynamic calibration of an instrument. The actual variation of input with time can be quite complex, but for calibration purposes, a simple form such as sinusoidal, step, and ramp signal input is used routinely for dynamic calibration of an instrument. Note that calibration of an instrument should be carried out routinely while comparing with a known standard to ensure validity in the performance of the measuring instrument. Based on the type of input variable to a measuring instrument, the general performance of an instrument can be divided into two categories: (i) static performance characteristics and (ii) dynamic performance characteristics, as discussed in the next section.

2.7.1 Static Performance Characteristics

i. Linearity

In a measuring instrument, it is desirable to have a linear relationship between output and measurand (input) as it can ensure accurate data reduction. A typical linear relationship between the static input (measurand) quantity (q_i) and output quantity (q_o) can be expressed as

$$q_o = Kq_i + C \qquad (2.1)$$

where K is the slope and C is the intercept. If the slope K remains constant, the instrument is assumed to have a linear relation between the measurand and output. The calibration data can be fitted to a curve. It is desirable to have a linear relationship between the measurand and output that can mimic calibration data as closely as possible. However, this condition is seldom met strictly in an actual situation. In an actual instrument, the maximum deviation of data from the linearity is specified in terms of the full scale of the instrument. For example, if linearity is specified as ±2% of the full scale, then actual data can vary from the ideal value within ±2% of the full scale. Two forms can specify the linearity of a measuring instrument: (a) independent and (b) proportional, as shown in Figure 2.3. Independent linearity means that the output will remain within two parallel lines set by percentage of linearity. For example, a 2% independent linearity as shown in Figure 2.3a indicates that the output data of an instrument will lie within the values set by parallel lines corresponding to ±2% of the full scale. In contrast, proportional linearity of the output data will vary depending on input magnitude but it will remain within a specified value of the full scale. For example, 2% proportional linearity as shown in Figure 2.3b indicates that variation of output data from the ideal increases with the magnitude of input but remains within ±2% of the full scale. Note that an instrument that possesses a higher level of linearity need not be an accurate one.

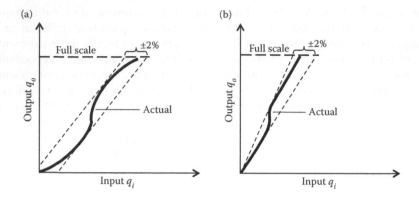

Figure 2.3

Two forms of linearity: (a) independent and (b) proportional.

ii. Static Sensitivity

The slope K of the relationship between static input (measurand) quantity (q_i) and output quantity (q_o), commonly known as the calibration curve, is termed as the static sensitivity, K, and expressed as

$$K = \left(\frac{dq_o}{dq_i} \right)_{q_{i,1}}$$

(2.2)

This static sensitivity is also known as *static gain*. It is a measure of change in output per unit change in measurand. If the calibration curve is linear then the static sensitivity (see Figure 2.4a) will be a constant for all ranges of input for an instrument. If the relationship between input and output is nonlinear, as shown in Figure 2.4b, the static sensitivity varies along the range and must be evaluated at a specific value of the input quantity.

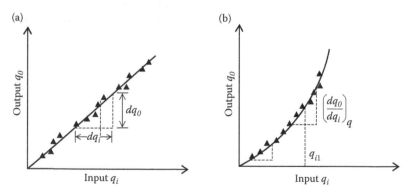

Figure 2.4

Static sensitivity: (a) linear and (b) nonlinear.

2. Elements of Measurement Systems

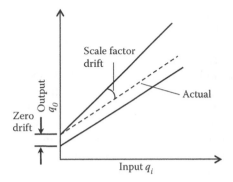

Figure 2.5

Illustration of zero drift and scale factor drift.

An instrument with higher sensitivity is desirable as it has a higher resolution with more accurate data. However, in some situations, higher sensitivity may cause more problems for the output of the instrument as it can interfere and modify the input in a negative way. For example, temperature is an interfering input in a pressure measurement that can cause a change in resistance and thus would drift the value of the output value even if change in pressure were zero. This is called *zero drift* and is shown in Figure 2.5. As well, during the measuring of a certain input, another interfering quantity called *scale factor drift* will hopefully affect a linear change in output, as shown in Figure 2.5. All these effects must be considered while calibrating an instrument.

iii. Repeatability

Generally, during calibration of an instrument and actual measurement, measurements are carried out several times for identical input. The output data must be scattered, as shown in Figure 2.6. The deviation of output data from its mean

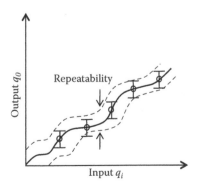

Figure 2.6

Illustration of a repeatability error.

2.7 Performance Characteristics of a Measuring Device

value in absolute or relative units is called repeatability. This is also known as precision error and considered a measure of the random variation in output during repeated measurement. In other words, it indicates the closeness of output data from its mean/ideal value when measurements are undertaken repeatedly for same input value under the same operating conditions. That means the same operator must be using the same instrument under the same ambient conditions for the same measuring input so that no other forms of errors except random errors are incurred during measurement. This process takes care of random errors of an instrument in the statistical sense. It can be expressed in terms of standard deviation σ, which indicates the measure of variation in the output for a given input. Generally, a repeated error can be expressed with respect to a full-scale value, as

$$\%e_R = \frac{2\sigma_{max}}{R_o} \times 100 \tag{2.3}$$

where R_o is the full-scale operating range and σ_{max} is the standard deviation. Note that repeated error represents the errors incurred only under controlled conditions during the calibration process. It does not represent the additional errors that might be incurred during actual measurement. It is important to mention the repeatability of any measured quantity to build the confidence of users. Sometimes it is also known as confidence level, which must be determined carefully for generation of authenticate data.

iv. Hysteresis

In order to ascertain the repeatability of data, sequential tests are carried out by increasing the input values followed by decreasing input values. It is often observed that input–output graphs would not coincide with each other for both sequential ascending and descending variation of the value of the measurand. A typical variation of an input–output graph is shown in Figure 2.7, which indicates

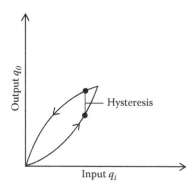

Figure 2.7

Illustration of the hysteresis effect.

noncoincidence of this relationship during sequential ascending and descending variations of input. The differences in output values between the upscale and downscale change of input during sequential testing is called the hysteresis error. For a particular value of a measurand, the difference between the upscale and downscale output value can be expressed as the hysteresis error:

$$e_H = q_{o,U} - q_{o,D} \tag{2.4}$$

Generally, the hysteresis error of an instrument is expressed in terms of maximum hysteresis error during calibration corresponding to the full-scale value, as given below:

$$\%e_H = \frac{e_{H,\max}}{R_o} \times 100 \tag{2.5}$$

The typical value of the hysteresis error for a pressure sensor is provided in Table 2.2. It must be kept in mind that hysteresis can occur only when the output of an instrument is dependent on the previous value indicated by the instrument.

The hysteresis error is attributed to internal friction, external friction, free-play/looseness, backlash of gears, elasticity of materials, linkage pivots, magnetic effects, thermal effects, and so forth. This kind of error can be minimized by proper design, selection of proper materials, proper fitting methods, and so forth.

v. Resolution

Resolution is the ability of a measuring instrument to sense a minimum change in the input value. In other words, it can be defined as the smallest measurable input change that produces definite perceptible change in output. On several occasions, the least count is being used in place of resolution, which is the smallest difference between two input values that can be observed on the output device. For

Table 2.2 Specification of a Typical Piezoresistive Pressure Sensor

Operation	Range
Input range	0–1000 mm of H_2O
Excitation	±10 V dc
Output range	0–5 V
Performance Parameters	
Linearity error	±1% FSO
Hysteresis error	±0.2% FSO
Sensitivity error	±0.2% reading
Thermal sensitivity error	±0.02/°C% reading
Thermal zero drift	±0.02/°C% FSO
Thermal range	0°C–50°C

Note: FSO = full-scale output.

example, 1 mm is the least count/resolution in a simple length-measuring scale. Generally, resolution can be expressed in terms of full-scale output or absolute unit. It must be noted that resolution indicates the smallest measurable change while hysteresis defines the smallest measurable input. It must not be construed that an instrument with a large hysteresis will have poor resolution.

vi. Range

Generally, an instrument is designed to measure a certain range of a measurand that can have a linear relationship with the output value. In other words, the range of an instrument indicates the linear operating span of the total scale. During calibration, input ranging from minimum to maximum value is applied that can have a linear relationship with the output value that can be detected and expressed easily. This limit is defined as the operating range of an instrument. The input range, R_i, can be expressed mathematically as

$$R_i = q_{i,\max} - q_{i,\min} \tag{2.6}$$

In a similar manner, output range is often known as the full-scale output (FSO) range, and R_o can be expressed as

$$R_o = q_{o,\max} - q_{o,\min} \tag{2.7}$$

Generally, it is not advisable to use any measuring instrument by extrapolation beyond its operating range as it will be not only erroneous but may damage the instrument itself. It is prudent to calibrate over the entire range of operating conditions, which will be helpful for carrying out error analysis.

vii. Accuracy

The accuracy of a measurement can be defined as the deviation of reading from the true value or standard value. During calibration if the input value is known exactly, it is known as true value. Generally, accuracy of an instrument is expressed as a percentage of the full-scale operating range. The accuracy indicates the ability of an instrument to provide a true value and is related to absolute error incurred by an instrument. In other words, accuracy is dependent on the random error, bias error, nonlinearity, reproducibility, zero drift, environmental effects, and so forth. The absolute error, e_A, can be defined as the difference between true value and indicated value, as expressed as

$$e_A = q_{o,T} - q_{o,\mathrm{m\,eas}} \tag{2.8}$$

The accuracy of an instrument can be expressed as

$$A = \left(1 - \frac{e_A}{q_{o,T}}\right)100 \tag{2.9}$$

2. Elements of Measurement Systems

where $q_{o,T}$ is the true value of measurement. Unfortunately, the true value is not known in most cases. Hence, the mean value of measured data is used to determine the accuracy of an instrument. It indicates the closeness of measurement to its true value.

viii. Precision and Bias Error

The precision of an instrument indicates the deviation of output data from its mean value when repeated. In other words, it indicates the closeness of output data from its mean value when measurements are undertaken repeatedly for same input value under same operating conditions. It is a measure of the random variation in output during repeated measurement. For example, let us consider the measurement of a known pressure of 100 kPa. If a Bourdon tube pressure gauge is used for this known pressure of 100 kPa, four readings are 98, 100, 102, and 104 kPa. The precision of this instrument would be ±3% as the maximum deviation from mean value (101 kPa) would be 3 kPa. However, the accuracy of this instrument would be 4%. Note that an instrument that can repeatedly indicate the same wrong values can be considered precise but it may not be accurate. The accuracy of an instrument can be improved by calibration but not beyond its precision level.

When the measured mean value differs from its true value, this deviation is called bias error. The difference between the true and mean values of repeated

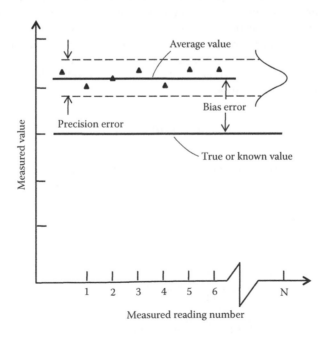

Figure 2.8

Illustration of precision and bias errors using calibrated data.

measurements is considered the bias error, as shown in Figure 2.8. Generally, bias error can be expressed with respect to true value:

$$\%e_B = \frac{q_{o,T} - q_{o,mean}}{q_{o,T}} \times 100 \tag{2.10}$$

In the above example pressure measurement using a Bourdon tube gauge, the bias error would be 1% with respect to the true value. Note that accuracy of an instrument can be influenced by both precision and bias errors.

2.7.2 Dynamic Performance Characteristics

In the above section, we discussed the static performance characteristics of a measuring instrument, which is valid if the measurand would not vary at all or vary slowly with time during the measurement process. On several occasions, sufficient time can be provided for the measuring system to attain a steady state so that it would not vary with time in order to carry out the static performance characteristics of an instrument. However, in many situations, the measurand will vary with respect to time, for which the dynamic performance of an instrument must be considered. Generally the measurand will be a function of time, which can be of various types. Four types of elementary inputs: (a) step, (b) ramp, (c) sinusoidal, and (d) impulse, as shown in Figure 2.9, are commonly used as an input for calibration purposes as they can be generated easily by a function generator. However, the actual dynamic input may differ from these elementary inputs. When a dynamic input is applied to an instrument, the difference between static output and dynamic output is considered at any instant of time. This difference is often known as absolute dynamic error, e_D, which can be expressed mathematically as

$$e_D(t) = q_{o,s}(t) - q_o(t) \tag{2.11}$$

where $q_{o,s}(t)$ is the static output and $q_o(t)$ is the dynamic output for a dynamic input $q_i(t)$. Note that the dynamic error will be dependent on the input, instrument response characteristics, amplification system, and display units. The dynamical behavior of a measuring instrument can be expressed mathematically as

$$a_n \frac{d^n q_o}{dt^n} + a_{n-1} \frac{d^{n-1} q_o}{dt^{n-1}} + \ldots\ldots + a_1 \frac{dq_o}{dt} + a_0 q_o$$

$$= b_n \frac{d^n q_i}{dt^n} + b_{n-1} \frac{d^{n-1} q_i}{dt^{n-1}} + \ldots\ldots + b_1 \frac{dq_i}{dt} + b_0 q_i = F_i(t) \tag{2.12}$$

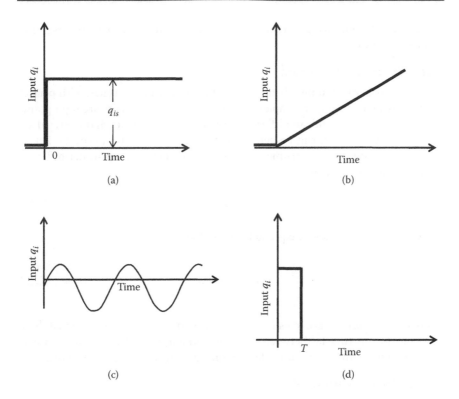

Figure 2.9

Four types of typical input to an instrument: (a) step, (b) ramp, (c) sinusoidal, and (d) impulse.

where a and b throughout are the physical constant, $q_i(t)$ is the dynamic input applied to measuring instrument, $q_o(t)$ is the dynamic output of measuring instrument, and $F_i(t)$ is the applied input to the instrument. Either D operator or the Laplace transformation method can obtain the solution to this kind of equation. The solution to the above differential equation can be obtained as

$$q_o(t) = q_{o,CF}(t) + q_{o,PI}(t) \tag{2.13}$$

where $q_{o,CF}$ is the complimentary function and $q_{o,PI}$ is the particular integral part of the solution. The complimentary part of the function corresponds to the natural characteristics of an instrument because it is the solution obtained when no input is applied. On the other hand, the particular integral part is the solution corresponding to the applied input to the instrument.

We have introduced the nth order differential equation of a system in the above paragraph. Note that order of an instrument can be designated by the order of the differential equation that describes the dynamic performance of an instrument. We will discuss zeroth-, first-, and second-order instruments in the following sections [1–6].

i. Zeroth-Order Instrument

The zeroth-order instrument is the simplest form of an instrument with perfect dynamic response. Some examples of zeroth-order instruments are light pointers, potentiometers, and amplifiers. The instrument can be characterized by static sensitivity, which can be determined by static calibration process. Let us consider a zeroth-order instrument that is governed by a zeroth-order differential equation as given below:

$$a_0 q_o = b_0 q_i = F_i(t) \qquad (2.14)$$

By simplifying the above equation, we can have

$$q_o = \frac{b_0}{a_0} q_i = K q_i \qquad (2.15)$$

where $K = b_0/a_0$ is the static sensitivity of a measuring instrument. Note that there would not be any time lag between the input and output no matter how the input varies with time. In other words, the dynamic response will be an ideal one.

ii. First-Order Instrument

The first-order instrument depicted by Equation 2.16 can expressed as

$$a_1 \frac{dq_o}{dt} + a_0 q_o = b_0 q_i \qquad (2.16)$$

The above equation can be rewritten as

$$\frac{a_1}{a_0} \frac{dq_o}{dt} + q_o = \frac{b_0}{a_0} q_i; \quad \Rightarrow \tau \frac{dq_o}{dt} + q_o = K q_i; \quad \Rightarrow (\tau D + 1) q_o = K q_i \qquad (2.17)$$

where $\tau = \dfrac{a_1}{a_0}$ is known as the time constant of an instrument as it has a dimension of time irrespective of the physical dimensions of a_1 and a_0, and K is the static sensitivity. The operational transfer function of the above equation can be written as

$$\frac{q_o}{q_i} = \frac{K}{(\tau D + 1)} \qquad (2.18)$$

The time constant is an important characteristic for measurement of a dynamic input signal as it indicates a measure of the speed of the instrument response. In order to appreciate the efficacy of this concept, let us learn how a first-order system responds to two kinds of input signals, namely (i) step function and (ii) periodic function.

i. *Step function input:* Generally, any sudden change of input such as pressure and temperature can be modeled mathematically as a step function as shown in Figure 2.9a. Input signal varies suddenly from a certain constant value to another constant value instantly. We are interested to know how the first-order instrument responds to this step input. For this purpose, we need to solve first-order equation (Equation 2.17) using an initial condition corresponding to the step function as given below:

$$q_i = 0, \quad \text{for } t \le 0;$$

$$q_i = q_{i,3}, \quad \text{for } t \ge 0; \tag{2.19}$$

The solution of Equation 2.17 for the above initial conditions becomes

$$q_o = Kq_{i,s} - Kq_{i,s}e^{-t/\tau} = Kq_{i,s}(1 - e^{-t/\tau}) \tag{2.20}$$

This solution indicates the time response of the first-order system. In other words, it indicates how it behaves with a step change input. The first part indicates the steady/static response while the second part indicates the transient/dynamic response. Generally, when time tends toward infinity (quite large), the dynamic term becomes almost zero and thus the systems attains a steady state. The variation of normalized output $q_o/Kq_{i,s}$ is plotted in Figure 2.10 with normalized time. Note that it varies from its initial value with time and attains a steady response value at a larger time ($t \to \infty$). The dynamic error in measurement at any time instant is the difference between steady and can be determined as

$$e_D = q_{i,s} - \frac{q_o}{K} = q_{i,s}e^{-t/\tau}; \; \frac{e_D}{q_{i,s}} = e^{-t/\tau} \tag{2.21}$$

The normalized dynamic error is plotted in Figure 2.9 with normalized time and varies from 100% error asymptotically to zero at a higher time ($t \to \infty$). When $t = \tau$, dynamic error becomes 36.8%; which means that for this condition, the system responds to 63.2% of the step input. The system responds to 99.3% of the step input only when $t = 5\tau$. The extent of response of the step input with normalized time is shown in Figure 2.10. In order to get a fast response, the time constant of the instrument must be reduced as much as possible. However in practice, a *rise time* is being used, which is often defined as the time required to respond to 95% of the step input value. This criterion may differ from instrument to

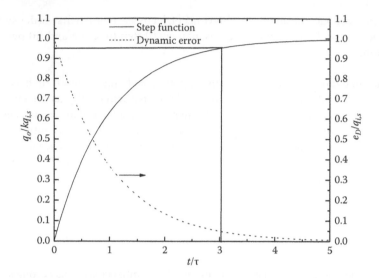

Figure 2.10

Response of step function and dynamic error with normalized time.

instrument. However, this rise time indicates the extent of the response of an instrument to a step input more quantitatively.

Example 2.1

The flame temperature is measured using a Pt-Pt/13% Rh thermocouple with a spherical bead diameter of 0.25 mm. When this thermocouple bead is suddenly introduced into the flame at 300 K, its temperature increases to a flame temperature of 1800 K, which can be modeled as a step function input. Assuming the average heat transfer coefficient to be 650 W/m² K, determine (a) the time constant of the thermocouple, (b) the time required to measure the flame temperature with an error of 5%, and (c) the time constant if the same thermocouple is used to measure oil temperature in which the average heat transfer coefficient is 5650 W/m² K. Take density of thermocouple material = 21,500 kg/m³, specific heat, $C = 135$ J/kg K.

Given: $d = 0.25$ mm, $T_F = 1800$ K, $T_i = 300$ K, $h = 650$ W/m² K, error in temperature = 0.05.

Assumption: The radiation and conduction effect of the conducting wire is neglected.

Solution: By striking an energy balance from the thermocouple bead with regard to the convection mode of heat transfer, we can have

$$mC\frac{dT(t)}{dt} = hA(T_F - T(t)) \tag{a}$$

where m is the mass of the thermocouple bead, C is the specific heat of the bead, and the convective heat transfer coefficient between the bead T_F is the flame temperature of the bead and environment A is the surface area of the bead. We know that the bead temperature at initial and infinity time becomes

$$t = 0; T = T_i \qquad\qquad (b)$$

$$t \rightarrow \infty; T = T_F$$

The energy balance equation can be rewritten as

$$\frac{mC}{hA}\frac{dT(t)}{dt} + T(t) = T_F \qquad\qquad (c)$$

1. The time constant, mC/hA, can be estimated as

$$\tau = \frac{mC}{hA} = \frac{\rho VC}{hA} = \left(\frac{21{,}500 \times \pi \times 0.00025^3}{6}\right) \times \frac{135}{650 \times \pi \times 0.00025^2} = 0.186 \text{ s}$$

2. By considering the initial and final bead temperatures, the solution for Equation a becomes

$$\frac{T(t) - T_i}{T_F - T_i} = (1 - e^{-t/\tau})$$

As the acceptable error in temperature is 5%, then the left-hand side of the above equation is equal to 0.95 as given below:

$$\frac{T(t) - T_i}{T_F - T_i} = 0.95 = (1 - e^{-t/0.186})$$

By solving the above expression, we get $t = 0.557$ s.

3. For an oil environment, the time constant, mC/hA, can be estimated as

$$\tau = \frac{mC}{hA} = \frac{\rho VC}{hA} = \left(\frac{21500 \times \pi \times 0.00025^3}{6}\right) \times \frac{135}{5650 \times \pi \times 0.00025^2} = 0.021 \text{ s}$$

Note that the time constant is small for an oil environment as heat transfer in a liquid phase is much higher as compared to a gas phase.

ii. *Periodic function input:* Periodic signals are encountered in engineering systems such as reciprocating pump flows and externally induced unsteady flames. When a periodic input is applied to a first-order instrument, the response of the instrument output with time is dependent on the frequency of the input. In order to understand the behavior of a first-order system, let us consider a simple sinusoidal periodic input:

$$q_i = A \sin \omega t \qquad (2.22)$$

The first-order equation, Equation 2.22, can be solved using an initial condition corresponding to a sinusoidal function:

$$q_i = 0, \quad \text{for } t < 0;$$

$$q_i = A \sin \omega t, \quad \text{for } t \geq 0; \qquad (2.23)$$

The general solution to the first-order differential equation (Equation 2.17) for the above sinusoidal periodic input is given by

$$q_o = Ce^{-t/\tau} + \frac{KA}{\sqrt{[1+(\omega\tau)^2]}} \sin(\omega t - \tan^{-1}\omega\tau) \qquad (2.24)$$

where C is the constant. Note that the first term in the above equation is the transient term, which decays with time and vanishes at a larger time $(t \to \infty)$. In other words, the transient response is important only during the initial period when input is applied. The second term is the steady response whose effects on output will remain as long as input is applied. Let us recast the above equation as

$$q_o = Ce^{-t/\tau} + B(\omega)\sin(\omega t + \psi(\omega))$$

where

$$B(\omega) = \frac{KA}{\sqrt{[1+(\omega\tau)^2]}}; \; \psi(\omega) = -\tan^{-1}(\omega\tau) \qquad (2.25)$$

Note that $B(\omega)$ represents the amplitude of the steady response that is dependent on the applied frequency ω, time response τ, and the amplitude of input A. The angle $\psi(\omega)$ is known as phase shift, which is dependent on the frequency of input. In order to understand these two terms, let us look at the relationship between the arbitrary sinusoidal input and output signals from a first-order instrument as shown in Figure 2.11. Note that there is a time lag between two signals and the amplitude ratio would not be unity. The form of the output signal is dependent on the value of the

2. Elements of Measurement Systems

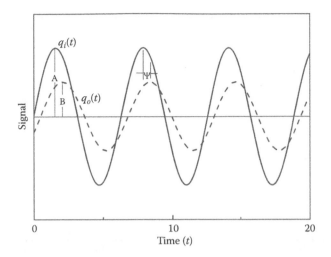

Figure 2.11

Relationship between a typical sinusoidal input and output.

frequency of the input signal as both B and ψ are frequency-dependent as per Equation 2.25. The steady response of the system to the applied input is known as the *frequency response*, which would approach to an ideal value when $\omega\tau$ is quite small.

In order to understand how the output signal is different from the input signal, we can define a term known as the *magnitude ratio, MR*, as

$$MR = \frac{B}{KA} = \frac{1}{\sqrt{[1+(\omega\tau)^2]}} \tag{2.26}$$

The variation of this magnitude ratio for a first-order system along with phase shift, $\omega\tau$, is plotted in Figure 2.12, which indicates the effects of the time constant and signal frequency of input on the output response. It can be seen from Figure 2.12 that the MR remains almost at unity value for a certain lower range of $\omega\tau$, indicating that the measurement system can transfer all input signals to output signals with almost no time lag. But at large values of $\omega\tau$, MR becomes quite a bit smaller than unity and the phase shift angle is quite high, resulting in a larger time delay that must be avoided for any sensible measurement system. Of course, any combination of ω and τ can produce the same result. For measurement of a signal with a high frequency, the system must have smaller time constant τ for achieving a higher MR. On the other hand, the larger time constant can suffice for measurement of a low-frequency signal. The dynamic error eD for a first-order system, which indicates how far the amplitude of an output signal differs from input for a given input frequency, is given by

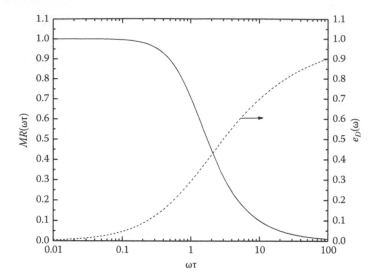

Figure 2.12

Variation of *MR* with ωτ.

$$e_D(\omega) = MR(\omega)^{-1} \tag{2.27}$$

It would not be possible to reproduce the input signal in any first measurement system because dynamic error e_D will have a certain finite value. This dynamic error e_D will be minimum when the first-order measurement system has a magnitude ratio equal to unity or around it with the input signal over an anticipated band of frequency. In an actual instrument, the dynamic error cannot be eliminated at all because perfect input signal cannot be reproduced. Therefore, it is important to use a certain range of magnitude ratio that can have a minimum dynamic error. It is customary to have a frequency bandwidth such that the MR must be greater than or equal to 0.707. Note that both the functions of $MR(\omega)$ and $\psi(\omega)$ characterize the frequency response of the measurement system to the input signal, which can be used as guidance in selecting an instrument for a particular measurement requirement. It should not be used for correcting the output signal as any deviations of a real system from ideal first-order behavior can lead to erratic errors.

Example 2.2

The temperature sensor used in a furnace can sense harmonic disturbances of 0.05 Hz with time constant of 3 s. Determine the time delay in the response of this sensor and the decrease in steady state amplitude response.

2. Elements of Measurement Systems

Given: $\tau = 3$ s, $\omega = .05$ Hz $= 0.31415$ rad/s.

Assumption: Harmonic disturbance.

Solution: The phase angle as per Equation 2.25 can be determined as

$$\psi(\omega) = -\tan^{-1}(\omega\tau) = -\tan^{-1}(0.31415 \times 3) = -48.11° = 0.84 \text{ rad}$$

Then the time delay would be determined as

$$\Delta t = \frac{\psi(\omega)}{\omega} = \frac{-0.84}{0.31415} = -2.67 \text{ s}$$

The decrease in the steady state amplitude response can be determined as

$$MR = \frac{1}{\sqrt{[1+(\omega\tau)^2]}} = \frac{1}{\sqrt{[1+(0.84)^2]}} = 0.7657$$

iii. Second-Order Instrument

Several instruments can have inertia, which affects the response of an input signal. For this kind of instrument, a second derivative term must be included in the model. Thus, this instrument can be modeled by a second-order differential equation that can expressed as

$$a_2 \frac{d^2 q_o}{dt^2} + a_1 \frac{dq_o}{dt} + a_0 q_o = b_0 q_i \tag{2.28}$$

The above equation can be rewritten as

$$\frac{a_2}{a_0} \frac{d^2 q_o}{dt^2} + \frac{a_1}{a_0} \frac{dq_o}{dt} + q_o = \frac{b_0}{a_0} q_i; \quad \Rightarrow \left(\frac{D^2}{\omega_n^2} + \frac{2\zeta D}{\omega_n} + 1 \right) q_o = K q_i \tag{2.29}$$

where $\zeta = \dfrac{a_1}{2\sqrt{a_0 a_2}}$ is known as the dimensionless damping ratio of instrument,

$\omega_n^2 = \dfrac{a_0}{a_2}$ is the undamped natural frequency (rad/s), and K is the static sensitivity.

The operational transfer function of the above equation can be written by using Equation 2.29 as

$$\frac{q_o}{q_i} = \frac{K}{\dfrac{D^2}{\omega_n} + \dfrac{2\zeta D}{\omega_n} + 1} \tag{2.30}$$

Examples of second-order instruments are diaphragm pressure transducers, piezoelectric sensors, force-measuring spring scales, and accelerometers. In order to appreciate the efficacy of this concept, let us learn how a second-order system responds to two kinds of input signals: (a) step function, and (b) periodic function, as given below.

   i. *Step function input:* As discussed earlier, any sudden change of input, such as pressure and temperature, can be modeled mathematically as a step function, as shown in Figure 2.10. We are interested to know how the second-order instrument responds to this step input. For this purpose, we need to solve the second-order equation, Equation 2.30, using the initial condition corresponding to the step function as given below:

$$q_i = 0, \quad \text{for } t < 0;$$

$$\frac{dq_o}{dt} = 0, \quad \text{for } t = 0; \tag{2.31}$$

$$q_i = q_{i,s}, \quad \text{for } t > 0;$$

Three distinct solutions of Equation 2.29 for the above initial conditions can be obtained as

For $0 \leq \zeta < 1$: Underdamped system

$$\frac{q_o}{Kq_{i,s}} = 1 - e^{-\zeta\omega_n t}\left[\frac{\zeta}{\sqrt{1-\zeta^2}}\sin\left(\omega_n t\sqrt{1-\zeta^2}\right) + \cos\left(\omega_n t\sqrt{1-\zeta^2}\right)\right] \tag{2.32}$$

For $\zeta = 1$: Critically damped system

$$\frac{q_o}{Kq_{i,s}} = 1 - e^{-\zeta\omega_n t}(1 + \omega_n t) \tag{2.33}$$

For $\zeta > 1$: Overdamped system

$$\frac{q_o}{Kq_{i,s}} = 1 - \left[\frac{\zeta+\sqrt{\zeta^2-1}}{2\sqrt{\zeta^2-1}}e^{\left(-\zeta+\sqrt{\zeta^2-1}\right)\omega_n t} + \frac{\zeta-\sqrt{\zeta^2-1}}{2\sqrt{\zeta^2-1}}e^{\left(-\zeta-\sqrt{\zeta^2-1}\right)\omega_n t}\right] \tag{2.34}$$

Note that this solution indicates the time response of the second-order system, which is dependent on the damping ratio, ζ. In order to understand the interesting features of the dynamic response of a second-order instrument to step input, the variations of $q_o/Kq_{i,s}$ using Equations 2.32 to 2.34 are plotted in Figure 2.13

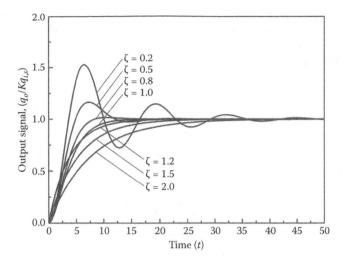

Figure 2.13

Response of step function along with error function.

for certain values of ζ. Note that for underdamped systems, a transient response is oscillatory in nature about its mean value. The oscillatory behavior known as ringing can be characterized in terms of period and ringing/natural frequency as

$$T_d = \frac{2\pi}{\omega_d}; \quad \omega_d = \omega_n\sqrt{1-\zeta^2} \tag{2.35}$$

where T_d is the *time period* and ω_d is the *ringing frequency*. These two parameters are independent of the input signal and are instead dependent on the characteristics of the measuring instrument. Note that for $\zeta = 0$, the system is displaced from its equilibrium by the free oscillation frequency. It can be observed in Figure 2.13 that with increased ζ, the oscillation gets reduced but its dynamic response is slowed down. The duration of dynamic response depends on the $\zeta\omega_n$ term. The system output responds back to the initial input value more quickly if the $\zeta\omega_n$ term is large. For all systems with $\zeta > 0$, the output attains steady state value at a larger time ($t \to \infty$). Hence, in a practical second-order system, we need to use rise time, which is defined as the time during which the output response is equal to 90% of the initial input. Note that this rise time can be reduced by decreasing the damping ratio, ζ. However, if the damping ratio is too small, then time required to reach ±10% of steady input value, known as the settling time, can be increased. This aspect can be observed in Figure 2.13 by comparing the output response for damping and ratio $\zeta = 0.2$ and $\zeta = 1.0$. Generally a damping ratio in the range from $\zeta = 0.6$ to 0.8 is preferred for a practical second-order instrument as it offers a compromise between rising and settling time. This range provides optimum frequency response with reasonable error.

ii. *Periodic function input:* As mentioned earlier, periodic signals are encountered in some engineering systems. Hence, let us determine how the second-order governing equation (Equation 2.29) behaves with a simple sinusoidal periodic input by applying the following initial condition:

$$q_i = 0, \quad \text{for } t < 0;$$

$$q_i = A \sin \omega t, \quad \text{for } t \ge 0; \tag{2.36}$$

The steady state solution to the second-order differential equation (Equation 2.29) for the above sinusoidal periodic input is given by

$$q_o = \frac{KA \sin[\omega t + \psi(\omega)]}{\sqrt{\left[1 - \left(\dfrac{\omega}{\omega_n}\right)^2\right]^2 + \left(\dfrac{2\zeta\omega}{\omega_n}\right)^2}} \tag{2.37}$$

$$\text{where, phase shift} = \psi(\omega) = -\tan^{-1} \frac{2\zeta(\omega/\omega_n)}{1 - \left[\dfrac{\omega}{\omega_n}\right]^2} \tag{2.38}$$

We can express the above equation in general form as

$$q_o = B(\omega)\sin[\omega t + \psi(\omega)] \tag{2.39}$$

where $B(\omega)$ is the amplitude of the above equation. Note that the amplitude of the output signal from a second-order instrument is dependent on the value of the frequency of the input signal. By using Equations 2.37 and 2.39, we can have an expression for magnitude ratio *MR* as given below

$$MR = \frac{B(\omega)}{KA} = \frac{1}{\sqrt{\left[1 - \left(\dfrac{\omega}{\omega_n}\right)^2\right]^2 + \left(\dfrac{2\zeta\omega}{\omega_n}\right)^2}} \tag{2.40}$$

In order to understand the characteristics of a second-order instrument subjected to sinusoidal input signal, the variation of MR is plotted in Figure 2.14 with ω/ω_n for several values of damping ratios. In a similar manner, the variation of phase shift $\psi(\omega)$ with ω/ω_n is plotted in Figure 2.15 for a wide range of damping ratios. It can be observed in Figure 2.14 that for zero damping ($\zeta = 0$), the MR increases with ω/ω_n and approaches infinity when $\omega/\omega_n = 1$, but the phase shift $\psi(\omega)$ jumps from zero to $-\pi$ like a step function. At this point resonance with a very large amount of amplitude occurs for a short duration when the forcing frequency

2. Elements of Measurement Systems

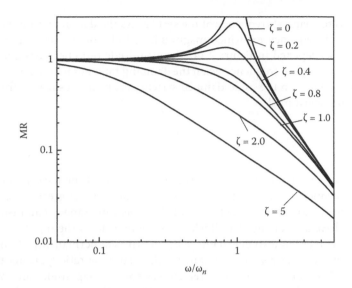

Figure 2.14

Variation of MR with ω/ω_n.

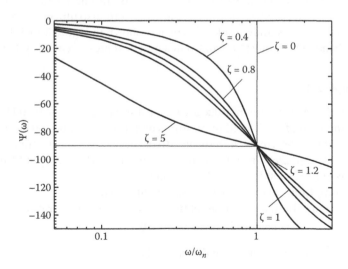

Figure 2.15

Variation of phase shift $\Psi(\omega)$ with ω/ω_n.

is equal to the natural frequency of the system. As the damping ratio increases, the range of frequency for which magnitude ratio becomes almost constant is enhanced, as depicted in Figure 2.14. The magnitude ratio becomes almost flat for a certain range of frequency when the damping ratio lies between 0.6 and 0.7. But the resonance in the underdamped system occurs at a resonance frequency that can be determined as

$$\omega_R = \omega_n \sqrt{1 - 2\zeta^2} \qquad (2.41)$$

Note that resonance frequency is the characteristic of the measurement system. As mentioned earlier, it can occur only when MR is greater than unity accompanied with significant phase shift. The operating range of an instrument must not have a resonance band, which not only distorts the input signal in a complex manner but also may damage the instrument permanently. In order to avoid resonance, the system is designed with a damping ratio ζ greater than or equal to 0.707. For an ideal measurement system the magnitude ratio must be around unity while the phase shift, $\psi(\omega)$, must be equal to zero for a wide range of input frequencies. Rarely will a real system meet this requirement for a restricted range of input frequency. Note that both MR and $\psi(\omega)$ remain around unity and zero value, respectively, at lower values of ω/ω_n. This region of frequency response during which this kind of ideal behavior is possible is known as the transmission band, as shown in Figure 2.15, and can be specified by frequency bandwidth or a certain range of dynamic error. For a second-order system, the transmission band, specified by 3 dB$\geq$MR$(\omega)\geq$–3 dB, is preferred. But magnitude ratio decreases with ω/ω_n accompanied with an increase in phase shift and becomes almost zero at large values of ω/ω_n. As a result, the instrument attenuates the input signal, which is undesirable. This region, shown in Figure 2.15, is known as the filter band, which must be avoided in a practical measurement system.

Example 2.3

A piezoelectric pressure transducer with a natural frequency of 3500 Hz and damping ratio of 0.4 is used to measure combustion chamber pressure in a rocket engine. Determine the resonance frequency of this pressure transducer. What will the phase shift and amplitude response be at a frequency of 1050 Hz?

Given: $\omega_n = 3500$ Hz, $\zeta = 0.4$, $\omega_1 = 1050$ Hz.

Assumption: Second-order measurement system.

Solution: The maximum amplitude point for damping ratio, $\zeta = 0.4$, and $\omega_n = 3500$ Hz can be obtained as $\dfrac{\omega_1}{\omega_n} \approx 0.8$

Then resonance frequency can determined as

$$\omega_1 = 0.8\omega_n = 0.8 \times 3500 = 2800 \text{ Hz}$$

At a frequency of 1200 Hz, we get the frequency ratio as

$$\frac{\omega_1}{\omega_n} = \frac{1050}{3500} = 0.3$$

By using Equation 2.38 we can determine the phase shift as

$$\text{phase shift} = \psi(\omega) = -\tan^{-1}\frac{2\zeta(\omega/\omega_n)}{1-\left(\dfrac{\omega}{\omega_n}\right)^2} = -\tan^{-1}\left[\frac{2\times0.4(0.3)}{1-(0.3)^2}\right] = -16.42°$$

Then by using Equation 2.40 the amplitude ratio, MR, would be determined as

$$MR = \frac{1}{\sqrt{\left[1-\left(\dfrac{\omega}{\omega_n}\right)^2\right]^2 + \left(\dfrac{2\zeta\omega}{\omega_n}\right)^2}} = \frac{1}{\sqrt{[1-(0.3)^2]^2 + (2\times0.4\times0.3)^2}} = 1.063$$

The dynamic error in its amplitude response becomes

$$1.063 - 1 = 0.063 = \pm6.3\%$$

Review Questions

1. What is accuracy? How is it different from precision?

2. What is hysteresis of an instrument? Can the accuracy of an instrument be affected by its hysteresis? Explain in terms of precision and bias error.

3. Why is it required to calibrate an instrument?

4. What is a primary standard? How is it different from a secondary standard?

5. Why are standards used while calibrating an instrument?

6. What is dynamic calibration? When is it required to carry out a dynamic calibration?

7. What is linearity? How is independent linearity different from proportional linearity?

8. What is static sensitivity in an instrument? How is nonlinear sensitivity different from linear sensitivity?

9. What is frequency response? How is it different from phase shift?

10. What is rise time?

11. What is steady state response?

12. What are zeroth- and first-order systems?

13. What are the factors that influence a time constant in a first-order system?

14. What is a second-order system? Illustrate using an example of how it is different from a first-order system.

Problems

1. In piston cylinder engines, cylinder pressure, temperature, and rpm of engines are measured using piezeoelectric pressure sensors, thermocouples, and optical tachometers, respectively. Identify the measurement system stages for these three instruments.

2. A piston cylinder system is used to generate pressure and calibrate it using a pressure transducer. The displacement of the piston changes the chamber pressure for its calibration. The following data is generated during the calibration process:

X(cm)	3.73	4.25	4.95	5.35	5.9
P(kPa)	0.5	0.6	1.0	1.5	2.5

Determine the static sensitivity at each x value and find the input value at which peak sensitivity occurs.

3. The following data is generated during the calibration of flow metering using change in displacement of a water manometer:

h(cm)	10	15	20	25	30
Q(LPM)	21.48	25.68	29.14	32.15	34.83

Plot the following data in appropriate format to fit the data by an expression, $Q = aX^b$. Determine coefficient a and b. Determine the static sensitivity at each X value and comment on it.

4. A piston cylinder system as described in Q2.2 is used to generate pressure and calibrate it using a pressure transducer. The following data is generated during the calibration process in the reverse direction:

X(cm)	5.9	5.35	4.95	4.25	3.73
P(kPa)	2.5	2.2	2.0	1.3	0.5

Plot the data in Q2.2 and the above table using a suitable scale. Determine the maximum hysteresis based on a full-scale range.

5. A thermometer with a time constant of 0.25 s at an initial temperature of 35°C is introduced suddenly to a furnace temperature of 1350 K. Determine time required to attain 95% of steady state value and temperature indicated by the thermometer at this instant of time.

6. A flame temperature is measured using a Pt-Pt/10% Rh thermocouple with a spherical bead diameter of 0.5 mm. When this thermocouple bead at 35°C is introduced suddenly into the flame at a gas temperature of 1500 K, assuming the average heat transfer coefficient to be 550 W/m² K, determine (a) the time constant of the thermocouple, (b) the time required to measure flame temperature with an error of 10%. The density of the thermocouple material = 21,000 kg/m³, and specific heat, $C = 133$ J/kg K.

7. A temperature sensor with a time constant of 11 s used for measurement of temperature in a furnace is subjected to slow harmonic disturbances of 0.02 Hz. Assuming it to be a first-order system, determine the time delay in the response of this sensor and decrease in steady state amplitude response.

8. A pressure sensor modeled as a first-order system has a phase shift of −35° at a certain frequency. If the frequency is decreased by a factor of 0.75, what will be the percentage of decrease in its amplitude?

9. A sensor when subjected to an input signal with 125 Hz is expected to operate within an amplitude response of ±5%. Assuming it to behave as a second-order system, determine the appropriate design parameter that can achieve this objective. Also determine natural frequency.

10. A pressure transducer with a natural frequency of 3500 Hz and damping ratio of 0.4 is used to measure the chamber pressure in a reactor. Determine the resonance frequency of this pressure transducer. What will the phase shift and amplitude response at a frequency of 1100 Hz be?

References

1. Beckwith, T. G., Marangoni, R. D. and Lienhard, J. H., *Mechanical Measurement*, Fifth Edition, Addison Wesley Longman, Inc., 1993.
2. Holman, J. P., *Experimental Methods for Engineers*, Sixth Edition, McGraw-Hill, Inc., New York, 1994.

3. Doebelin, E. O., *Measurement System Application and Design*, Fourth Edition, McGraw-Hill Publishing Co., New York, 1990.
4. Halliday, R. and Resnick, D., *Fundamentals of Physics*, Third Edition, John Wiley & Sons, Inc., New York, 1988.
5. Thomson, W. T., *Theory of Vibration with Applications*, Second Edition, Prentice Hall, Englewood Cliffs, NJ, 1981.
6. Figliola, R. S. and Beasley, D. E., *Theory and Design for Mechanical Measurements*, Third Edition, John Wiley & Sons, Inc., New York, 2000.

3

Data Acquisition

Data is a precious thing and will last longer than the systems themselves.

Time Berners-Lee

3.1 Introduction

In Chapter 2, we discussed instrument systems that can produce a certain output when the measurand is acted upon. In this chapter we will learn how to collect the requisite measurement data in a proper manner and process it to garner appropriate understanding of the phenomena that occur during a process or to control the process. In other words, we will be discussing data acquisition and data handling in terms of how the integrity of data must be maintained properly as it is likely to become flawed due to the adaptation of improper procedures. The sampling of data must be carried out properly to circumvent the loss of signal that can lead to misinterpretation of actual phenomena during a process. The data can be acquired by a person or number of persons who can take readings from a single or number of instruments manually and record on the observation data sheet/notebook. For example, a combustion engineer can look at the needle of a pressure gauge and read its value in order to recode it into a data log book for further analysis. In earlier days, this was the usual procedure of acquiring data. In modern times, experimental data is acquired electronically using analog-to-digital converters that can collect large amounts of data at very high speeds as compared to manual modes. The objective of this chapter is to expose the readers to microprocessor-based data acquisition systems and data processing systems that can be used for measurement of not only combustion systems but other systems as well.

3.2 Data Acquisition System

The main function of a data acquisition is to collect, transmit, and process the acquired data in a desired form such that it can be analyzed easily. Of course the data must be recorded properly and stored in suitable form for subsequent presentation and analysis purposes. As most of the data signals generated by the instrument are analog in nature, they must be converted into digital signals. Analog signals are continuous both in amplitude and time while digital signals are noncontinuous both in amplitude and time. It is quite cumbersome to use an analog signal directly and there are ample chances for the acquired signal to be lost during transmission, so analog-to-digital conversion is used as a part of the data acquisition system.

The main component in a data acquisition system is the sensor, which senses variation in the physical quantity and converts most of all forms of signals (mechanical, chemical, magnetic, etc.) into an electrical form. This electrical signal may be acquired in various forms, such as voltage current resistance and frequency, either in analog or digital form. We have already discussed the various kinds of transducers in Chapter 2. Readers are urged to revisit/recall the basic concepts of the various transducers and their responses before proceeding further. A typical data acquisition system consists of a sensor–transducer, signal conditioner, signal conversion, data processing, data storage, and display units, as shown in Figure 3.1. It can be broadly divided into three parts: (i) input, (ii) signal converter, and (iii) output. We can observe in Figure 3.1 that the input stage is comprised of a transducer for sensing change in physical variables, an input circuit, and a signal conditioner. But for conditioning the signal, amplifiers, filters, and so forth are used routinely. The data transmission unit helps to transfer

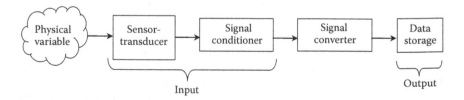

Figure 3.1

Schematic of a general data acquisition system.

the data to display unit without any loss or contamination. Note that data can be transmitted over a short distance or a long distance. As an analog signal is quite difficult to transmit over a longer distance, it is converted into digital form using an analog-to-digital (A/D) card. The output stage is meant to display and store the data in an appropriate form that can be used for further analysis and investigation. The data can be displayed with the help of a digital panel meter (DPM) or oscilloscope or chart recorder, printer, or computer. The microprocessor-based data acquisition system is popular in modern times as it can collect, store, display, and analyze input data rapidly and accurately with the least cost.

In most practical systems, the number of experimental variables will be more than one and hence a data acquisition system discussed for one variable cannot be used directly. If we want to acquire several pieces of data simultaneously we will have to employ several replicas of a single-channel data acquisition system. However, the cost of doing this would be quite exorbitant if the number of channels is more than a certain critical value, and it would be cumbersome to handle and maintain so many individual systems. Therefore, it is better to have a single data acquisition system that can acquire and analyze multiple channels. For this purpose, a scanner/programmer is used whose main function is to sample a data channel rapidly in sequence so that one conversion and output device can be used at any instance in time. The schematic of a typical multichannel data acquisition system is shown in Figure 3.2, which contains a data scanning system (data scanner/multiplexer) in addition to three main components: (i) input, (ii) signal converter, and (iii) output, as in a single-channel data acquisition system. The system can be programmed to acquire data or the desired range of data in a particular order. Note that the number of transducers employed depends on the number of variables to be considered for any sets of measurement. On certain occasions, data may be acquired at certain regular intervals of time for which a digital timer is included in the multichannel data acquisition system. This feature can be used for controlling some systems through the data acquisition system. As well, an

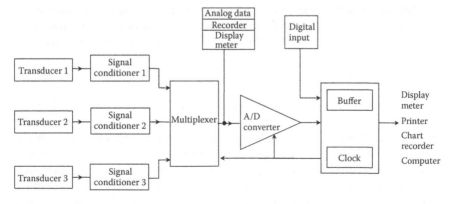

Figure 3.2

Schematic of a multichannel data acquisition system.

intermediate signal condition system is provided in some data acquisition systems for several functions, including voltage-frequency conversion, analog-to-digital conversion, data filtering, amplification, and harmonic analysis. This kind of multichannel data acquisition system is quite flexible and cost effective. Some of the data acquisition systems in earlier days had built-in microprocessors for data analysis and storage, but in recent times, these can be integrated easily with personal computers (PCs). The acquired data can be stored in a PC and can be fed easily to a printer/plotter for data display and analysis. Several standard software packages are also available that can be used to acquire and manipulate measured data for further analysis.

In order to select a proper data acquisition system, the following important factors should be considered judiciously: (i) resolution and accuracy, (ii) number of channels, (iii) sampling rate per channel, (iv) signal conditioning requirement of each channel, and (v) cost. Note that resolution of any measurement variable is dictated by the overall desired accuracy of the system. The resolution is chosen around four to five times of the desired accuracy. The resolution of any measurement is also dependent on the rate at which a sample is acquired. The resolution is decreased with a higher measurement rate and lower sampling frequency. Hence, a specific resolution desired in any experiment has to be decided with the proper understanding of the requirements and constraints of the data acquisition system. The number of channels to be employed for measurement is dependent on the overall number of bits within a data acquisition system. Apart from this, sampling rate is also dependent on the overall number of bits within a data acquisition system and the desired resolution. For example, if the number of channels, N, are to be used for acquiring data at K sampling per second with resolution of S, then the bit rate of the data acquisition system will be approximately equal to 3 NKS bits/s. Increasing either of these variables can enhance the overall conversion rate of a data acquisition system. Several signal conditioning features, such as ratiometric conversion, high-resolution conversion, range biasing, automatic gain switching, and logarithmic conversion can be included during the selection of a data acquisition system. Several noise reduction features, such as filtering, an integrating converter, and digital processing, can be considered for identifying a suitable data acquisition system during the design of an experimental setup.

3.3 Signal Conditioning Systems

Most data that is being acquired contains spurious data known as noise in some form or another, which is inherent in any measurement system as per the second law of thermodynamics. Hence it is important to eliminate or reduce noise in the measured data for improving its quality by conditioning the input signal. Analog data can be handled using either piecewise linear or smooth series approximations. Several types of filters, amplifiers, and lock-in-amplifiers are also used for signal conditioning. It is important to have prior knowledge about the frequency range of the expected data for choosing proper signal conditioning systems. For example, if atmospheric temperature is to be measured by a thermocouple sensor, then it is obvious that the atmospheric temperature would not change quickly. Hence high-frequency noise in the voltage of a thermocouple sensor can be easily eliminated without any distortion of the input signal.

3.3.1 Signal Amplifiers

One of the basic requirements of a signal conditioning system is to amplify the signal properly. Signals from any instrument can be quite weak on several occasions and need to be amplified for driving the output device used in data manipulation. Hence low-level signal must be amplified to a standard level with a suitable amplifier (Figure 3.3) before processing them further. As well, on several occasions the impedance of a transducer may not match with the operating device. Buffer amplifiers are used to overcome the mismatching problem of impedance. For example, in the case of a piezoelectric transducer used for pressure measurements, a signal level may not be too low but the impedance may be too high for the current to flow. In order to overcome this problem a current amplifier is used to convert the current signal into voltage form. Therefore, signal conditioning amplifiers must be chosen properly depending on the type of data that will be used. Several types of amplifiers exist, such as a single-stage buffer amplifier, potentiometer, instrumentation amplifier, isolation amplifier, integration, differentiation amplifier, current-voltage amplifier, and lock-in amplifier.

Most modern amplifiers are based on solid state devices or integrated circuits. In this section we will discuss the overall performance of amplifiers. Most modern amplifiers have a feedback loop. Let us consider a typical amplifier with a feedback loop as shown in Figure 3.2. In this amplifier when the feedback loop is not operational, input voltage E_i is considered to be amplified to output voltage E_o with a gain of G. The signal gain G can be expressed as

$$G = \frac{E_o}{E_i} \tag{3.1}$$

The signal gain in most instruments can range between 1 to 1000. Of course one can achieve higher gain for certain instruments when the signal is too weak to be handled. Note that the term gain G is also used for representing the attenuation of voltages in instruments and hence the value of G can be less than 1.

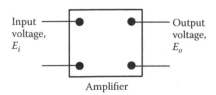

Input voltage, E_i — ⬤ ⬤ — Output voltage, E_o

Amplifier

Figure 3.3

Schematic of a typical amplifier.

Gain is expressed more often in the algorithm scale in the decibel (dB) unit, as given below:

$$G_{dB} = 20\log_{10}\left(\frac{E_o}{E_i}\right)$$ (3.2)

If the gain factor G happens to be 10, then G_{dB} would be 20 dB and an amplifier with G of 10,000 would be equal to 80 dB. As mentioned earlier for an attenuated signal, decibel gain G_{dB} will be negative. Note that the main function of an amplifier is to enhance/attenuate a signal but it also affects the original signal in a number of ways, including frequency distortion, phase distortion, common-mode effect, and source loading. We know that amplifiers are used for signals covering a certain range of frequencies. The characteristics of a typical amplifier—frequency response and phase response—are shown in Figure 3.4. Gain remains constant between frequencies f_{c1} and f_{c2}, which is known as bandwidth. The upper and lower frequencies f_{c1} and f_{c2}, respectively, are known as cutoff frequencies, which are defined as the frequencies beyond which gain is reduced at least by 3 dB. Most of the practical amplifiers may not have the same gain for a wider range of frequencies. For example, an amplifier may have a gain of 40 dB for 1 kHz but 10 dB at 50 kHz. Hence care must be taken to select a proper amplifier, which must have the same gain for the range of frequencies of the signal to be amplified. An amplifier with a narrow bandwidth is likely to distort the shape of the signal, as shown in Figure 3.5. Note that a square wave is distorted by the amplifier, which is often called frequency distortion. During amplification of a signal one has to worry about the distortion of the input signal due to a change of phase although the gain of the amplifier remains almost constant over the

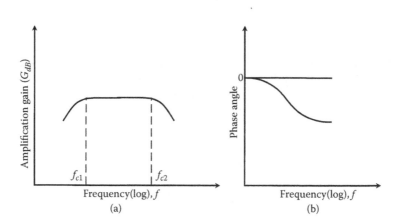

Figure 3.4

Typical characteristics of an amplifier: (a) frequency response and (b) phase response.

3. Data Acquisition

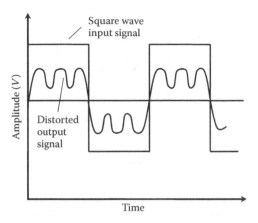

Figure 3.5

Frequency distortion of a square wave signal.

requisite bandwidth. Let us consider the input voltage E_i to the amplifier that can be expressed mathematically as

$$E_i = V_i \sin(2\pi f t) \tag{3.3}$$

where V_i is the amplitude and f is the frequency of the sine wave. Then the output signal E_o would be

$$E_o = GV_i \sin(2\pi f t + \phi) \tag{3.4}$$

where ϕ is the phase angle difference between input and output. The phase response of an amplifier is represented in a plot of phase angle and logarithm of frequency, as shown in Figure 3.4b. Both amplitude and frequency response as shown in Figure 3.4 is known as the Bode plot, which depicts the dynamical characteristics of any signal. For a simple periodic signal like sine or cosine waveforms, phase distortion may not pose any serious problems, but for complicated signals, it is important to avoid/minimize phase distortion during the amplification of signals.

Let us understand two other problems—input and output loading—that are encountered during amplification of a signal. Of course the input voltage may either be generated by the sensor or by another signal conditioning device. If this input voltage from the source (sensor or any other signal conditioning device) gets changed when an amplifier is connected to this device, this change is known as the input loading problem. Similarly when the output voltage of an amplifier is changed, it is called the output loading problem. In order to appreciate the loading problems, let us consider a simple model as depicted in Figure 3.6 in which the sensor can be considered as a voltage generator with E_g and a resistance R_g. In a similar way, we can model the input voltage to the amplifier E_i with an input

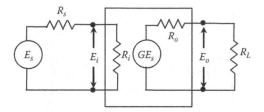

Figure 3.6

Model for the loading problem of an amplifier.

resistance R_i and the output voltage to the amplifier E_o with an output resistance R_o and loading resistance R_L, as shown in Figure 3.6. Note that E_g is the voltage of the sensor (voltage generator) only when it is not connected to the amplifier. But when the sensor is connected to the amplifier, the sensor output voltage is not the same as E_g because there will be voltage drop due to the presence of two resistors, R_g and R_i, in the input circuit. In other words, load will be exerted on the sensor/source by the amplifier. Similarly, output loading will be placed on the amplifier itself. Let us analyze it further to determine how the loading problems can be minimized. For this purpose, we can express the amplifier input voltage E_i in terms of E_g as

$$E_i = \frac{R_i E_g}{R_g + R_i} \tag{3.5}$$

Similarly, we can express the amplifier output voltage E_o in terms of E_g as

$$E_o = \frac{R_L G E_i}{R_o + R_L} \tag{3.6}$$

Combing Equations 3.5 and 3.6, we can express E_o in terms of E_g and various resistances as

$$E_o = \frac{R_L}{R_o + R_L} \frac{R_i}{R_g + R_i} G E_g \tag{3.7}$$

We can see from the above expression that when $R_L \gg R_o$ and $R_i \gg R_g$, Equation 3.7 can be approximated as

$$E_o \approx G E_g \tag{3.8}$$

This no-loading effect can be achieved by choosing high values of input resistance R_i and low values of output resistance R_o. This is known as the ideal amplifier, which has infinite input resistance/impedance and zero value of output resistance/impedance.

3. Data Acquisition

Example 3.1

A pressure sensor has an open circuit output voltage of 65 mV with a resistance of 150 Ω. This sensor with input resistance of 1.5 MΩ is connected to an amplifier with a gain of 25 and a resistance of 50 Ω. Determine the output voltage and loading input error for the loading resistance of 2.5 kΩ.

Given: $E_g = 65$ mV, $R_g = 150$ Ω, $R_i = 1.5$ MΩ, $G = 25$, $R_o = 50$ Ω, $R_L = 2.5$ kΩ.

To Find: Loading error ε_L.

Solution: The pressure transducer can be modeled as a 65-mV voltage generator in series with a resistance of 150 Ω, as shown in Figure 3.6. By using Equation 3.8, we can find out the output voltage for $R_L = 2.5$ kΩ as given below:

$$E_o = \frac{R_L}{R_o + R_L}\frac{R_i}{R_g + R_i}GE_g = \frac{2500}{50 + 2500} \times \frac{1500000}{150 + 1500000}10 \times 65 = 624.759\,\text{mV}.$$

The loading error ε_L can be determined as

$$\varepsilon_L = \frac{E_{o,\text{ideal}} - E_o}{E_{o,\text{ideal}}} \times 100 = \frac{650 - 624.759}{650} \times 100 = 3.88\%.$$

Note that although this error is acceptable it can be reduced further by increasing the input resistance/impedance of the amplifier or loading resistance.

3.3.2 Operational Amplifier

An operational amplifier, often known as an op-amp, is a dc differential amplifier with a very high gain in the order of 10^4 to 10^6. It can operate from dc voltage to ac with a certain higher frequency in the order of 1.0 MHz. It is generally constructed as low cost solid-state integrated circuits packed in a compact container, which can be represented schematically by a triangle symbol as shown in Figure 3.7. The input voltages—noninverting (E_p: positive voltage) and inverting (E_n: negative voltage)—can be applied to the input terminal, while the output voltage is obtained from a single terminal. Note that output from the noninverting voltage input is in phase with the input while the inverting input is 180° phase with the input voltage. Besides these terminals, two additional power terminals are provided for adjusting certain characteristics. This kind of amplifier can be used for several applications, such as summer, integrator, voltage comparator with suitable feedback networks, function generator, filters, and impendence transformer. It is used extensively for measurements. Most operational amplifiers have the following characteristics:

- High-input impedance (MΩ to teraΩ)
- Higher gain (10^4–10^6)
- Low output impedance (even to fraction of Ω with feedback)

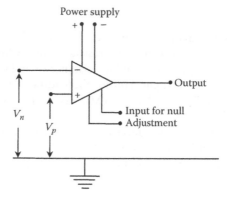

Figure 3.7

Schematic of a typical operational amplifier.

- Fast response/slew rate (several volts per μs)
- Effective in rejecting common-mode inputs

Several IC op-amps are used in instruments, and the most popular op-amp is the μA741, whose differential input resistance is 300 kΩ. However, most operational amplifiers do not ensure a differential amplifying property. As a result it would not match properly with transistors and resistors and other elements. Hence, provisions such as null adjustment input (see Figure 3.7) are provided to balance the unwanted offset voltage. Another limitation of op-amps is that they have finite common-mode rejection ratios (CMRRs). The CMRR of a typical op-amp varies from 60 to 100 dB, which is typically 10^3 to 10^6 smaller than the differential gain. Besides this, it also suffers from thermal drift. However, most modern op-amps have improved thermal stability, CMMR, and offset voltage, of course with an added cost.

Generally, negative feedback loops are used in most operational amplifiers as even a slight input voltage difference can cause an amplifier to become saturated due to higher gain value. In order to overcome this problem, a connection is made between the inverting terminal and the output terminal. As a result, any increase in output voltage E_o can be fed back to inverting input and thus would result in reducing the voltage difference. Hence, saturation is avoided as the input voltage difference will be almost zero ($E+ ≈ E-$). Op-amps have high input impedance, so the input current i_+ is almost zero (pA: picoamp). As the loading is quite small, output current i_- is almost zero (pA: picoamp). Let us now consider an inverting amplifier as shown in Figure 3.8a in which the output voltage relative to ground is opposite to that of the input voltage. In this case impedance is low and is in the order of 100 Ω as compared to a noninverting op-amp (in the order of 100 MΩ). As it has negative feedback, $E+$ and $E-$ are equal to zero. The inverting currents draw no current ($i_1 = i_2$). Hence by applying Ohm's law to R_1 and R_2, we have

3. Data Acquisition

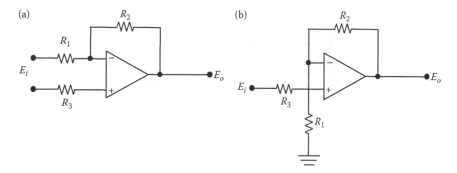

Figure 3.8

Schematic of operational amplifier configurations: (a) inverting amplifier and (b) noninverting amplifier.

$$i_1 = \frac{E_i - 0}{R_1}; \quad i_2 = \frac{0 - E_o}{R_2} \tag{3.9}$$

We know that $(i_1 = i_2)$, so then we have

$$G = \frac{E_o}{E_i} = -\frac{R_2}{R_1} \tag{3.10}$$

It can be shown that the resistor R_3 can be made approximately equal to

$$R_3 \approx -\frac{R_1 R_2}{R_1 + R_2} \tag{3.11}$$

In the case of a noninverting amplifier as shown in Figure 3.8b, the feedback voltage is connected to the negative terminal, resulting in a closed/feedback loop configuration and then the signal input is provided to a positive terminal. In order to find an expression for G, we will analyze this circuit that includes R_1, R_2, and R_3, noting that the input current i_+ is almost zero (pA: picoamp). As the loading is quite small, output current i_- is almost zero (pA: picoamp), as discussed above. Hence by applying Ohm's law to R_1 and R_2, we can find out the output voltage E_o. This can be related to the inverting terminal voltage/input voltage as

$$G = \frac{E_o}{E_i} = \frac{R_1 + R_2}{R_1} = 1 + \frac{R_2}{R_1} \tag{3.12}$$

Gain G is a function of only R_1 and R_2 but does not depend on actual resistor R_3. Typical values of R_1 and R_2 in this kind of noninverting amplifier are in the range of 1 kΩ to 1 MΩ. Let us understand how the gain G of this kind of amplifier is dependent on the frequency. A typical frequency response of the amplifier

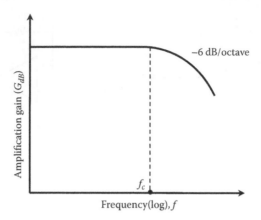

Figure 3.9

Schematic of the frequency response of an operational amplifier.

is shown in Figure 3.9, which indicates that gain G remains almost constant until a certain cutoff frequency f_c. This cutoff frequency f_c is defined as threshold frequency at which gain is declined at a rate of 6 dB/octave. Note that an octave is a doubling of frequency above f_c during which gain is declined by 6 dB. As discussed earlier, the range of frequency from $f = 0$ to f_c is known as the bandwidth of this amplifier. It is desirable to have constant high gain with a certain bandwidth but the phase angle between input and output, φ, is dependent strongly on frequency. For a noninverting amplifier, the phase angle can be related with frequency as

$$\varphi = -\tan^{-1}\left(\frac{f}{f_c}\right) \tag{3.13}$$

When f is equal to f_c, the phase angle becomes $-45°/-0.7853$ rad, which indicates that output lags behind the input by about 1/8 of a cycle. As the variation of the phase angle is almost linear, the signals within bandwidth can have modest phase distortion.

Example 3.2

In a noninverting op-amp-based amplifier with a gain of 15, the resistances R_1 and R_2 happen to be 5 and 75 kΩ, respectively. If the cutoff frequency is 100 kHz, determine the resistance R_2 and phase shift for a periodic wave with a frequency of 15 kHz.
   Given: $R_1 = 5$ kΩ, $G = 15$, $f_c = 100$ kHz, $f = 15$ kHz.
   To Find: phase shift φ.

3. Data Acquisition

Solution: The gain for a noninverting op-amp-based amplifier can be expressed by Equation 3.12 as

$$G = \frac{E_o}{E_i} = 1 + \frac{R_2}{R_1}; \quad 15 = 1 + \frac{R_2}{5}; \quad \Rightarrow R_2 = 70 \, k\Omega$$

The phase shift for a periodic wave with a frequency of 15 kHz can be determined as

$$\varphi = -\tan^{-1}\left(\frac{f}{f_c}\right) = -\tan^{-1}\left(\frac{15}{100}\right) = -9.48°$$

This indicates that output lags behind the input signal by 2.6% of this periodic cycle, which may be acceptable.

3.3.3 Differential Amplifier

A differential amplifier amplifies the difference between two voltages but does not amplify any particular voltage. This kind of amplifier is quite useful for amplification and measurement of small signals without contaminating them with external noises. For example, any small signal from an instrument may get affected by a line voltage of 230 V at 50 Hz, which can be avoided by using a differential amplifier. In this amplifier, two input and one output terminals as in an op-amp are required whose output will be proportional to the difference between two input voltages. In order to have better performance, this type of amplifier must have well-matched internal components with the ability to adjust the differential voltages. Note that when different voltages are applied to two input terminals, the input is known as a differential mode voltage, but when the same voltage with respect to ground is applied to two input terminals, this input is called a common mode voltage. An ideal instrumentation amplifier would not produce any output to this common mode voltage but would produce a certain output for differential mode voltage. A real amplifier will produce a finite output to both modes of voltages. Of course the response to the differential mode voltage is quite higher as compared to the common mode voltage, so it is important to know the relative response to differential and common mode voltages. This measure of relative response is often referred to as CMRR, as defined below:

$$CMMR = 20\log_{10}\left(\frac{G_{diff}}{G_{cm}}\right) \tag{3.14}$$

where G_{diff} is the gain for a differential mode voltage between the input terminals and G_{cm} is the gain for a common mode voltage applied to both input terminals. A typical CMMR of an amplifier is around 120 dB.

3.4 Signal Filtering System

We have discussed above how the signal sensed by an instrument may pick up unwanted signals known as noise, which can distort the true signal and thus may produce intolerable errors in measurement. The noise signal can distort a true signal both in the frequency and amplitude components. One way of overcoming this corruption of signals is to filter a signal while sensing the measurand by the instrument or during the conditioning of the measured signal. As we know, filtering is the process of attenuating noise while permitting the desired dynamic signal generated due to the measurand. Depending on the type of signals, filters can be broadly divided into two categories: (i) analog and (ii) digital filters. An analog filter is generally used when we need to control the frequency content of the signal. Antialiasing filters can be used to eliminate a signal by using Nyquist frequency criteria even before sampling. We will discuss the Nyquist frequency criteria when we discuss sampling methods. Keep in

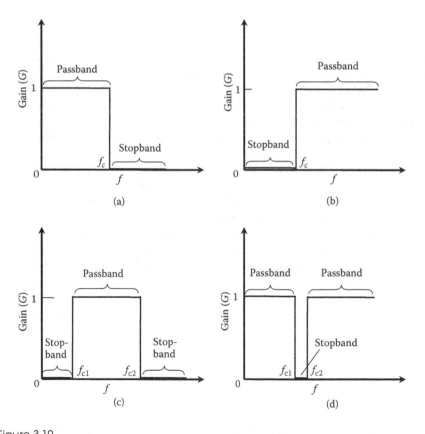

Figure 3.10

Ideal characteristics of four types of filters: (a) lowpass, (b) highpass, (c) bandpass, and (d) notch.

mind that some modern data-acquisition cards have analog filters for avoiding aliasing of a virgin signal with noises from other sources, while digital filtering is carried out using software with the help of mathematical tools. In addition to the above classifications, filters can be classified based on the range frequency qualitatively as (a) lowpass, (b) highpass, (c) bandpass, and (d) band-reject filters, as shown in Figure 3.10. These filters are differentiated by frequency of signal. In the case of a highpass filter, transmission of a high-frequency signal with little or no attenuation is allowed above a certain cutoff value of frequency, f_c, while for a lowpass filter, a signal is allowed to transmit below a certain cutoff value. A bandpass filter permits a range of frequency to transmit for further processing. In contrast, the band-reject filter allows the passage of all frequencies except a certain selected narrow frequency band. The sharp cutoff shown in ideal filters as depicted in Figure 3.10 cannot be achieved in any real filter. Let us consider the variation of magnitude ratio (gain) with frequency ratio (f/f_c) for a typical lowpass filter as shown in Figure 3.11. Note that there is a certain transition band over which the gain decreases with frequency. This rate of transition is termed as filter roll-off/cutoff expressed in decibels per octave. There will also be a phase shift between the input and output signals. In Figure 3.11b the characteristic of a typical bandpass filter is shown in which the frequency bandwidth between the lower and upper skirt is defined with respect to the average/center frequency between lower frequency, f_l, and upper frequency, f_u. Similar terms are also useful for both bandpass and band-reject filters. It must be kept in mind that signals get attenuated at a higher rate as the filter frequency moves away from the cutoff frequency although there may be discontinuities in attenuation at the transition frequency. Although filters manage to remove the noise in the outside of the bandwidth matched to the signal spectrum, noise cannot be really removed inside the filter bandwidth.

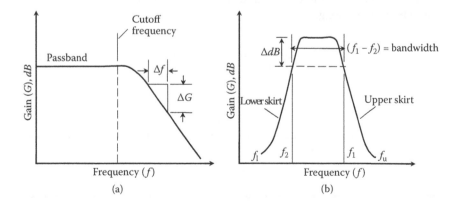

Figure 3.11

Gain versus frequency ratio for typical lowpass: (a) bandpass and (b) filters.

3.4.1 Analog Filters

Analog filters are designed and developed using various electrical/electronic circuits consisting of resistors, capacitors, inductors, operational amplifiers, and so forth that condition the continuous signals. Analog filters can be broadly divided into two categories: (i) active and (ii) passive filters. In the case of passive filters, passive elements—resistance, R, capacitance, C, and inductance, L—are used to filter signals. In contrast, active filters use powered components such as amplifiers (e.g., op-amps) to filter analog signals. Several kinds of analog electronic filters have been designed and developed for various applications of measurement systems. We will restrict our discussion to the simplest lowpass and highpass passive filters comprised of a single resistor, capacitor, and inductor, as shown in Figure 3.12. Let us consider a first-order resistance-capacitance (RC) lowpass analog filter as shown in Figure 3.12a. The capacitor allows the high-frequency currents but blocks the low-frequency current of the input voltage, E_i, and the high-frequency current component of the input signal is short-circuited. Let us consider the input voltage, E_i, to this filter circuit that can be expressed mathematically as

$$E_i = V_i \sin(2\pi ft) \tag{3.15}$$

In order to find out the frequency characteristics, we need to determine the output voltage of this lowpass filter, E_o. Note that current drawn at the output is almost zero, so the same current will pass through both the resistor and capacitor. Hence we can have

$$I = \frac{E_i - E_o}{R} = C\frac{dE_o}{dt} \tag{3.16}$$

By simplifying Equation 3.16 and substituting the expression for E_i in terms of V_i and frequency f, we get

$$RC\frac{dE_o}{dt} + E_o = V_i \sin(2\pi ft) \tag{3.17}$$

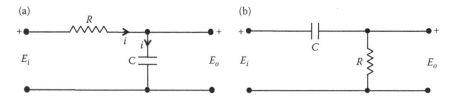

Figure 3.12

First-order RC filters: (a) lowpass and (b) highpass.

3. Data Acquisition

By solving the above ordinary differential equation, we get

$$E_o = A_1 e^{-t/RC} + \frac{V_i}{\sqrt{1+(2\pi RCf)^2}} \sin(2\pi ft + \phi) \qquad (3.18)$$

where phase angle

$$\phi = -\tan^{-1}(2\pi RCf) \qquad (3.19)$$

The first term on the right-hand side of Equation 3.4 indicates the transient response and the second term represents the steady state response. For filtering, we will be interested in the frequency response corresponding to the steady state portion, and then Equation 3.18 can be rewritten as

$$E_o = V_o \sin(2\pi ft + \phi) = \frac{V_i}{\sqrt{1+(2\pi RCf)^2}} \sin(2\pi ft + \phi) \qquad (3.20)$$

In order to characterize the performance of a filter, we need to use a cutoff frequency, f_c, as

$$f_c = \frac{1}{2\pi RC} \qquad (3.21)$$

By using Equations 3.18 and 3.20, the frequency response of this lowpass filter can be expressed in terms of the cutoff frequency, f_c, as

$$\frac{V_o}{V_i} = \frac{1}{\sqrt{1+(f/f_c)^2}} \qquad (3.22)$$

The phase response can be expressed in terms of the cutoff frequency, f_c, as

$$\phi = -\tan^{-1}(f/f_c) \qquad (3.23)$$

By using Equation 3.22, we can plot the frequency response in log form in Figure 3.13a, which is commonly known as a Bode plot. Note that amplitude attenuates from 0 to −40 dB when the frequency ratio varies over several orders of magnitude. The cutoff frequency is chosen corresponding to −3 dB, which means that the gain is reduced by 70.7%. For this condition, signal power is reduced by half at the cutoff frequency, f_c. The response of this lowpass filter is quite flat when the signal frequency is below the cutoff frequency, f_c. It can be observed from this figure that the bandpass response is attenuated to the rejection band region gradually with an increase in frequency. Apart from attenuating the amplitude,

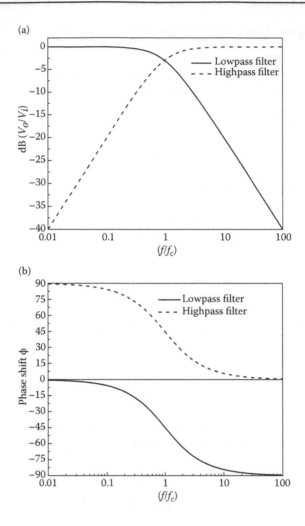

Figure 3.13

(a) Frequency response and (b) phase response for (i) lowpass and (ii) highpass RC filters.

this filter causes an increase in phase shift with an increase in signal frequency, as shown in Figure 3.13b. The output signal lags behind the input signal by 45° at a –3-dB response point.

Let us consider a highpass filter that can be obtained easily by interchanging the capacitor with the resistor, as shown in Figure 3.12b. As a result, the capacitor will be passing the signal with high frequencies while blocking the low-frequency

3. Data Acquisition

signal. By carrying out a similar analysis using a lowpass filter we can get almost the same results under a steady state condition, as given below:

$$E_o = V_o \sin(2\pi ft + \phi) = \frac{V_i(2\pi RCf)}{\sqrt{1+(2\pi RCf)^2}} \sin(2\pi ft + \phi) \tag{3.24}$$

where phase angle, ϕ, becomes positive for a highpass filter as given below:

$$\phi = 90° - \tan^{-1}(2\pi RCf) \tag{3.25}$$

As in the case of a lowpass filter, we need to use a cutoff frequency, f_c, in order to characterize the performance of a highpass filter, as given below:

$$f_c = \frac{1}{2\pi RC} \tag{3.26}$$

By using Equations 3.25 and 3.26, the frequency response of this highpass filter can be expressed in terms of the cutoff frequency. f_c, as

$$\frac{V_o}{V_i} = \frac{(f/f_c)}{\sqrt{1+(f/f_c)^2}} \tag{3.27}$$

The phase response can be expressed in terms of the cutoff frequency, f_c, as

$$\phi = 90° - \tan^{-1}(f/f_c) \tag{3.28}$$

In a similar way, by using Equation 3.16, we can plot the frequency response in log form in Figure 3.13a. Note that amplitude gets amplified from 0 to −40 dB when the frequency ratio varies over several orders of magnitude. The cutoff frequency is chosen corresponding to −3 dB, which means that the gain is reduced by 70.7%. The response of this highpass filter is quite flat when the signal frequency is beyond the cutoff frequency, f_c. Apart from attenuating the amplitude, this filter causes an increase in phase shift with a decrease in the signal frequency, as shown in Figure 3.13b. The output signal is ahead of the input signal by 45° at the −3 dB response point. Because of their simplicity, first-order RC filters are being used widely in eliminating noises in input signals although they have relatively slow roll-off. Besides these simple RC filters, both lowpass and highpass filters can be designed using a single resistor and inductor (RL) pair. Unfortunately, they are not preferred in practical applications because they are too bulky and large, particularly when the operation range of frequency exceeds 100 kHz. Besides this it is quite difficult to have optimal value of inductance in RL filter. Note that resistors in these filters make a circuit dissipative in nature and hence make the circuit inefficient. In order to achieve sharper cutoff characteristics, a number of simpler filters can be used in series. For n number of RC filters in series, the total frequency response can be expressed as

$$\frac{V_o}{V_i} = \frac{1}{\sqrt{1+[(f/f_c)^2]^n}}$$

(3.29)

Total phase response can be expressed in terms of the cutoff frequency, f_c, as

$$\phi_{total} = \sum_{i=1}^{n} \phi_i$$

(3.30)

Note that the current drawn by one filter can alter the performance of the preceding filter. Generally a voltage follower is used as a buffer between each successive filter. We will not be discussing analog filters any further; however, interested readers can refer to the following books [4,5] for advanced complex filters that can be used in practical measuring instruments.

Example 3.3

A pressure sensor senses a signal of 45 mV at 2500 Hz while incurring 10-mV noise at a frequency of 50 Hz. In order to reduce the 50-Hz noise, a highpass filter is used at a cutoff frequency of 500 Hz. Determine the output voltage and % error due to the noise in the filtered signal.

Given: V_i = 45 mV; V_{in} = noise voltage = 10 mV, f_c = 500 Hz, f = 2500 Hz, f_n = 50 Hz.

To Find: V_o, % error in V_o due to noise.

Solution: The output voltage of the filtered signal at a cutoff frequency of 500 Hz can be determined by using Equation 3.27 as

$$\frac{V_o}{V_i} = \frac{(f/f_c)}{\sqrt{1+(f/f_c)^2}} = \frac{(2500/500)}{\sqrt{1+(2500/500)^2}} = 0.98; \quad V_o = 45 \times 0.98 = 44.1\,\text{mV}$$

Similarly, the output voltage of the filtered noise signal at a cutoff frequency of 500 Hz can be determined by using Equation 3.27 as

$$\frac{V_{on}}{V_{in}} = \frac{(f/f_c)}{\sqrt{1+(f/f_c)^2}} = \frac{(50/500)}{\sqrt{1+(50/500)^2}} = 0.099; \quad V_{on} = 10 \times 0.099 = 0.99\,\text{mV}$$

The error in amplitude due to noise, which is about 22.22% in the raw signal is being reduced to only 2.2%. Note that although the noise is reduced drastically, it reduces the original signal by only 2%. If it is not acceptable, then one may use a complex filter to avoid the noise in the signal.

3.4.2 Active Filters

In order to enhance the performance characteristics of passive filters, operational amplifiers are used because they have high-frequency gain characteristics. The basic components of a generic active filter is shown in Figure 3.14. Passive filter networks consisting of resistors, capacitors, and so forth are connected to an operational amplifier that provides power and ensures higher impedance characteristics. These filters can provide higher output current without deterioration in its performance as the lower output impedance is achieved. As a result, active filters are used whenever steep roll-off, adjustable cutoff frequencies, arbitrary flat passband, and so forth are required. In recent times, they are being used extensively in practical applications, which have ushered in a great deal of development in active filter design. We will be covering just the basics of active filters. Interested readers can refer to books on active filter design and applications [4,5].

The schematic of typical first-order, single-stage, active filters, such as (a) lowpass, (b) highpass, and (c) bandpass, are shown in Figure 3.15. Let us consider a lowpass active filter as shown in Figure 3.15a, which is comprised of passive filter components along with a single op-amp. By carrying out a similar analysis that was discussed previously for a passive filter, it can be shown that the cutoff frequency can be related to resistance R_2 and capacitance C_2 as given below:

$$f_c = \frac{1}{2\pi R_2 C_2} \tag{3.31}$$

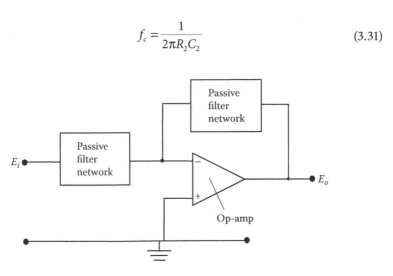

Figure 3.14

Schematic of a typical active filter.

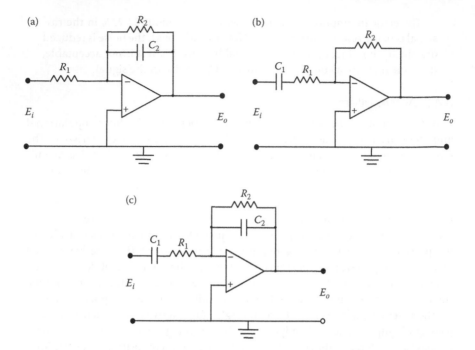

Figure 3.15

Schematic of first-order active filters: (a) lowpass, (b) highpass, and (c) band pass.

Similarly, for a first-order and single-stage highpass active filter, the cutoff frequency can be related to resistance R_2 and capacitance C_2 as given below:

$$f_c = \frac{1}{2\pi R_1 C_1} \tag{3.32}$$

The gain for a highpass filter can be related to frequency ratio and static sensitivity as given below:

$$G = \frac{R_2}{R_1} \frac{(f/f_c)}{\sqrt{1+[(f/f_c)^2]^{1/2}}} \tag{3.33}$$

where R_2/R_1 is the static sensitivity ratio for active filters as the operational amplifier is being used. Note that the filter signals with a cutoff ratio of 80 dB/octave and more than 60-dB attenuation in the rejection band can be achieved using modern active filters. Generally, higher-order active filters become more compact as they are designed using integrated circuits.

3. Data Acquisition

3.4.3 Digital Filters

Digital filters are quite convenient to use, particularly when analog signals are converted successfully into digital form. We will learn briefly how analog signals can be converted into digital form by proper sampling in Section 3.5. But in digital filters, the samples undergo certain operations to enhance/reduce certain aspects of the signals. First, an analog signal from an instrument has to be digitized and manipulated mathematically and subsequently reconstructed as a new analog signal. Digital filters are available in most virtual lab software (e.g., LabVIEW), and are used extensively in modern laboratories as they are quite convenient. Although digital filters are quite effective for eliminating noise from the sampled signals, they cannot remove its effects completely. During the digital filtering process, the sampled time domain signal is converted into a frequency domain using the Fourier transformation method and the signal amplitude is multiplied by the desired frequency response depending on the type of filters used (lowpass, bandpass, highpass, etc.). Subsequently the signal is transferred back to the time domain by using the inverse Fourier transformation method.

A simple digital filter can be conceived as a moving average filter that can be employed to eliminate broadband noise. Generally a digital filter replaces the actual data points with average data point values. In order to illustrate this aspect further, let us consider a forward-moving averaging smoothing scheme that can be represented mathematically as

$$\bar{V}_i = (V_i + V_{i+1} + \ldots + V_{i+n})/(n+1) \tag{3.34}$$

where $\bar{V}_i$ is the average value that is determined and used in place of V_i and $(n + 1)$ is the number of successive values used for averaging. In a similar way, a backward-moving averaging smoothing scheme as given below is being used for filtering data digitally:

$$\bar{V}_i = (V_{i-n} + \ldots + V_{i-1} + V_i)/(n+1) \tag{3.35}$$

On some occasions, the center-weighted-moving averaging smoothing scheme as given below is used for filtering data digitally:

$$\bar{V}_i = (V_{i-n} + \ldots + V_{i-1} + V_i + V_{i+1} + \ldots + V_{i+n})/(2n+1) \tag{3.36}$$

These filtering schemes for data analysis can be implemented easily even in a spreadsheet program as well as in specialized software. Note that for any data, when filtering is light in nature, three terms should be used, but when heavy filtering is called for, more than 10 terms should be considered.

3.5 Sampling and Data Acquisition

We need to convert the analog signal to a digital form as it is convenient to handle measured signals from instruments. As discussed earlier, analog signals are continuous both in amplitude and time while digital signals are discrete both in amplitude and time. For converting a continuous signal to a discrete signal, we will have to sample the measured analog signal properly so that it will contain all the information in the original signal. In other words, the process of sampling must be done so that all information in the discrete signal must be present in the digital form.

Let us consider a sinusoidal analog signal, $E(t)$, that varies with time t, as shown in Figure 3.16. This continuous signal can be converted into its discrete form with the help of an on/off switch that can remain closed or open for the same time interval, Δt. For this example, we can take a time interval of 10 s, and then the number of times the discrete data, N, will be acquired by the data acquisition system will be 20. These discrete data points are plotted and resemble very closely the sinusoidal analog signal in terms of amplitude and frequency. But if the signal is not sampled properly, the discrete data points forming the digital signal may not represent the analog signal. The extent by which a digital signal represents the analog signal will depend on (i) the frequency content, (ii) the time interval, and (iii) the total sampling time of the measured signal. Note that the actual signal will be quite complex in nature, unlike the sinusoidal analog signal shown in Figure 3.16. As the complex dynamical signal can be well represented by a Fourier series, then any complex dynamical signal can be converted into its

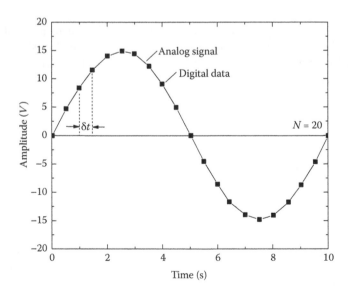

Figure 3.16

Sinusoidal analog signal and its digital form.

3. Data Acquisition

equivalent digital form using a discrete Fourier transformation (DFT). Therefore, a Fourier analysis can be invoked for converting an analog signal into its equivalent discrete form. However, it is necessary to determine the frequency content of the signal during the digital conversion of the time-dependent analog signal. Let us consider a periodic signal of 5 Hz, as shown in Figure 3.17, in which amplitude $E(t)$ varies for a time interval of 2 s. We need to convert this analog signal to its discrete form by sampling data at a certain time interval, Δt. In other words, the data from the analog signal is to be sampled at a certain frequency, which is known as the sampling frequency, f_s. Then the sampling frequency is inversely related to time interval, Δt, between two sampling frequencies ($f_s = 1/\Delta t$). The questions arise, what would the time interval be between two discrete points? Would the sampling frequency be equal to or higher/lower than the frequency of the signal? In order to illustrate this point further let us sample the data from this analog signal at three sampling frequencies: (i) $f_s = 6$ Hz, (ii) $f_s = 25$ Hz, and (iii) $f_s = 100$ Hz. Note that the sampling interval decreases with an increase in the sampling frequency. The variations of this discrete data are plotted in Figure 3.17b, c,

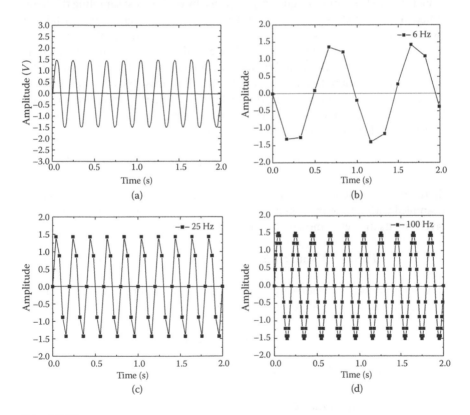

Figure 3.17

Sampling of a signal at three frequencies: (a) Analog signal at 5 Hz, (b) digital data at $f_s = 6$ Hz, (c) digital signal at $f_s = 25$ Hz, and (d) digital signal at $f_s = 100$ Hz.

and d, respectively. It can be observed that the discrete data with a lower sampling rate preserves similar variation but distorts the actual analog signal with a lower frequency than that of the original one. In other words, the sampling rate during the conversion of an analog signal to its discrete form is a very critical accurate representation of the actual analog signal. In order to ensure the proper representation of an actual analog signal, we need to choose the proper sampling frequency by invoking the Nyquist sampling theorem, which states that an analog signal can represented by a set of discrete data provided that the sampling rate be greater than or equal to twice the highest frequency present in the signal. This can be represented mathematically as

$$f_s \geq 2 f_{max} \tag{3.37}$$

where f_{max} is the maximum frequency of the continuous signal. This expression provides a criterion for a minimum sampling rate. It is interesting to observe in Figure 3.17 that the distortion of the output signal when data is sampled at a lower frequency than the sample frequency prescribed by the Nyquist sampling theorem. Although the discrete signal looks similar to the actual analog signal, it takes a lower frequency. This phenomena of constructing two similar signals from a given set of data values is known as *aliasing*. The actual signal from an instrument is quite complex, containing several frequency components, unlike single frequency considered in the example of a sine wave. For handling this type of complex signal, the DFT can be used for converting the analog signal into its digital form in which the alias phenomena may likely occur. Let us consider a simple voltage signal that can be depicted by one of the Fourier series as given below:

$$E(t) = V \sin(2\pi f t + \phi(f)) \tag{3.38}$$

where f is the frequency and ϕ is the phase shift. Let us consider that the signal is sampled with a time increment of Δt that covers N time increments over one waveform. Its discrete time signal can be expressed as

$$E(n\Delta t) = V \sin(2\pi f n \Delta t + \phi(f)); n = 1,2,3,......(N-1) \tag{3.39}$$

As it is a periodic waveform, we can have the same amplitude at certain different frequencies that can be expressed mathematically as

$$\sin(t) = \sin(t + 2\pi f p) \tag{3.40}$$

where p is any integer. We can rewrite Equation 3.38 as

$$V \sin(2\pi f n \Delta t + \phi(f)) = V \sin(2\pi f n \Delta t + 2\pi p + \phi(f))$$
$$= V \sin(2\pi n \Delta t (f + m/\Delta t) + \phi(f)) \tag{3.41}$$

where $m = p/n = 0, 1, 2.....$ is another integer. The above equation clearly indicates that for any value of Δt, the frequency f and $f + m/\Delta t$ must have a similar wave form. Note that the false frequency in the form $f + m/\Delta t$ for any value of integer m is known aliasing frequency of original frequency f. But we can avoid the problems of aliasing if we use the Nyquist sampling theorem and when $m \leq 1$. Similar arguments can hold for complex periodic, aperiodic, and even random waveforms. We can also extend a similar argument for a complex signal by considering the general form of the Fourier series as given below:

$$E(n\Delta t) = \sum_{i=1}^{\infty} V_i \sin(2\pi n\Delta t(if + im/\Delta t) + \phi_i(f)) \tag{3.42}$$

The above equation also indicates aliasing phenomena as in Equation 3.41. The resolution of the DFT also plays an important role in simulating the measured signal. In general the Nyquist frequency that represents a folding frequency for an aliasing frequency can be expressed as

$$f_N = \frac{f_s}{2} = \frac{1}{2\Delta t} \tag{3.43}$$

All of the actual frequencies above the f_N in the analog signal can appear as the aliasing frequencies when the sampling frequency is less than f_N. Therefore, it is very important and essential to convert an analog signal into its digital form while avoiding the problems that arise due to aliasing. This can be easily alleviated by following the Nyquist sampling theorem. Another way of avoiding aliasing is to filter the signal before carryout analog-to-digital conversion. For example, a sound signal from any source may contain a signal harmonic beyond 20 Hz. If we are interested in analyzing this signal, naturally it would be prudent to filter the signal beyond 20 Hz before converting it into digital form. It can be concluded that the digital signal can only represent a true waveform of the actual measured signal provided proper sampling rate and resolution of the DFT are chosen for A/D conversion. We will discuss the basic aspects of A/D and D/A conversion in Section 3.6.

Example 3.4

A periodic wave with a frequency of 120 Hz is being sampled at 180 Hz. Determine the maximum frequency that can be captured by a discrete signal. What is the frequency of sample by which the data can be extracted correctly from this wave?

   Given: $f = 120$ Hz, $f_s = 180$ Hz.
   To Find: f_N, f_s.

Solution: Note that the Nyquist frequency represents the maximum frequency that can be captured by a discrete signal while sampling at $f_s = 180$ Hz. Then, by using Equation 3.43 the maximum frequency can be determined as

$$f_N = \frac{f_s}{2} = \frac{180}{2} = 90 \text{ Hz}$$

It is clear that the signal with a frequency of 120 Hz can be captured if it is sampled at 180 Hz as it does not satisfy the Nyquist criteria. Hence the signaled must be sampled at

$$f_{s,c} = 2f_N = 2f = 2 \times 120 = 240 \text{ Hz}$$

It is important to satisfy the Nyquist criterion for deciding about the sampling rate which can ensure minimum distortion in original signal while acquiring in digital form.

3.6 A/D and D/A Conversion

We have already discussed the basic need to convert an analog signal to its digital form and vice versa. Remember that the first requirement of converting an analog signal to a digital signal is to avoid noise during its transmission as digital signals are immune to spurious signals. In this section, we will briefly discuss how to convert an analog signal to its digital form and vice versa. Note that digital signals are discrete in both amplitude and time. A binary numbering system in terms of either 0 or 1 is used to represent the digital measuring signals and transmit them for signal processing and data display. We discussed previously that an open or closed switch using different voltage levels can be employed to generate digital signal from an analog signal. Note that the smallest unit of information in the binary system is known as a *bit* or *digit*. These bits can be thought of as on/off switches that are used to transmit both logical and numerical information. By combining these bits, we can represent any integer number. The collection of these bits known as a *word* can be used to represent any kind of numerical information. A word length may range from 4 bits to 32 bits. Note that a word consisting of 8 bits is known as a *byte*. The number of bits in a word can decide the size of the maximum number that a word can hold. For example, an 8-bit word can represent a number from 0 to 255 as a combination of two bits (0, 1) and can be represented by 2^8. The binary number system used in computer and A/D cards has a base of 2. Any positive integer number can be represented as a series of the power of 2 as

$$a_n \times 2^n + a_{n-1} \times 2^{n-1} + \ldots\ldots a_i \times 2^i + \ldots\ldots a_1 \times 2^1 + a_0 \times 2^0 \qquad (3.44)$$

where a_i represents an ith bit/digit. The most significant bit (MSB) is the a_n and the least significant bit (LSB) is a_0. For an 8-bit word, the MSB is 128 while the LSB is 0. This method is known as the straight binary code. Note that a bit can have either 0 or 1 as a logic circuit and is used to convert any number into a binary number. For example, if we want to convert a decimal number 155 into binary, we can express it in the following form:

$$155 = 1 \times 128 + 0 \times 64 + 0 \times 32 + 1 \times 16 + 1 \times 8 + 0 \times 4 + 1 \times 2 + 1 \times 2$$

$$= 1 \times 2^7 + 0 \times 2^6 + 0 \times 2^5 + 1 \times 2^4 + 1 \times 2^3 + 0 \times 2^2 + 1 \times 2^1 + 1 \times 2^0 \quad (3.45)$$

Hence, the digit 155 can be represented as the 8-bit binary number 10011011. In order to produce this as an electrical signal corresponding to this number we need to use eight off-on switches that are connected in parallel. All bits required to form an 8-bit word are available simultaneously.

3.6.1 Digital-to-Analog Conversion

Digital-to-analog (D/A) conversion occurs when a digital binary word is converted into an analog voltage. In order to illustrate the processes involved during conversion of a digital signal into its analog form, let us consider an 8-bit digital device, as shown in Figure 3.18, which consists of an amplifier and eight binary-weighted resisters with a common summing point. In this device, an operational

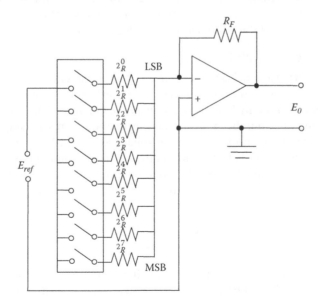

Figure 3.18

An 8-bit digital-to-analog converter.

amplifier is connected to eight resisters in parallel that can act as an inverter. The register with the MSB will have a definite value. For example, the switch 7 has the MSB as it provides the largest voltage to the amplifier. In contrast, switch 0 would have the LSB as it provides the lowest voltage to the amplifier input. Note that a stable reference voltage, E_{ref}, must be applied to the network of registers. Generally the switches in the integrated circuit are activated depending on the position of the bits in the digital signal. It will allow the digital signals to convert its analog equivalent. At each successive bit the register is doubled and hence the resister associated with the LSB would have a value of $2^{n-1} R$. The output voltage of the amplifier, E_o, can be represented by

$$E_o = -E_{ref} R_f \left(\frac{1}{R} + \frac{1}{2R} + \ldots + \frac{1}{128} \right) \tag{3.46}$$

where R_f is the resister that can be varied to produce the desired output voltage. To see how this D/A device works, we will consider our previous example of an 8-bit word 10011011. The sum of the inverse register would be equal to

$$\frac{1}{R} \left(\frac{1}{128} + 0 + 0 + \frac{1}{16} + \frac{1}{8} + 0 + \frac{1}{2} + \frac{1}{1} \right) = \frac{1}{R} 1.6953125 \tag{3.47}$$

Both reference voltage E_{ref} and R_f can be varied to have the desired output range. The full-scale reading of a typical D/A converter is within ±5 V dc. Higher resolutions can be obtained by using a higher bit D/A converter.

3.6.2 Analog-to-Digital Conversion

The analog value of a signal can be converted into its digital form using an analog-to-digital (A/D) converter. This process of converting an analog signal to its digital form is commonly known as *quantization*. Quantization takes place in a discrete manner one number at a time. During quantization, the sample voltages are rounded up or down to quantization voltage levels, E_q. Let us consider that the analog signal varies between minimum voltage E_{min} and maximum voltage E_{max}, as shown in Figure 3.19, which is to be quantized between 0 to 10 V dc into q digits by an A/D converter. The quantization interval can be expressed as

$$\Delta E_q = \frac{E_{max} - E_{min}}{q - 1} \tag{3.48}$$

Quantization produces an error as we will be approximating the actual voltage to digitized voltage, known as quantization error, ε_q, which is expressed as

$$\varepsilon_q = E_q - E_i \tag{3.49}$$

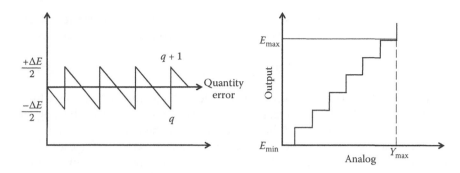

Figure 3.19

Schematic quantization of an analog signal.

The E_i is above the halfway point between two levels q and $q + 1$, which is rounded up to E_{q+1}. If it is below halfway, it is rounded down to E_q, as shown in Figure 3.19. The maximum quantization error, $\varepsilon_{q,max}$, can be expressed as

$$\varepsilon_q^{max} = \pm\frac{\Delta E_q}{2(E_{max} - E_{min})} = \pm\frac{1}{2(q-1)} \tag{3.50}$$

The quantization error reduces when the quantization level q increases, which depends on the number of bits (digits) being used in the card. Because the relationship between E_q and E_i is characterized by a series of discrete steps, the resolution of an A/D converter can be defined as the smallest voltage change that can be registered. In other words, it can be expressed as the fraction of a signal that can be expressed in terms of full-length scale voltage, which is given by

$$E_R = \frac{E_{FSR}}{2^M} \tag{3.51}$$

where M is the number of bits. For example, an 8-bit A/D converter with an E_{FSR} of 10 V can detect a minimum signal of 0.0391 V. If we want to have higher resolution then we need to use a higher bit A/D converter card, as shown in Table 3.1.

Table 3.1 Resolution of Bits

Bits (M)	E_q (V)	SNR (dB)
4	0.625	24
8	0.039	48
12	0.0024	72
16	0.15×10^{-3}	96
32	2.328×10^{-9}	193

Note: SNR = signal-to-noise ratio.

Another way of specifying resolution is the signal-to-noise ratio (SNR), which relates to the power of a signal. SNR is often expressed in terms of decibels as

$$SNR(dB) = 20\log(2^M) \qquad (3.52)$$

Various values of SNR are provided in Table 3.1 for a range of bits from which it can be observed that SNR increases with a higher bit A/D converter. If we again go back to the example of an 8-bit A/D converter, remember that it can have a resolution of 0.0391 V. Suppose that this A/D converter receives voltage less than this value, then it can detect and change this value to zero. On the other hand, if it can receive a higher voltage than the full-scale voltage E_{FRS}, then it cannot indicate the correct value. For example, for full-scale voltage $E_{FRS} = 10$ V, the input signal of 11 V would only be accepted as 10 V. In this situation, the A/D converter incurs a *saturation error*, which must be avoided. For this purpose a signal conditioning system is used to adjust the input voltage appropriately. Note that an A/D converter, like any other instrument, is subjected to *conversion errors* such as hysteresis, linearity, sensitivity, and repeatability. The settling time, signal noise during analog sampling, temperature effects, excitation power fluctuations, and so forth, are some of the main causes of conversion error. This conversion error will be dependent on the method adopted for A/D conversion. Hence, several types of A/D converters are being designed and developed. We restrict our discussion in the next section to two commonly used A/D converters: (i) successive approximation and (ii) ramp. For more types of A/D converters, interested readers can refer to advanced books on A/D conversions [4,5].

3.6.3 Successive Approximation A/D Converter

This is the most commonly used A/D converter that uses a trial and error method for determining input voltage to be converted into its digital form. A schematic of an 8-bit successive approximation A/D converter is shown in Figure 3.20 that consists of an amplifier and eight binary-weighted resisters having a common summing point voltage comparator, controlled logic with a clock, and a storage register. In this device, an operational amplifier is connected to eight resisters in parallel that can convert the register content into an analog voltage. Note that a stable reference voltage needs be fed into the switch network as discussed above with regard to a D/A converter. The output from this switch network is supplied to the operational amplifier. Subsequently, the voltage output from this amplifier is supplied to the voltage comparator, which is used to compare the input voltage with internally generated voltage. Beginning with the smallest register (MSB), this A/D converter guesses the successive binary values to narrow down the appropriate representation of the input analog voltage. Let us look at how the A/D conversion takes place sequentially:

i. The value of the MSB is set to 1 while all other bits are set to 0. This generates an equivalent value of E_q at the D/A converter output.
ii. If $E_q > E_i$ (initial voltage) at the comparator, E_q becomes low and the value of the MSB is reset to 0.

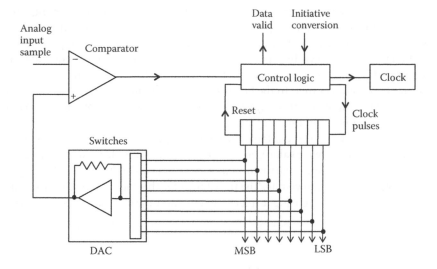

Figure 3.20

Schematic of a successive approximation A/D converter.

    iii. If $E_q < E_i$ (initial voltage) at the comparator, E_q becomes high and the value of MSB is reset to 1.

    iv. The second MSB is then set to 1 and steps from (i) to (iii) are repeated.

    v. This process is repeated until it reaches the LSB value.

    vi. Finally, the register value provides the quantization value of E_i.

Note that the quantization process mentioned above goes on successively using one clock tick per bit. For example, an 8-bit successive approximation A/D converter with 0–10 V is used to convert an input voltage of $E_i = 7.5$ V. Note that the converter has a resolution of 7.375 V. The final register counts 10101011 (to be counted), which represents 7.357 V. Because it differs from the input voltage of $E_i = 7.5$ V, the quantization error becomes 0.143 V. Note that this error can be reduced easily by using a higher bit A/D converter. Then the digital information placed on the memory can be used for manipulation of digital signal. This type of successive approximation A/D converter has the advantage of rapid conversion time. For example, an 8-bit A/D converter will take around 2 µs to perform the conversion and therefore it is employed when conversion speed is important (of course at reasonable cost). But it incurs an error in the digitized value if the input signal changes during the bit-by-bit process. In order to avoid this problem, a sample hold amplifier is used that can ensure a higher scan rate.

Ramp A/D Converter

A ramp A/D converter is preferred when high accuracy and a low-level input signal (<1 mV) is to be digitized. In this converter, a linear reference voltage signal, E_{ref} is used to differentiate the analog input signal for converting into its binary form. The input signal is digitized with time steps that increase by factors of 2^M, where M is the number of bits of the A/D device. A typical ramp A/D converter is shown in Figure 3.21 that consists of an analog comparator, ramp function generator, switch, comparator, control logic, counter, clock, buffer register and digital output. In the beginning, the reference voltage signal is set to zero. Then the input signal voltage E_i is fed into the integrator for a fixed period of time t. The output of the integrator increases with time in proportion to the slope of the input signal and attains the value of E_q at end of the time interval. At this time a fixed reference value is applied to the integrator. Because the reference voltage has the opposite polarity to that of an input signal, it reduces the output of the integrator at a known rate. The time taken for the output to reach zero value indicates the measure of the input voltage signal. This time interval can be measured

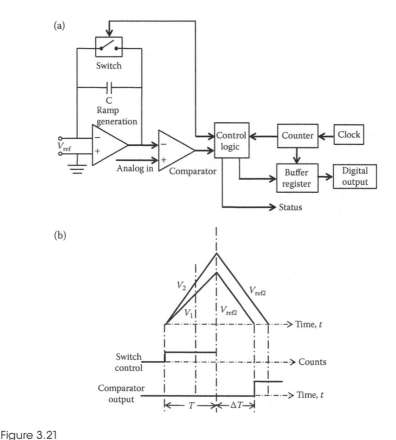

Figure 3.21

Schematic of a dual-ramp A/D converter: (a) ramp A/D converter and (b) dual-slope voltage integrator.

3. Data Acquisition

with the help of a digital counter. The same resister and capacitor set is used for both processes, so the input voltage E_i can be related to reference voltage E_{ref} as

$$E_i = E_{ref} \frac{t_m}{t} \tag{3.53}$$

where t_m is the time interval between two steps. Note that accuracy of the dual-ramp converter is dependent on the exactness of the reference voltage and stability of the time counter during conversion. The dual-ramp converter has the ability to reject a 50-Hz line frequency noise, of course when time period t is an even multiple of this frequency. Although a dual-ramp converter is quite inexpensive it acquires data at a slow rate. The maximum conversion time t_c of a dual-ramp converter can determined as

$$t_C = 2 \frac{2^M}{CS} \tag{3.54}$$

where CS is the clock speed. For example, for an 8-bit A/D converter with a 1-MHz clock speed, a dual-ramp convertor needs around 0.256 ms for a conversion. Therefore, this type of converter can be used satisfactorily in applications for combustion system as it changes slowly.

Example 3.5

A 12-bit A/D card is used to measure temperature from a combustion system using a Pt-Pt10%Rh thermocouple. It is essential to have 1°C resolution at 1000°C. Determine the maximum full-scale voltage to be covered by this card.

Given: A 12-bit A/D card, resolution = 1°C at 1000°C.
To Find: E_{FSR}, f_s.
Solution: The sensitivity of this thermocouple at 1000°C can be obtained by considering the millivolt data from Table 5.3.

$$S = \frac{11.951 - 9.587}{200} = 0.0118 \text{ mV/°C}$$

Hence, at a resolution of 1°C, this A/D card must have a resolution of 0.0118 mV. Therefore, we can have

$$0.0118 = \frac{E_{FSR}}{2^{12}} \Rightarrow E_{FSR} = 0.0118 \times 2^{12} = 48.33 \text{ mV}$$

Hence, a temperature of 1750°C can be measured easily as it has a voltage of 18.61 mV at a maximum temperature of 1750°C.

3.7 Data Storage and Display

In order to analyze and understand physical phenomena, we need to store experimental data and display it in a format that is amenable to the user. Therefore the data presentation system that can be comprised of both storage and display is the last but not the least element in any measurement system. Its main function is to communicate the measured value to the human observer who can decipher the physical phenomena. Some instruments may not have the ability to store data for future use. For example, an ordinary glass thermometer may measure temperature but may not store the displayed values itself for further analysis. Of course a person can store these values easily in a notebook if obtaining the measurement involves a slow process. But if the process is very fast—for example, pressure fluctuations during combustion instability phenomena—it will be very difficult for a person to write down the pressure data. It is therefore very important to store data in an appropriate storage device because most processes are too fast to be recorded by human observation. As well, the amount of data generated by a device may be quite large. For example, measurements of pressure, temperature, flow rate, and so forth in a combustor are quite vast, and this data must be stored for further quantitative analysis. Data is also recorded and stored in certain situations in a continuous manner for future analysis, as in cases of failures that occur in rocket engines and aircraft.

Several data storage and display systems are being devised and used in modern times for measurement systems. These can be broadly classified into two categories: indicator (display) or recorder (storage), and each category is classified further into analog and digital, as shown in Figure 3.22. In the case of an analog indicator (display), a pointer scale as in a Bourdon tube pressure gauge is used routinely. But in this pointer indicator, an observer has to interpolate the data lying between two scale marks using his or her instant judgment, which may incur error in the measured data. In order to avoid this problem it is important to use a digital indicator. There are two categories of digital indicators: (i) small-scale alphanumeric units and (ii) large-scale display units. The small-scale alphanumeric display devices are suitable for indicating the measured values of a single variable. LEDs or LCDs are used routinely for this purpose. In LED, semiconductor diodes are used to display the data as they emit electromagnetic radiation over a certain band of wavelength when subjected to positive voltage. The response of LEDs to changes to input current must be quite fast to display even time-variant experimental data; for example, turn-on and turn-off can occur in 10 ns. But it is quite difficult for an observer to properly record and then store this fast-varying data. For displaying a lot of data simultaneously, an LCD system is used. An LCD system does not emit light like an LED but rather uses incident light from other sources to display experimental data. A thin layer of liquid crystal is sandwiched between two thin transparent metal electrodes for forming an LCD display unit for a measurement system. Interestingly, an LCD display needs less power than an LED

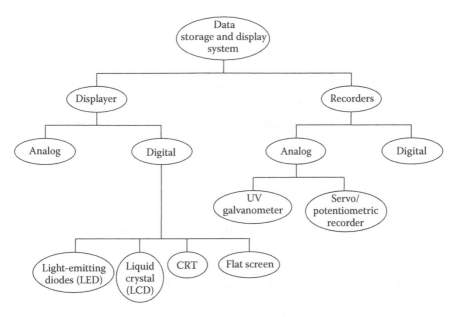

Figure 3.22

Classification of data storage and display systems.

array, but an LCD is not visible in dimly lit surroundings unlike an LED display that can be displayed in a dark environment.

For displaying data in a large-scale display, a cathode ray tube (CRT) is used that consists of a visual display unit (monitor) along with a keyboard. This can be used for storing data temporarily, which is known as a storage oscilloscope. As the CRT is not preferred in modern times due to its bulkiness and fragility, several kinds of flat screen display devices such as liquid crystal, electroluminescence, gas plasma discharge, and vacuum fluorescence are replacing it. Note that in all these units, the data is displayed in the form of images using a matrix of pixels. Both alphanumeric and graphic forms can be displayed on these flat screens. Other devices such as compact discs (CDs), magnetic cores, magnetic disks, and various kinds of semiconductor devices are also being used for storing experimental data.

Experimental data can be recorded using both analog and digital means, as shown in Figure 3.22. An analog chart recorder is directly connected to an analog voltage source such as an amplifier or deflection bridge. The ultraviolet (UV) galvanometer chart recorder is a delicate device with low impendence and highly sensitive high bandwidth that is used for measuring high-speed data in laboratories. In contrast, a closed-loop servo or potentiometric recorder is a rugged device with high impendence and a high accuracy bit with low bandwidth that is used mostly in industrial setups. Several types of alphanumeric printers, such as dot-matrix, laser, and inkjet, are also used to record digital data.

Questions

1. Why is a data acquisition system required for meaningful measurement in combustion systems?

2. What are various constituents of a typical data acquisition system? Describe each component both pictorially and verbally.

3. What are the kinds of amplifiers used in measurement systems?

4. What is an operational amplifier? Explain its components and their respective functions.

5. What is an inverting amplifier and how is it different from a noninverting amplifier?

6. What is an instrumentation amplifier? You may refer other resources to answer this question.

7. Why does one need to use filters while acquiring data? What are the kinds of filters used in measurement systems?

8. What is the difference between active and passive filters?

9. Why is a digital signal preferred for transmitting and conditioning of signals?

10. What is aliasing? How can it be achieved?

11. Define the resolution for an A/D converter with an example.

12. How is an analog signal quantized? Explain using an example.

13. What are the standard methods used for data storage in modern times?

Problems

1. Determine the gain G and G_{dB} when 10 mV is being amplified to produce an output voltage of 1.2 V.

2. An amplifier with a gain of 100 dB is to be used to amplify input voltage of 1.2 mV. Determine its output voltage.

3. A transducer is being modeled as a 75-mV voltage generator in series with a resistance of 450 Ω. This transducer with an input resistance of 2.1 MΩ is connected to an amplifier with a gain of 15 and resistance of 45 Ω. Determine the loading input error for a loading resistance of 3.0 kΩ. What will be the reduction in loading error if loading resistance is decreased to 1.0 kΩ?

4. A pressure sensor has an open-circuit output voltage of 85 mV with a resistance of 350 Ω. This sensor with an input resistance of 2.5 MΩ is

connected to an amplifier with a gain of 20 and a resistance of 65 Ω. Determine the output voltage and loading resistance if the loading input error must be restricted to 2%.

5. A transducer contaminated by 50-Hz noise has to respond to oscillations up to 3 Hz. A first-order lowpass Butterworth filter is used to reduce the 50-Hz noise. The resistance R_1 is equal to 12 kΩ. Determine the amplitude of noise that can be reduced with this filter.

6. A transducer senses a signal of 65 mV at 250 Hz while incurring 15-mV noise at a frequency of 10 Hz. In order to reduce the 10-Hz noise, a highpass filter is used at a cutoff frequency of 50 Hz. Determine the output voltage and % error in it due to noise in the filtered signal. Determine the phase change for the output voltage.

7. A transducer senses a signal of 65 mV at 10 Hz while picking 15-mV noise at a frequency of 200 Hz. In order to reduce noise at 200 Hz, a lowpass filter is used at a cutoff frequency of 50 Hz. Determine the output voltage and % error in it due to noise in the filtered signal.

References

1. Holman, J. P., *Experimental Methods for Engineers,* Sixth Edition, McGraw-Hill, Inc., New York, 1994.
2. Doebelin, E. O., *Measurement System Application and Design,* Fourth Edition, McGraw-Hill Publishing Co., New York, 1990.
3. Vasudevan, K., *Digital Communications and Signal Processing,* Universities Press, Hyderabad, India, 2010.
4. Jones, B. E., *Instrumentation Measurement and Feedback,* Tata-McGraw-Hill, New Delhi, India, 1989.
5. Bentley, J. P., *Principles of Measurement Systems,* Pearson India Ltd., New Delhi, India, 2009.

4

Data Analysis

Data has to be analyzed using one's own judgment and intelligence for deciphering its true meaning.

D. P. Mishra

4.1 Introduction

In Chapter 3, we discussed microprocessor-based data acquisition systems and data processing systems that are used for measurement of not only combustion systems but other systems as well. Recall that we also discussed data acquisition and data handling in terms of how the integrity of data must be maintained properly as it can lead to misinterpretation of actual phenomena during a process. In this chapter we will learn how to properly analyze the measured data in order to have a better understanding of the phenomena during a process or to control the process. The objective of this chapter is to teach readers how to analyze experimental data and identify and estimate various types of experimental errors. Various types of statistical tools are discussed to analyze errors in experimental data that can be used for the measurement of combustion as well as other measurement systems. In order to develop a physical understanding of the subject, various types of graphical analysis are covered. Some commonly used curve-fitting techniques are also covered briefly, as well as a short introduction to report writing.

4.2 Critical Analysis of Experimental Data

We need to analyze obtained experimental data critically to find out certain patterns that can be generalized to test an existing hypothesis or carve out a new hypothesis. As mentioned earlier, data analysis is the process of synthesizing and analyzing experimental data with the intent to unravel useful information and derive meaningful conclusions. But before deriving any conclusions from experimental data, it is essential to ascertain the quality of the data produced during experimentation. Recall that we try to measure the value of a physical quantity. In order to assess the quality of experimental data we will have to determine how the measured data deviates from the actual value of the physical quantity being measured. We will have to use our common sense in consistently recognizing certain patterns while acquiring the measured data. For example, temperature will increase with the addition of heat to a system. If some data point defies this common sense, it may be eliminated before carrying out any further data analysis. But if more data in an experiment defy common sense consistently, we need to review the entire experimentation procedure diligently. After examining the experimental data for consistency using common sense or previous theories, we need to carry out statistical analysis, particularly when experiments are repeated several times, which will help us to determine the level of confidence. Once confidence in the acquired experimental data is obtained, it is important to determine the uncertainty level that can be stated along with the experimental data. It must be emphasized that the experimental data must be visualized properly to reveal inherent concepts, phenomena, patterns, or hypotheses lying hidden in the experimental data. For this purpose, it is customary to have a hypothesis based on either previous theory or antithesis. Certain physical nondimensional terms are used for deriving a better understanding of experimental data. Once we understand the experimental data, it is equally important and essential to develop a semi-empirical relationship for future use in design and development of products and procedures. It is also important to present the findings of any experimental endeavor in the form of a suitable report. In this chapter, readers will be made aware of various tools namely error analysis, statistical analysis, uncertainty analysis, regression analysis, and report writing.

4.3 Experimental Error Analysis

Errors are likely to occur during the process of experimentation regardless of how careful an experimenter is. Generally measurement error can be defined as the difference between the measured value and true value of a physical quantity. Although this definition of error provides a way to ascertain the quality of the data, it would not be helpful to quantify the experimental error in any measurement system as we can never determine the true value of a physical quantity without measuring it. That does not mean that we cannot ascertain the extent of errors incurred during any measurement; rather, we can determine and express them in terms of an uncertainty level. This experimental uncertainty can be conceived of as the possible value of the errors incurred during experimentation. Before we delve into the extent of these errors, we need to understand the types of errors that are likely to occur in the measurement of any physical quantity. Therefore, it is important and essential to identify the types of errors incurred during the experimentation. The specific causes of errors can vary from one observation to another. Even in an experiment, errors may come from several sources. There might be a gross error in an instrument or experimental setup or an error that might occur due to a faulty experimental procedure, such as a fault in a reading scale, which may result in an error in measurement. As well, personal judgment differs from one person to another in taking the reading from the same instrument under the same conditions. These kinds of errors can be detected easily if several experimentalists acquire the same measurement data. Even then it is difficult to have error-free data unless certain standard values are known. Generally, personal bias in collecting certain data can lead to errors in measurement when the data will be interpreted by the observer himself or herself to prove a certain hypothesis. Of course this can be avoided either by having another person collect the data who is not involved in the process of testing this hypothesis or theory, or by collecting the data using an unbiased system such as a data acquisition system and then scrutinizing this data using the statistical analysis method. Measurement errors can occur due to backlash, friction, or hysteresis in an instrument. These kinds of errors can be avoided during the calibration of an instrument, which must be carried out before initiating any serious experimental work. Besides these errors, there are certain built-in errors that can occur due to incorrect design, fabrication, or poor maintenance of equipment. If an instrument has a certain limitation of resolution, errors may incur that can be avoided by using an instrument with better resolution. If detected, these errors can be recertified easily to improve the quality of the data. However, an instrument may have a fixed error for some unknown reason, which will result in the same or a similar error every time measurements are repeated. For example, if a mercury-in-glass thermometer is used to measure the temperature of hot fluid, it may measure a lower temperature than the actual fluid temperature, due to heat loss from the stem of thermometer. If it happens to read fluid temperature by a 2% higher value than its true value, then on repeating this measurement using the same instrument, the entire set of measured values will be offset by 2% from its true value. This kind of error is known as a *fixed error*, which is also sometimes referred to as bias or *systematic error*. There is another kind of error known as a *random (precision) error* in which measured value varies from true value by a different amount even when it is repeated using the same instrument. This kind of error may be caused by random fluctuations in the instrument either due to mechanical, electrical, or electronic noises that are basically beyond the control

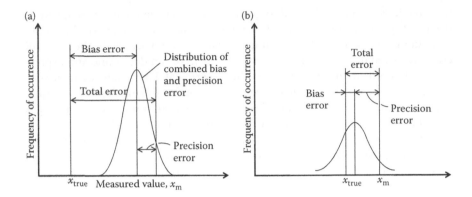

Figure 4.1

Schematic of the types of errors in a typical measurement system: (a) a bias error greater than a precision error and (b) a bias error less than a precision error.

of the experimenter. For example, vibration in an instrument can cause the reading of the measurement to vary from the true value. On repetition, the measured value varies slightly from its true value, creating a distribution around it. Most times, the random errors may follow a certain statistical distribution. In such situations, the extent of the errors can be quantified with the help of statistical analysis methods, which will be discussed briefly in subsequent sections. In contrast, the statistical method cannot be used to determine the extent of the fixed error as it is fixed and does not have any distribution.

In a practical situation, bias and random errors occur simultaneously. Hence, total errors for a particular measurement can be expressed as the sum of fixed (bias) errors and random (precision) errors, as depicted in Figure 4.1. In this case, a random error is higher than a bias error. In some other situations, a bias error is higher than a random error, and in other circumstances, a bias error is in the same order as a random error. Data that has small random errors is termed *high-precision data*. When both fixed and random errors are small, then it is termed *high-accurate data*.

4.4 Uncertainty in Experimental Data

We learned in the previous section that no measurement is perfectly accurate. Hence, it is important to ascertain the extent of inaccuracies in data. In other words, there will be uncertainty in experimental data that is mainly caused by two types of errors: (i) bias (systematic) and (ii) random (precision). The extent of inaccuracies in data can be expressed in terms of uncertainty. An uncertainty should not be construed as an error as it is the difference between true value and the recorded value. Note that an error is a fixed number and cannot be considered as a statistical variable. On the other hand, an uncertainty is a possible value of the error that can vary over a certain range around the average of measured data. The uncertainty in any measurement system can vary a great deal depending on the circumstances under which experiments are conducted. Its causes cannot be identified precisely as it is random in nature. Rather, the uncertainty can be conceived of as a histogram of values and thus can be determined using the statistical method as it is inherently a statistical variable. Ideally, the uncertainty level in an experiment is estimated from the repeated data. The data can be obtained from experiments in two ways: by (i) single-sample and (ii) multiple-sample. When a whole set of data is obtained using a single instrument, then this is termed single-sample data. In this case certain uncertainties may not be identified by repeating the experiment because the same instrument is used for obtaining the same data. For example, a thermometer is used to measure the temperature of fluid. If the measurements are carried out several times, the same error inherent in this thermometer will be recorded no matter how many times readings are taken. On the other hand, if more than one thermometer is used to measure the temperature of the same fluid, then it is possible to identify the uncertainty in measurements. This kind of data is known as multiple-sample data, which is rarely considered in actual practice due to its prohibitive cost and time commitment. Hence, we will restrict our discussion to single-sample data analysis in this chapter. Readers are encouraged to refer to others' works [1,4,5] for advanced data analysis.

In order to estimate uncertainty in experimental results we will devote this section to discussing the precise uncertainty analysis method as enumerated in the work of Kline and McClintock [1]. Recall that uncertainty expresses the extent of error in any set of measurements. It is the possible value of error that might be incurred during any measurement. But the total error, δ, is usually expressed in terms of two components—bias error, β, and random error, ε, as

$$\delta_i = \beta + \varepsilon_i \tag{4.1}$$

where the subscript, i, is the ith measurement. Note that the bias error, β, is fixed and systematic in nature while the random error, ε, is variable, which differs for each measurement. When the number of measurements, N, is quite large and approaches infinity the data would follow a normal distribution, as shown in Figure 4.2. Then the systematic/bias error, β, can be determined by taking the difference between the mean value of the N readings, X_m, and true value, X. In contrast the random error would cause the readings to vary around the mean value. Let us consider an example of temperature measurements of fluid by a thermometer. A histogram of these temperature measurements is shown in Figure 4.3. We will discuss the statistical point of view further in the next section. For the time being, let us understand how to specify the error in the measured data. Note that the thermometer is biased by 0.5°C and the true value of the fluid happens to be 39.5°C (see Figure 4.3). The average temperature turns out to be 39°C. Unfortunately, in a practical situation, the true value of a measurement, X, would not be known. Therefore, it would not be

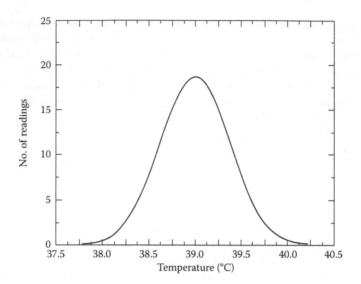

Figure 4.2

A typical normal distribution of experimental data.

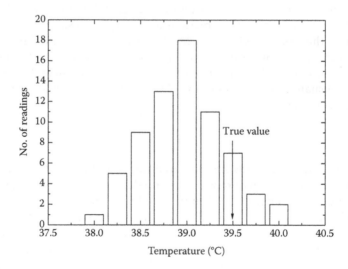

Figure 4.3

Histogram of temperature readings of fluid by a thermometer.

possible to determine the exact bias and random errors for a set of measurements. However, based on measurements, we can express the value of data, X, that can lie within a certain interval as

$$X = X_m \pm U \qquad (4.2)$$

4. Data Analysis

where X_m is the mean value of the N readings and U is the uncertainty in the value of X that expresses a reasonable limit for error for a given parameter. For example, the temperature measured in the above example can be expressed as

$$T = 39 \pm 1°C \qquad (4.3)$$

The above expression states that the temperature data is within $\pm 1°C$ with odds of 20:1. In other words, the probability of this measurement within error of $\pm 1°C$ is 19 in 20 measurements. This kind of statement can be made by the experimenter after conducting a particular experiment several times. This idea of degree of confidence in uncertainty specification while studying the specific heat of ammonia was introduced in 1930 by P. H. Myers and his colleagues at NBS (National Bureau of Standard, USA) in a humorous manner in their research paper:

> We think our reported value is good to one part in 10,000; we are willing to bet our own money at even odds that it is correct to two parts in 10,000; furthermore if by any chance our value is shown to be in error by more than one part in 1000, we are prepared to eat our apparatus and drink the ammonia.

We will learn more how this uncertainty, U, can be estimated by carrying out a statistical analysis in the next section. However, let us learn how to carry out an uncertainty analysis, as illustrated in Section 4.5.

4.5 General Uncertainty Analysis

Let us consider a set of measurements for which we need to determine the uncertainty in each primary measurement that can be expressed with the same odds. The desired results of the experiments can be estimated from these measurements. For result, R, let $X_1, X_2, X_3, \ldots\ldots X_i, \ldots\ldots\ldots, X_N$, be the independent variables in experimental measurements and $U_1, U_2, U_3, \ldots\ldots U_i, \ldots\ldots\ldots, U_N$, be the uncertainties in these independent variables. Then the uncertainty in result, U_R, can be expressed [1] as

$$U_R = \left[\left(\frac{\partial R}{\partial X_1} U_1 \right)^2 + \left(\frac{\partial R}{\partial X_2} U_2 \right)^2 + \ldots\ldots\ldots + \left(\frac{\partial R}{\partial X_i} U_i \right)^2 + \ldots\ldots\ldots \left(\frac{\partial R}{\partial X_N} U_N \right)^2 \right]^{\frac{1}{2}}$$

$$(4.4)$$

where the results, R, can be expressed as

$$R = R(X_1, X_2, X_3 \ldots\ldots\ldots\ldots X_i, \ldots\ldots\ldots X_N) \qquad (4.5)$$

The effect of error in a measured individual variable, X_i, on R can be estimated by an analogy to the derivative of a function. The change in R can be related to the small changes in X as

$$\delta R = A_1 \delta X_1 + A_2 \delta X_2 + A_3 \delta X_3 + \ldots\ldots\ldots + A_N \delta X_N = \sum_{i=1}^{N} A_i \delta X_i \qquad (4.6)$$

The partial derivative obtained from Equation 4.6 with the ith variable can be determined as

$$\frac{\partial R}{\partial X_i} = A_i \qquad (4.7)$$

Note that Equation 4.7 can be exact only if any change in each X_i is infinitesimal. If it is not, then it will be just an approximation. For each variable of the measured data, δX_i is known as uncertainty in the variable, which can be denoted by U_i. It must be appreciated that each term in the right-hand side of Equation 4.6 can be either positive or negative as it represents the most probable error for each variable. Therefore, if we will sum up all the contributions of uncertainty from each dependent variable in Equation 4.7, it would not represent the true representative value of uncertainty in any measurement because some of the terms would cancel each other, which may lead to a zero value of uncertainty U in certain situations. But we can estimate the maximum possible uncertainty U_{max} in R by taking note of all terms in the right-hand side of Equation 4.8:

$$U_{max} = \sum_{i=1}^{N} \left| \frac{\partial R}{\partial X_i} U_i \right| \qquad (4.8)$$

But the above method of estimating error will be erroneous, as it would not represent the true value of the uncertainty because it will never be possible for all terms in Equation 4.8 to be positive simultaneously. Therefore, let us consider the root mean square of uncertainty in each term of Equation 4.8 for estimating the uncertainty, which can represent the error more accurately as

$$U_R = \left[\sum_{i=1}^{N} \left(\frac{\partial R}{\partial X_i} U_i \right)^2 \right]^{\frac{1}{2}} = \left[\sum_{i=1}^{N} (A_i U_i)^2 \right]^{\frac{1}{2}} \qquad (4.9)$$

In certain situations, the result function, R, of Equation 4.9 can take the form of a product of the respective primary variables raised to exponents, and can be expressed as

$$R = X_1^{A_1} X_2^{A_2} \ldots\ldots X_i^{A_i} \ldots\ldots\ldots X_N^{A_N} \qquad (4.10)$$

By performing a partial derivative of Equation 4.10, we have

$$\frac{\partial R}{\partial X_i} = X_1^{A_1} X_2^{A_2} \dots\dots A_i X_i^{A_i-1} \dots\dots X_N^{A_N} \tag{4.11}$$

Dividing Equation 4.11 by Equation 4.10, we obtain

$$\frac{1}{R}\frac{\partial R}{\partial X_i} = \frac{A_i}{X_i} \tag{4.12}$$

Substituting the above relation in Equation 4.9, we get

$$\frac{U_R}{R} = \left[\sum \left(\frac{A_i U_i}{X_i} \right)^2 \right]^{\frac{1}{2}} \tag{4.13}$$

The above relation for fractional uncertainty in the results, $\frac{U_R}{R}$, can be used only when the result takes the form of a product of the respective primary variable raised to exponents (Equation 4.10). It must also be appreciated that the uncertainly propagation in the results, U_R, depends on the square of the uncertainty in each independent variable. In other words, the largest uncertainty of one variable or more will dictate the actual value of total uncertainty. Therefore, the uncertainty of other variables with smaller values can be neglected easily. Let us now consider one example for the additive function of error to illustrate how the uncertainty analysis can be used for practical purposes.

Example 4.1

A venturi flow meter is used to measure steady gas flow rate, which can be expressed in terms of pressure change across as per the following expression:

$$\dot{m} = C_D A \sqrt{\left[\frac{2P_1}{RT_1}(P_1 - P_2) \right]}$$

where C_D is the discharge coefficient, A is the flow area, P_1 and P_2 are the upstream and downstream pressure across the venturi flow meter, T_1 is the upstream temperature, and R stands for gas constant used in ideal gas law. Determine the uncertainty in the mass flow rate for the data of the following variables: $C_D = 0.98 \pm 0.02$, $A = 50 \pm 0.2$ mm^2, $P_1 = 150 \pm 5$ kPa, $T_1 = 25 \pm 0.2$ °C, $\Delta P = P_1 - P_2 = 15 \pm 0.05$ kPa.

Given: $C_D = 0.98 \pm 0.02$, $A = 50 \pm 0.2$ mm^2, $P_1 = 150 \pm 5$ kPa, $T_1 = 25 \pm 0.2$ °C, $\Delta P = P_1 - P_2 = 15 \pm 0.05$ kPa.

To Find: Uncertainty, $U_{\dot{m}}$.

Solution: The mass flow rate is dependent on several variables: C_D, P_1, T_1, ΔP, A, which can be expressed as

$$\dot{m} = f(C_D, A, P_1, T, \Delta P) \tag{a}$$

The derivatives for each variable can be obtained from the expression of the mass flow rate as

$$\frac{\partial \dot{m}}{\partial C_D} = A \sqrt{\left[\frac{2P_1}{RT_1} \Delta P \right]} \tag{b}$$

$$\frac{\partial \dot{m}}{\partial A} = C_D \sqrt{\left[\frac{2P_1}{RT_1} \Delta P \right]} \tag{c}$$

$$\frac{\partial \dot{m}}{\partial P_1} = 0.5 A C_D \sqrt{\left[\frac{2}{RT_1} \Delta P \right] P_1^{0.5}} \tag{d}$$

$$\frac{\partial \dot{m}}{\partial T_1} = -0.5 A C_D \sqrt{\left[\frac{2P_1}{R} \Delta P \right] T_1^{-3.2}} \tag{e}$$

$$\frac{\partial \dot{m}}{\partial \Delta P} = 0.5 A C_D \sqrt{\left[\frac{2P_1}{RT_1} \right] \Delta P^{-0.5}} \tag{f}$$

By substituting these derivatives in Equation 4.9 and dividing it by Equation a, we have

$$\frac{U_{\dot{m}}}{R} = \left[\left(\frac{U_{C_D}}{C_D} \right)^2 + \left(\frac{U_A}{A} \right)^2 + \left(\frac{U_{P_1}}{2P_1} \right)^2 + \left(\frac{U_{T_1}}{2T_1} \right)^2 + \left(\frac{U_{\Delta P_1}}{2\Delta P_1} \right)^2 \right]^{\frac{1}{2}} \tag{g}$$

By substituting the values in the above equation, we can determine uncertainty in the mass flow rate as

$$\frac{U_{\dot{m}}}{R} = \left[\left(\frac{0.02}{0.98} \right)^2 + \left(\frac{0.2}{50} \right)^2 + \left(\frac{5}{2 \times 150} \right)^2 + \left(\frac{0.2}{2 \times 25} \right)^2 + \left(\frac{0.04}{2 \times 15} \right)^2 \right]^{\frac{1}{2}}$$

$$= [4.165 \times 10^{-4} + 1.6 \times 10^{-6} + 2.77 \times 10^{-4} + 1.6 \times 10^{-5} + 1.77 \times 10^{-6}]^{\frac{1}{2}} = 0.0208 \tag{h}$$

The uncertainty in the mass flow rate is ±2.08%, which is mainly influenced by the uncertainty in C_D and inlet pressure. Hence, efforts must be made to reduce uncertainties in C_D and inlet pressure if possible.

4.5.1 Applications of the Uncertainty Analysis

The uncertainty analysis has several applications, some of which are:

- It can be used to carry out a thorough examination of the experimental procedure for identification of potential sources of errors
- It helps in recognizing the need for using improved instruments and experimental procedures for achieving the desired output accuracy
- It helps in identifying the instrument and/or experimental procedure that controls accuracy
- Using this analysis method, experimental cost can be minimized for requisite output accuracy
- It can be used to carry out a feasibility analysis of an experiment for achieving certain results with the desired accuracy
- It provides the appropriate basis for ascertaining the validity of both experimental and computational data pertaining to certain hypotheses or testing of certain components
- It provides the basis for guarantees of accuracy in the testing of large equipment such as power plants, spacecraft, and so forth
- It allows the design of probes and experimental procedures for a minimum uncertainty level
- It serves as a check against making mistakes by using grossly erroneous data with larger uncertainty levels
- It provides an integrated grasp on how to run a given experiment within the available constraints

Besides these above applications, there may be several other currently undiscovered uses that can be explored by researchers of the future.

4.6 Statistical Analysis of Data

In the previous section, we learned about bias and precision errors and also about the degree of confidence in error estimation. As precision errors are random in nature, it is important to learn how to determine the quantitative value of scatter in data and its degree of confidence. The quantitative measure of scatter in data can be determined by carrying out a statistical analysis, which we will discuss in this section.

In order to illustrate the efficacy of this method, let us consider our previous example of temperature data using a thermometer, which is plotted in a histogram as shown in Figure 4.3. The data varies within a certain limit around the mean value of 39°C with some data being higher and others being lower. Of course, these are finite data points. However, if a large amount of data is considered such that it almost approaches infinity, the histogram would become smoother and resemble a bell-shaped curve, as shown in Figure 4.2. This is an ideal situation when an infinite amount of data will be available as it is quite cumbersome and costly to generate and handle such large amounts of data. In a real situation, we do not have the luxury of generating a plethora of data within the constraints of time and resources. Hence, we will have to make do with samples of data composed of a finite number of data taken from the parent population.

4.6.1 Normal Error Distribution

When a series of experimental data is obtained, there will be variations in the readings due to errors that are random in nature. For an infinite (large) number of data points, their distribution can follow a smooth curve that is known as the normal/Gaussian distribution. The normal/Gaussian distribution mimics most of the experimental and numerical data and is used extensively in real situations provided a large number of data is available. Let us consider N data points, obtained for a particular variable X. The fraction dN/N of the population of observations whose value lies in the range of X and $X + dX$ can be expressed as

$$P(X) = \frac{dN}{N} = \frac{1}{\sigma\sqrt{2\pi}} e^{-(X-X_m)^2/2\sigma^2} dX \tag{4.14}$$

where $P(X)$ is probability density, X_m is the mean reading of data set (population) that is the most probable data, and σ is the standard deviation that specifies the distribution of the data. The true error, ε, is specified by the deviation, d, as stated earlier, which is given by

$$\varepsilon = X - X_m \tag{4.15}$$

From a set of experimental data, we need to determine the mean as each set of data will be different from each reading. For N number of readings of its variable denoted by X_i, the arithmetic mean X_m can be determined by

$$X_m = \frac{1}{N} \sum_{i=1}^{N} X_i \tag{4.16}$$

The deviation d_i of the ith variable from its mean value can be estimated as

$$d_i = X_i - X_m \tag{4.17}$$

Note that the average deviation of all readings becomes zero, as given below:

$$\bar{d_i} = \frac{1}{N} \sum_{i=1}^{N} d_i = \frac{1}{N} \sum_{i=1}^{N} (X_i - X_m) = 0 \tag{4.18}$$

4. Data Analysis

However, the average of the absolute values of the deviation is not necessarily zero. The standard deviation, σ, is basically the root mean square of the deviation, which is expressed as

$$\sigma = \lim_{N \to \infty} \sqrt{\left[\frac{1}{N} \sum_{i=1}^{N} (X_i - X_m)^2 \right]}$$

(4.19)

Note that the square of the standard deviation that indicates the measure of the spread of a data set (population) is known as variance or biased standard deviation because it is valid only for a large number of samples. However, on several occasions it may not be possible to use several data sets. For general validity of data, it is desirable to have at least 20 data points to estimate standard deviations reliably. But for a small number of data, the unbiased standard deviation, also known as the *precision index* (PI), can be determined by

$$PI = \sqrt{\frac{1}{N-1} \sum (X_i - X_m)^2}$$

(4.20)

Note that the factor $N - 1$ is used instead of N as in Equation 4.19. This unbiased standard deviation can be used only when the population of the sample is known.

The variation of the probability density is plotted in Figure 4.4 for three cases of standard deviation, $\sigma = 0.25, 0.5$, and 1.0 for which the mean, X_m, is equal to 10.

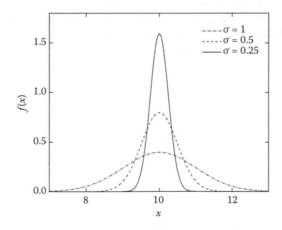

Figure 4.4

Normal distribution plot for $\sigma = 0.25, 0.5,$ and 1.0.

The maximum probability density occurs at $X = X_m$ whose value can be determined by using Equation 4.21 as

$$P(X_m) = \frac{dN}{N} = \frac{1}{\sigma\sqrt{2\pi}}$$ (4.21)

It can be observed from this figure that the probability of X values being closer to the mean value, X_m, increases with the decrease in the standard deviation, σ. In other words, scatter in the data of X becomes larger with an increase in the standard deviation, σ, leading to a larger precision error. Hence, $P(X_m)$ can be considered as the measure of the precision of the data. If we can express Equation 4.21 in a normalized form then the total area under this curve becomes unity, which can be expressed mathematically as

$$\int_{-\infty}^{+\infty} P(X)\, dX = 1$$ (4.22)

The probability of the data has to be integrated between plus and minus infinity. But for a practical case, we need to determine the probability of data that will fall within a certain range around the mean value in a normal distribution, which can be expressed as

$$P(\Delta X) = \int_{X_m - \Delta X}^{X_m + \Delta X} \frac{1}{\sigma\sqrt{2\pi}} e^{-(X - X_m)^2/2\sigma^2}\, dX$$ (4.23)

Note that it would not be possible to determine this integral in a closed form. Even if its value can be tabulated for a range of ΔX, then we need to have a table for each pair of (X_m, σ) values. In order to avoid this cumbersome problem of handling an infinite number of tables, we can normalize the integral such that a single table will be sufficient to determine its integral. Let us define the normalized deviation, η, from its mean value, X_m

$$\eta = \frac{X - X_m}{\sigma}$$ (4.24)

By substituting this normalized variable in Equation 4.23, we have

$$P(\eta) = \frac{1}{\sqrt{2\pi}} \int_{-\eta_1}^{\eta_1} e^{-\eta^2/2}\, d\eta$$ (4.25)

where $\eta_1 = X_1/\sigma$. The value of the probability, $P(\eta)$, can be represented as the area under the normal distribution of the curve, as shown in Figure 4.5, between $-\eta_1$ and η_1, which is known as a two-trailed probability. As this normal distribution

4. Data Analysis

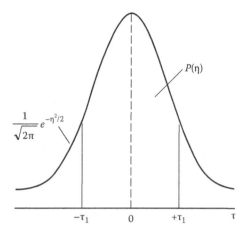

Figure 4.5

Graphical representation of the probability $P(\eta)$.

is symmetric, the probability of a set of data is equal to 1/2 $P(\eta)$, which is known as a single-trailed probability. The values of a two-trailed probability $P(\eta)$ for η varying from 0 to 5 are given in Table 4.1, which can used for its evaluation.

4.6.2 Student's *t* Distribution

We have considered that we can estimate the true value of a particular piece of data by taking an arithmetic mean of a set of experimental data. But we have not asked how good this arithmetic mean that represents the true value is. Of course, to answer this question, one has to conduct a set of measurements and find a new arithmetic mean. Generally we may find that this new arithmetic value may not be same as the previous mean value. Therefore it is difficult to resolve this problem unless we otherwise acquire a large amount of data. This problem can be alleviated to some extent if we can consider the mean standard deviation, σ_m, which is defined as

$$\sigma_m = \frac{\sigma}{\sqrt{N}} \tag{4.26}$$

where σ is the standard deviation of a set of measurements and N is the number of measurements in a set. This expression indicates that the random component of uncertainty in measurements can be reduced by acquiring more experimental data and averaging them. It can be applied for experimental data even if its distribution would not be Gaussian in nature. The method of determining the mean standard deviation would not be satisfactory, particularly when the number of data points is quite low ($N < 10$), as in engineering. In order

Table 4.1 Values of a Single-Trailed Normal Distribution Probability $P(\eta)$ from 0 to ∞

η	0.00	0.01	0.02	0.03	0.04	0.05	0.06	0.07	0.08	0.09
0.0	.0000	.0040	.0080	.0120	.0160	.0199	.0239	.0279	.0319	.0359
0.1	.0398	.0438	.0478	.0517	.0557	.0596	.0636	.0675	.0714	.0753
0.2	.0793	.0832	.0871	.0910	.0948	.0987	.1026	.1064	.1103	.1141
0.3	.1179	.1217	.1255	.1293	.1331	.1368	.1406	.1443	.1480	.1571
0.4	.1554	.1591	.1628	.1664	.1700	.1736	.1772	.1808	.1844	.1879
0.5	.1915	.1950	.1985	.2019	.2054	.2088	.2123	.2157	.2190	.2224
0.6	.2257	.2291	.2324	.2357	.2389	.2422	.2454	.2486	.2517	.2549
0.7	.2580	.2611	.2642	.2673	.2704	.2734	.2764	.2794	.2823	.2852
0.8	.2881	.2910	.2939	.2967	.2995	.3023	.3051	.3078	.3106	.3133
0.9	.3159	.3186	.3212	.3238	.3264	.3289	.3315	.3340	.3365	.3389
1.0	.3413	.3438	.3461	.3485	.3508	.3531	.3554	.3577	.3599	.3621
1.1	.3643	.3665	.3686	.3708	.3729	.3749	.3770	.3790	.3810	.3830
1.2	.3849	.3869	.3888	.3907	.3925	.3944	.3962	.3980	.3997	.4015
1.3	.4032	.4049	.4066	.4082	.4099	.4115	.4131	.4147	.4162	.4177
1.4	.4192	.4207	.4222	.4236	.4251	.4265	.4279	.4292	.4306	.4319
1.5	.4332	.4345	.4357	.4370	.4382	.4394	.4406	.4418	.4429	.4441
1.6	.4452	.4463	.4474	.4484	.4495	.4505	.4515	.4525	.4535	.4545
1.7	.4554	.4564	.4573	.4582	.4591	.4599	.4608	.4616	.4625	.4633
1.8	.4641	.4649	.4656	.4664	.4671	.4678	.4686	.4693	.4699	.4706
1.9	.4713	.4719	.4726	.4732	.4738	.4744	.4750	.4756	.4761	.4767
2.0	.4772	.4778	.4783	.4788	.4793	.4798	.4803	.4808	.4812	.4817
2.1	.4821	.4826	.4830	.4834	.4838	.4842	.4846	.4850	.4854	.4857
2.2	.4861	.4864	.4868	.4871	.4875	.4878	.4881	.4884	.4887	.4890
2.3	.4893	.4896	.4898	.4901	.4904	.4906	.4909	.4911	.4913	.4916
2.4	.4918	.4920	.4922	.4925	.4927	.4929	.4931	.4932	.4934	.4936
2.5	.4938	.4940	.4941	.4943	.4945	.4946	.4948	.4949	.4951	.4952
2.6	.4953	.4955	.4956	.4957	.4959	.4960	.4961	.4962	.4963	.4964
2.7	.4965	.4966	.4967	.4968	.4969	.4970	.4971	.4972	.4973	.4974
2.8	.4974	.4975	.4976	.4977	.4977	.4978	.4979	.4979	.4980	.4981
2.9	.4981	.4982	.4982	.4983	.4984	.4984	.4985	.4985	.4986	.4986
3.0	.4987	.4987	.4987	.4988	.4988	.4989	.4989	.4989	.4990	.4990
3.1	.4990	.4991	.4991	.4991	.4992	.4992	.4992	.4992	.4993	.4993
3.2	.4993	.4993	.4994	.4994	.4994	.4994	.4994	.4995	.4995	.4995
3.3	.4995	.4995	.4995	.4996	.4996	.4996	.4996	.4996	.4996	.4997
3.4	.4997	.4997	.4997	.4997	.4997	.4997	.4997	.4997	.4997	.4998
3.5	.4998	.4998	.4998	.4998	.4998	.4998	.4998	.4998	.4998	.4998
3.6	.4998	.4998	.4999	.4999	.4999	.4999	.4999	.4999	.4999	.4999
3.7	.4999	.4999	.4999	.4999	.4999	.4999	.4999	.4999	.4999	.4999
3.8	.4999	.4999	.4999	.4999	.4999	.4999	.4999	.4999	.4999	.4999
3.9	.5000	.5000	.5000	.5000	.5000	.5000	.5000	.5000	.5000	.5000
4.0	.5000	.5000	.5000	.5000	.5000	.5000	.5000	.5000	.5000	.5000

to overcome this problem, an amateur statistician, W. S. Gasset (1876–1937), under the pseudonym "Student," proposed the following distribution of quantity t as given below:

$$t = \frac{X_m - X_i}{PI/\sqrt{N}} \tag{4.27}$$

Note that the standard deviation is replaced by the PI. The probability distribution of t-statistics will be similar in nature to that of Gaussian distribution. Therefore, the confidence level, Δ, can be estimated by using the t distribution, given below as

$$\Delta = \frac{t\,PI}{\sqrt{N}} \tag{4.28}$$

Table 4.2 provides the value of Student's t distribution for different degrees of freedom. This Student's t distribution becomes the normal distribution when the amount of data tends toward infinity (large N).

Example 4.2

A series of experiments is carried out to calibrate a turbine flow meter. Under the same conditions, measurement of the flow rate is repeated seven times and the values are 453, 458, 449, 443, 460, 439, and 444 m³/s. Determine the mean value, precision index, and 95% confidence interval for this mean value.

Given: $Q = 453, 458, 449, 443, 460, 439$, and 444 m³/s, $N = 7$.

To Find: X_m, X_m with 95% confidence.

Solution: Because the sample size is less than 30, we can use Students's t distribution to determine the confidence level. For this let us first calculate the mean and precision index of the data given in this example as

$$X_m = \frac{453 + 458 + 449 + 443 + 460 + 439 + 444}{7} = 449.43 \tag{a}$$

$$\sigma = \left(\frac{1}{6} \left[\begin{array}{l} (453-449.43)^2 + (458-449.43)^2 + (449-449.43)^2 + (443-449.43)^2 \\ + (460-449.43)^2 + (439-449.43)^2 + (444-449.43)^2 \end{array} \right] \right)^{0.5} = 7.94 \tag{b}$$

Table 4.2 Values of Two-Tailed Probability $P(\eta)$

Degree of Freedom ν	$t_{0.20}$	$t_{0.10}$	$t_{0.05}$	$t_{0.02}$	$t_{0.01}$	$t_{0.002}$	$t_{0.001}$
1	3.078	6.314	12.71	31.82	63.66	318.3	637
2	1.886	2.920	4.303	6.965	9.925	22.230	31.6
3	1.638	2.353	3.182	4.541	5.841	10.210	12.92
4	1.533	2.132	2.776	3.747	4.604	7.173	8.610
5	1.476	2.015	2.571	3.365	4.032	5.893	6.869
6	1.440	1.943	2.447	3.143	3.707	5.208	5.959
7	1.415	1.895	2.365	2.998	3.499	4.785	5.408
8	1.397	1.860	2.306	2.896	3.355	4.501	5.041
9	1.383	1.833	2.262	2.821	3.250	4.297	4.781
10	1.372	1.812	2.228	2.764	3.169	4.144	4.587
11	1.363	1.796	2.201	2.718	3.106	4.025	4.437
12	1.356	1.782	2.179	2.681	3.055	3.930	4.318
13	1.350	1.771	2.160	2.650	3.012	3.852	4.221
14	1.345	1.761	2.145	2.624	2.977	3.787	4.140
15	1.341	1.753	2.131	2.602	2.947	3.733	4.073
16	1.337	1.746	2.120	2.583	2.921	3.686	4.015
17	1.333	1.740	2.110	2.567	2.898	3.646	3.965
18	1.330	1.734	2.101	2.552	2.878	3.610	3.922
19	1.328	1.729	2.093	2.539	2.861	3.579	3.883
20	1.325	1.725	2.086	2.528	2.845	3.552	3.850
21	1.323	1.721	2.080	2.518	2.831	3.527	3.819
22	1.321	1.717	2.074	2.508	2.819	3.505	3.792
23	1.319	1.714	2.069	2.500	2.807	3.485	3.768
24	1.318	1.711	2.064	2.492	2.797	3.467	3.745
25	1.316	1.708	2.060	2.485	2.787	3.450	3.725
26	1.315	1.706	2.056	2.479	2.779	3.435	3.707
27	1.314	1.703	2.052	2.473	2.771	3.421	3.690
28	1.313	1.701	2.048	2.467	2.763	3.408	3.674
29	1.311	1.699	2.045	2.462	2.756	3.396	3.659
30	1.310	1.697	2.042	2.457	2.750	3.385	3.646
32	1.309	1.694	2.037	2.449	2.738	3.365	3.622
34	1.307	1.691	2.032	2.441	2.728	3.348	3.601
36	1.306	1.688	2.028	2.434	2.719	3.333	3.582
38	1.304	1.686	2.024	2.429	2.712	3.319	3.566
40	1.303	1.684	2.021	2.423	2.704	3.307	3.551
42	1.302	1.682	2.018	2.418	2.698	3.296	3.538
44	1.301	1.680	2.015	2.414	2.692	3.286	3.526
46	1.300	1.679	2.013	2.410	2.687	3.277	3.515
48	1.299	1.677	2.011	2.407	2.682	3.269	3.505
50	1.299	1.676	2.009	2.403	2.678	3.261	3.496
55	1.297	1.673	2.004	2.396	2.668	3.245	3.476
60	1.296	1.671	2.000	2.390	2.660	3.232	3.460

(continued)

4. Data Analysis

Table 4.2 (Continued) Values of Two-Tailed Probability $P(\eta)$

Degree of Freedom ν	$t_{0.20}$	$t_{0.10}$	$t_{0.05}$	$t_{0.02}$	$t_{0.01}$	$t_{0.002}$	$t_{0.001}$
65	1.295	1.669	1.997	2.385	2.654	3.220	3.447
70	1.294	1.667	1.994	2.381	2.648	3.211	3.435
80	1.292	1.664	1.990	2.374	2.639	3.195	3.416
100	1.290	1.660	1.984	2.364	2.626	3.174	3.390
150	1.287	1.665	1.976	2.351	2.609	3.145	3.357
200	1.286	1.653	1.972	2.345	2.601	3.131	3.340

From Student's t distribution, for degree of freedom $\nu = N - 1 = 6$, $t_{95} = 2.447$, and then the interval can be determined as

$$\Delta = \frac{t_{95}PI}{\sqrt{N}} = \frac{2.447 \times 7.94}{\sqrt{7}} = 7.34 \tag{c}$$

Note that the estimated interval for this value has to be increased for enhancing confidence in the mean value of the given data.

4.7 Data Visualization and Analysis

We have already learned in Chapter 3 how to analyze our desired data obtained by conducting experiments to determine its errors, precision, and general validity. Now we need to understand whether this hard-earned data can provide us with a certain physical understanding of the scientific principles. For this purpose, we need to visualize this data, which will help us to discern the underlying hidden principles of science so that we can prove or disprove our hypothesis. A simple but quick analysis can be carried out while generating the scientific data itself. It may take the form of theoretical analysis or scrutinizing/matching of data with certain physical principles or comparing it with other experimental/theoretical data. The validity of the data must be verified in a rigorous manner by comparing it with certain known models so that confidence can be built with regard to your methodology of generating data. Once this preliminary check is done, it is important to analyze the data to decipher the underlying scientific principles which can be carried out by data visualization. *Data visualization* is the study of the visual representation of data, which will help us in deciphering the information embedded in an abstract form in terms of certain variables. The main goal of data visualization is to extract the information clearly and effectively through graphical means that can be communicated to others easily. Care must be taken to convey ideas effectively, which will provide better insight into intricate processes in a more intuitive way. The data can be visualized through conventional formats such as tables, histograms, pie charts, and bar graphs. However, one has to be careful to visualize the data with the specific objective of extracting specific information with adequate attention. As well, if you need to convey a message to your readers effectively, sometimes you may have to adopt a different kind of data visualization. In fact, data can be visualized in profound, creative, and fascinating ways that utilize the imagination of researchers and experimenters.

4.8 Graphical Analysis

Graphical analysis also helps not only to analyze the experimental data but also helps to convey intricate aspects of features more effectively as compared to tabular form. As it has been rightly said, one single graph or picture can communicate more effectively than an entire well-written page. Graphs are quite useful for developing theoretical relationships that can be used for designing and developing engineering systems and in validating theoretical models against experimental data. It must also be kept in mind that graphical analyses cannot be fruitful unless there is a sound understanding of the physical phenomena within a particular experiment. There should not be an excess use of graphs because this can confuse and overwhelm the experimenters as well as the readers. It is therefore essential to choose the proper graphical format in order to best present and display the underlying physical phenomena.

4.8.1 Choice of Graphical Format

Several graphical formats, such as bar charts, pie charts, column charts, x–y plots, contour maps, and x–y–z plots are being used for presenting data in better way. Among all the graphical formats, the x–y plot is the most popular among experimental investigators because it is easy to plot and interpret data. There are several variations in this simple x–y plot that can be chosen depending on the type of data involved and the nature of the information that comes out of the raw data.

Some simple types of graphs are shown in Figure 4.6. We can see that Figure 4.6a depicts emission index data from swirl and TVC burners in a bar chart, which makes it easier to visualize the differences between the two burners for three emission species (CO, NO_x, and HC). Figure 4.6b indicates the variation of a viscosity index with temperature data which is cris-crossing each other and cannot be revealed easily when it is plotted using bar chart. Data may be presented as a volume (barrel) chart, as shown in Figure 4.6c, but this format is not preferred by researchers. The particulate emission data presented in the pie chart (Figure 4.6d) can also be represented well in a bar chart but not in x–y plots.

The basic philosophy is that one has to represent data in a graphical form in a way that the reader can easily and immediately understand and appreciate it, and researchers can easily decipher the underlying physical principles. Let us consider general guidelines that may be followed for making effective graphs:

i. The graph must be designed such that readers can easily understand and interpret the data.
ii. The proper type of axes must be chosen (e.g., linear, semi-log, or log-log) to facilitate the presentation of physical principles.
iii. Axes must be labeled clearly with proper names and symbols of the quantity along with the respective units (SI format is preferable).

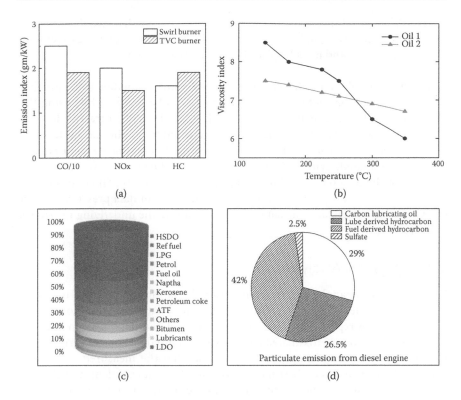

Figure 4.6

Four types of simple graphs: (a) bar chart, (b) x–y plot, (c) volume chart, and (d) pie chart.

    iv. Axes must be numbered with proper check marks for significant numerical digits. An optimum number of check marks must be used without cluttering the axes.

    v. It is advisable to include zero in the axes. If it is important to present a small range of data, it is better to include zero while breaking the axes (see x-axis of Figure 2.8).

    vi. Error bars for data points must be placed to indicate uncertainty levels.

    vii. Scales and proportions must be chosen judiciously depending on the relative importance of the variations in the data.

   viii. When several different sets of data are presented in a single graph, it is advisable to use different prominent symbols for each variable that is defined by the legend on the graph. In order to show trends, if curves are plotted, different lines such as solid, dashed, dash-dot, and so forth should be used to distinguish them. If they are placed close to each other, it is important to use labels for identifying them. Experimental data is always accompanied with symbols along with curves while theoretical curves are traditionally drawn using lines.

ix. A bare minimum of text must be placed on the graph to avoid confusion.

x. The title of the graph must precisely represent its characteristics as closely as possible and must be placed just below each graph.

Generaly, graphs and charts are generated using computer software that chooses its axes range automatically. But one has to be careful while using the chosen ranges for an axis by the software. It might be quite attractive to have good-quality colorful graphs but it is more important to use the various features of the software along with your own common sense to best uncover and explain the trends and underlying principles of intricate phenomena. Therefore, one should plot the data on a chart or graph keeping in mind the intentions of the experimenter. Several permutations and combinations of data presentation can be explored to understand the data and communicate the underlying principles to the readers.

4.9 Regression Analysis

We use regression analysis to find a functional relationship between independent and dependent variables within experimental data. This functional relationship will be representing the average value of the data. The anticipated relationship can be approximated either by polynomial or Fourier series form. Generally it is assumed in regression analysis that the measured dependent variable follows normal distribution around the independent variable. A relationship between experimental data can be obtained by using regression analysis so that the curve-fitted line will be around the precision error of the experimental data. In other words, the main aim of fitting a curve to the experimental data by regression analysis is to average out the precision error so that this fitted curve will follow the central tendency of the scattered data. Several methods of regression analysis can be carried out for fitting experimental data. The most common form of regression analysis is the least-square method, which is described below.

4.9.1 Least-Square Regression Method

The least-square regression method can be applied for curve-fitting experimental data that must have a higher level of precision compared to dependent variables. Suppose we have a set of experimental data, $y_1, y_2, \ldots \ldots y_N$ (N set of experimental data), which can be expressed in terms of independent variables x in the form of $y = f(x)$. Then, by using regression analysis we can curve-fit the experimental data with an mth order polynomial curve fit as

$$y_c = a_0 + a_1 x + a_2 x^2 + a_3 x^3 + \ldots \ldots \ldots + a_m x^m \qquad (4.29)$$

where $a_0, a_1, a_2, \ldots a_m$ are the coefficients of the polynomials, y_c is the value of curve fitted data obtained using an mth order polynomial equation for a set of experimental data. In case of regression analysis, the sum of the squares of the deviations between the actual experimental data and curve-fitted data is to be minimized. By this method, we need to determine the coefficients of polynomials $a_0, a_1, a_2, ... a_m$ of Equation 4.29. The sum of the squares of the deviation D for all values of experimental data $y_i(y_1, y_2, ...y_N)$ can be expressed as

$$D = \sum_{i=1}^{N}(y_i - y_{i,c})^2 = \sum_{i=1}^{N}[y_i - (a_0 + a_1x + a_2x^2 + ... + a_mx^m)]^2 \quad (4.30)$$

In the least-squares analysis, the sum of the squares of the deviation D are minimized to obtain the coefficients of polynomials $a_0, a_1, a_2, ... a_m$. For this purpose the above equation is differentiated with each coefficient and this partial derivative equates to zero as given below:

$$\frac{\partial D}{\partial a_0} = \frac{\partial}{\partial a_0}\left(\sum_{i=1}^{N}[y_i - (a_0 + a_1x + a_2x^2 + ... + a_mx^m)]^2\right) = 0 \quad (4.31)$$

$$\frac{\partial D}{\partial a_1} = \frac{\partial}{\partial a_1}\left(\sum_{i=1}^{N}[y_i - (a_0 + a_1x + a_2x^2 + ... + a_mx^m)]^2\right) = 0 \quad (4.32)$$

$$\frac{\partial D}{\partial a_m} = \frac{\partial}{\partial a_m}\left(\sum_{i=1}^{N}[y_i - (a_0 + a_1x + a_2x^2 + ... + a_mx^m)]^2\right) = 0 \quad (4.33)$$

The above $(m + 1)$ equations can be solved simultaneously to determine the m number of regression coefficients. Let us consider an example to illustrate how to use this method for determining the regression coefficients.

Example 4.3

The following temperature data along with distance given below follows a linear relationship. Determine the regression coefficients by using the least-squares method.

n	x_i (cm)	T_i (K)
1	1	500
2	2	650
3	2.5	740
4	3.0	880

Given: Independent variable = x, dependent variable = T, $N = 4$, linear relationship.

To Find: a_0, a_1.

Solution: Let us assume a linear polynomial in the form, $T_c = a_0 + a_1 x$. In order to determine the regression coefficients a_0, a_1, we need to minimize the sum of the squares of the deviation, D, as

$$\frac{\partial D}{\partial a_0} = -2\left(\sum_{i=1}^{N} y_i - (a_0 + a_1 x_i) \right) = 0 \qquad\qquad \text{(a)}$$

$$\sum_{i=1}^{4} T_i = 2a_0 + a_1 \sum_{i=1}^{4} x_i \qquad\qquad \text{(b)}$$

$$\frac{\partial D}{\partial a_1} = -2\left(\sum_{i=1}^{N} [y_i - (a_0 + a_1 x_i)] x_i \right) = 0 \qquad\qquad \text{(c)}$$

$$\sum_{i=1}^{4} T_i \sum_{i=1}^{4} x_i = a_0 \sum_{i=1}^{4} x_i + a_1 \sum_{i=1}^{4} x_i \sum_{i=1}^{4} x_i \qquad\qquad \text{(d)}$$

By solving Equations b and d simultaneously, we have

$$a_0 = \frac{\displaystyle\sum_{i=1}^{4} x_i \sum_{i=1}^{4} x_i T_i - \sum_{i=1}^{4} x_i^2 \sum_{i=1}^{4} T_i}{\left(\displaystyle\sum_{i=1}^{4} x_i\right)^2 - 2\sum_{i=1}^{4} x_i^2} \qquad\qquad \text{(e)}$$

$$a_1 = \frac{\displaystyle\sum_{i=1}^{4} x_i \sum_{i=1}^{4} T_i - 2\sum_{i=1}^{4} x_i T_i}{\left(\displaystyle\sum_{i=1}^{4} x_i\right)^2 - 2\sum_{i=1}^{4} x_i^2} \qquad\qquad \text{(f)}$$

Let us evaluate the summed quantities individually as they appear in the above equation.

n	x_i (cm)	T_i (K)	x_i^2	$x_i T_i$
1	1	500	1	500
2	2	550	4	1100
3	2.5	580	6.25	1450
4	3.0	600	9.0	1800
	$\sum x_i = 8.5$	$\sum T_i = 2230$	$\sum x_i^2 = 20.25$	$\sum x_i T_i = 4850$

By substituting the above values in Equations e and f, we get

$$a_0 = \frac{\sum_{i=1}^{4} x_i \sum_{i=1}^{4} x_i T_i - \sum_{i=1}^{4} x_i^2 \sum_{i=1}^{4} T_i}{\left(\sum_{i=1}^{4} x_i\right)^2 - 4\sum_{i=1}^{4} x_i^2} = \frac{8.5 \times 4850 - 20.25 \times 2230}{8.5^2 - 4 \times 20.25} \qquad (g)$$

$$= \frac{-393.25}{-8.75} = 449.429$$

$$a_1 = \frac{\sum_{i=1}^{4} x_i \sum_{i=1}^{4} T_i - 4\sum_{i=1}^{4} x_i T_i}{\left(\sum_{i=1}^{4} x_i\right)^2 - 4\sum_{i=1}^{4} x_i^2} = \frac{8.5 \times 2230 - 4 \times 4850}{8.5^2 - 4 \times 20.25} = \frac{-445}{-8.75} = 50.857 \quad (h)$$

The curve-fitted relation for this experimental data is given by

$$T_c = 449.429 + 50.857x. \qquad (i)$$

The plot of this curve-fitted relationship along with data is shown in Figure 4.7, which indicates a linear relationship. We need to evaluate how good the curve-fitting for this data is by considering the correlation coefficient as discussed below.

Correlation coefficient: We need to determine how the actual data is related to the regressed (curve-fitted) data by defining a correlation coefficient r as given below:

$$r = \left(1 - \frac{\sigma_{y,x}^2}{\sigma_y^2}\right)^{0.5} \qquad (4.34)$$

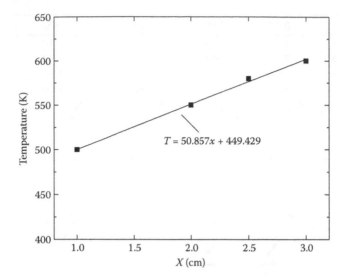

$T = 50.857x + 449.429$

Figure 4.7

Linear regression analysis curve along with its data points.

where σ_y is the standard deviation of y and $\sigma_{y,x}$ is the standard deviation of y, x as given below:

$$\sigma_y = \left(\frac{\sum_{i=1}^{N}(y_i - y_m)^2}{N-1} \right)^{0.5} \qquad (4.35)$$

$$\sigma_{y,x} = \left(\frac{\sum_{i=1}^{N}(y_i - y_{i,c})^2}{N-m} \right)^{0.5} \qquad (4.36)$$

Note that y_i is the actual data and y_{ic} is the curve-fitted data for the same value of an independent variable and m is number of coefficients in the polynomial. For the above example $N - m = N - 2$ as we have used two coefficients in the linear equation, which can basically eliminate two degrees of freedom from the set of data. Note that for a perfect fit, $\sigma_{y,x}$ must be equal to 0 and the correlation coefficient r becomes equal to 1. On the other hand, if $r = 0$, then it indicates that the fitted curve to the data is quite poor. Therefore it is expected that the correlation coefficient r must be closer to 1 to ensure better curve-fitting. But even when the correlation coefficient r is closer to the unity value, it may not ensure good curve-fitting, so it is advisable to place the actual data and curve-fitted line together in a plot. For this reason, the actual data of the example is plotted along with a least-squares fit to indicate that this curve fit matches with the experimental data reasonably and can be improved further by considering more data within this range.

4.9.2 Multivariable Regression Analysis

We learned in the section above how we can apply regression analysis for single independent variables. In real situations, we may have several independent variables that might be affecting the dependent variables. Therefore it is important for us to learn about multivariable regression analysis. The least-squares method discussed above can be used for multivariables. For linear relationships, the polynomial equation for multivariables can be of the following form:

$$y_c = a_0 + a_1 x_1 + a_2 x_2 + a_3 x_3 + \ldots\ldots + a_n x_n \tag{4.37}$$

where a_0, a_1, a_2, ... a_n are the coefficients of the polynomials, x_n are the independent variables, and y_c is the value of the curve-fitted data obtained using the above linear equation for the set of experimental data. In the case of regression analysis, the sum of the squares of the deviations is between the actual experimental data and curve-fitted data. By this method, we need to determine the coefficients of the linear equation, a_0, a_1, a_2,...a_n, of Equation 4.37. The sum of the squares of the deviation D for all values of the experimental data $y_i(y_1, y_2, \ldots\ldots y_N)$ can be expressed as

$$D = \sum_{i=1}^{N} (y_i - y_{i,c})^2 = \sum_{i=1}^{N} [y_i - (a_0 + a_1 x_1 + a_2 x_2 + \ldots\ldots + a_n x_n)]^2 \tag{4.38}$$

In the least-squares analysis, the sum of the squares of the deviation D are minimized to obtain the coefficients of the polynomials a_0, a_1, a_2,...a_n. For this purpose the above equation is differentiated with each coefficient and this partial derivative equates to zero, as

$$\frac{\partial D}{\partial a_0} = \frac{\partial}{\partial a_0} \left(\sum_{i=1}^{N} [y_i - (a_0 + a_1 x_1 + a_2 x_2 + \ldots\ldots + a_n x_n)]^2 \right) = 0 \tag{4.39}$$

$$\frac{\partial D}{\partial a_1} = \frac{\partial}{\partial a_1} \left(\sum_{i=1}^{N} [y_i - (a_0 + a_1 x_1 + a_2 x_2 + \ldots\ldots + a_n x_n)]^2 \right) = 0 \tag{4.40}$$

$$\frac{\partial D}{\partial a_n} = \frac{\partial}{\partial a_n} \left(\sum_{i=1}^{N} [y_i - (a_0 + a_1 x_1 + a_2 x_2 + \ldots\ldots + a_n x_n)]^2 \right) = 0 \tag{4.41}$$

The above $(n + 1)$ equations can be solved simultaneously to determine the n number regression coefficients and a_0. Note that this multivariable regression analysis can be used for an exponential form as

$$y_c = a_0 a_1^{x_1} a_2^{x_2} \ldots\ldots a_{1n}^{x_n} \tag{4.42}$$

where a_0, a_1, a_2, ... a_n are the coefficients (constants) that can be determined by using the least-squares analysis. By taking the logarithm of the above equation, we get

$$\log y_c = \log a_0 + x_1 \log a_1 + x_2 \log a_2 + \ldots\ldots + x_n \log a_n \tag{4.43}$$

This is similar to the linear multivariable equation that can fit the experimental data. The calculation procedure for determining correlation coefficients of the multivariable regression equation is provided in detail in advanced texts such as [3,4]. Advanced calculators or computers are used routinely for multivariable regression analysis as it is quite cumbersome to carry out such involved calculation by hand. The most common software is the Excel® package, which has the ability to carry out multivariable and exponential regression analyses.

4.10 Report Writing

In order to present an accurate representation of their experiments, researchers need to develop good oral and written communication skills. Therefore, it is important to present these efforts in the form of a report or record so that it can be communicated to others. For this it is equally important for the experimenter to develop good communication abilities, both written and oral. Of course good writing and good speaking abilities are very essential in all walks of life. During the process of presenting the results of any experimental endeavor, we have ample chances to deepen our thought processes and understanding about our experimental work, which may in turn help us to improve our experimental skills further. As humans, we all want to reach perfection and although it is elusive, we strive to achieve it through deliberation and discussion with the help of experimentation, writing, and presentation.

Several types of scientific reports and records are used for the dissemination and propagation of scientific knowledge. We will be restricting our discussion to only three types of written reports, which are

    i. Laboratory reports
   ii. Technical reports
  iii. Research papers

       a. *Laboratory Reports.* Laboratory reports are written for a limited group of readers; namely, teachers, supervisors, and fellow investigators who are familiar with the experimental work. On several occasions, the experimenter can read out the report either to himself or others for improving his understanding and augment the experimental procedure to improve the methodologies. Note that laboratory reports must be as short as possible while not compromising any information that will help the experimenter to visualize and reconstruct the process or phenomenon aptly. All pertinent information and data must be included so that anyone, including the experimenter, can refer back to his or her report at any time. The reports must include background, objective, experimental setup, results, and discussion. Guidelines for laboratory reports are as follows:

Principle: The basic underlying principle of each experiment is to be stated and stressed in clear and lucid terms and must cover all the details.

Objectives: The main objectives of the experiment must be presented along with allied aims. In other words, they will state what the experiment demonstrates.

Background: A theoretical or physical explanation is to be included, containing all equations to make the necessary calculations. This section can also highlight the rationale behind the experiment.

Apparatus: A detailed description of all apparatus and instrumentation necessary to conduct the experiment is presented in a precise manner so that it can be replicated easily.

Procedure: A detailed step-by-step procedure should be included so that anyone can execute this procedure easily. All precautions for the experiments must also be included.

Results and discussion: The results should be presented in tabular and graphical forms for all quantities to be measured and calculated along with a brief discussion. Wherever necessary, typical results, figures, and sketches should be presented to clarify the quantities being measured.

Uncertainty analysis: A sample calculation for an uncertainty analysis should be provided. Errors in each measurement must be estimated using statistical tools.

References: A collection of easily accessible references should be included so that the readers have access to information for further studies.

Notation: A listing of symbols used in the report should be provided.

b. *Technical Reports.* A technical report is generally quite exhaustive, encompassing all the facts, data, and information pertinent to large projects that might consist of several sets of experiments. Several experimenters might be involved in a big research endeavor, for which a technical report must be written in a very comprehensive manner. The main purpose of the project must be stated very clearly and experimental methods and results must be presented in a precise but exhaustive manner as the readers may not be fully acquainted with the intricacies of the experimental work. A technical report must also contain full details about the experimental setup and procedures and be presented in a lucid manner so that other researchers can replicate the experiment without any difficulties. Generally, the following format is used in preparing technical reports: (i) title, (ii) table of contents, (iii) list of figures, (iv) introduction, (v) objectives, (vi) experimental setup, (vii) experimental procedures, (viii) results and discussion, (ix) conclusions, and (x) references. Details about technical report writing can be found in [5–7].

c. *Research Papers.* Generally, research papers that contain experimental work are meant to be presented at conferences or published in journals. These technical documents undergo scrutiny by reviewers and editors who recommend them for acceptance or rejection. Professional writing skills must be learned in order to produce a research paper. The writer must keep in mind the constraints of space, deadlines, and the number of words, limits on figures, graphs, tables, and so forth, while preparing a research paper. Hence, it is critical to develop the ability to highlight problems, statements, and results along with important conclusions in a precise but convincing manner. Generally, the following format is used in preparing research papers: (i) title, (ii) table of contents, (iii) list of figures, (iv) introduction, (v) objectives, (vi) experimental setup, (vii) experimental procedures, (viii) results and discussion, (ix) conclusions, and (x) references. Note that the sequence and style may differ from one journal to another. Details about writing research papers can be found in [6,7].

It must be kept in mind that the main purpose of any technical writing is to convey the facts and finding to the readers, so care must be taken to communicate with clarity, precision, and of course with utmost completeness. The nontechnical prose in technical writing must be more precise and specific than in a literary piece of work. Therefore, a technical writer must understand his or her material thoroughly before beginning and must organize his or her writing properly beforehand. The results must be presented in a logical order: (i) experimental methods, (ii) experimental results and discussion, and (iii) conclusions. Before initiating the writing process of any technical document, the experimenter must assemble his or her experimental data and carry out a thorough analysis on it with specific tangible conclusions. It makes sense to prepare brief notes on the results and discussion before beginning any serious writing. As mentioned earlier, it is essential to first prepare a detailed outline of what will be discussed before beginning to write. Sentences in technical documents must not be too short or too long or they will hinder the efficacy of communication. Correct form, conventions, and nomenclature should be used to enhance the effectiveness of communication. Generally, formal style is used when writing a technical document, but in recent times, a more mixed style has become more accepted. Interested readers may refer to [6,7] for more on technical writing.

Review Questions

1. What is accuracy? How is it different from precision?

2. What is the hysteresis of an instrument? Can the accuracy of an instrument be affected by its hysteresis? Explain in terms of precision and bias error.

3. Why is it required to calibrate an instrument?

4. What is a primary standard? How is it different from a secondary standard?

5. Why are standards being used while calibrating an instrument?

6. What is dynamic calibration? When is it required to carry out a dynamic calibration?

7. What is linearity? How is independent linearity different from proportional linearity?

8. What is static sensitivity in an instrument? How is nonlinear sensitivity different from linear sensitivity?

9. What is frequency response? How is it different from phase shift?

10. What is rise time?

11. What is a steady state response?

12. What are zeroth- and first-order systems?

13 What are the factors that influence a time constant in a first-order system?

14. What is a second-order system? Illustrate how it is different from a first-order system using an example.

Problems

1. A bomb calorimeter used by an engineer for measuring the heating value of a biodiesel value provides the following readings: 45237, 46315, 48219, 44345, 47547, 48936, 43974, 44347, and 45946 kJ/kg. Determine the mean value and the standard deviation (precision index).

2. The power consumption in an ignition coil in a combustion system can be calculated by measuring voltage and current as

$$\text{Voltage } V = 120 \pm 1.5 \text{ V; current } I = 15 \pm 0.05 \text{ A}$$

Determine the maximum possible errors and also the uncertainty level in the calculation of power.

3. The pressure in a manometer is expressed as the height of a liquid column. The density of manometric fluid is 950 kg/m³. The maximum pressure that can be measured by this manometer with an accuracy of 2% is 1500 Pa. If the uncertainty of a reading scale in a manometer is 0.02 mm, determine the allowable uncertainty in the density of the manometric fluid.

4. An orifice flow meter is used to measure the steady flow rate of a water flow rate through a pipe. During this experiment, the flow discharge coefficient C_D of an orifice can be obtained by collecting and measuring the water flow rate through this orifice in a certain time interval under a constant manometric head. The flow discharge coefficient C_D can be expressed in terms of mass flow rate, water head, and other variables as per the following expression:

$$C_D = \frac{\dot{m}}{A\rho\sqrt{\Delta h}}$$

where C_D is the discharge coefficient, A is the flow area, and Δh is the pressure head across the venturi flow meter. Determine the uncertainty in discharge coefficient C_D with a 95% confidence value and maximum possible error for the data of the following variables: $m = 59 \pm 0.05$ kg, diameter of tube, $d = 50 \pm 0.2$ mm, pressure head, $\Delta h = 120 \pm 5$ cm, density $\rho = 999 \pm 0.1\%$ kg/m³.

5. In a combustor, a Pt-Pt/10%Rh thermocouple is used to measure gas temperature when wall temperature, T_w, is 800 K. Due to radiation heat loss, the actual temperature of the hot gas is expected to be lower than the measured surface temperature $T_s = 1950$ K. In order to correct this measurement data, the following relationship is being used:

$$\Delta T_{correction} = \frac{\varepsilon}{h}\sigma\left(T_s^4 - T_w^4\right)$$

where σ is the Stefan-Boltzmann constant, ε is the emissivity of the thermocouple bead, and h is the heat transfer coefficient. Determine the temperature correction and uncertainty in this correction for the data of the following variables: $h = 59 \pm 5$ W/m²-K, the emissivity, and $\varepsilon = 0.8 \pm 0.1$ mm.

6. A viscometer used by an engineer for measuring the viscosity of a biodiesel value provides the following readings: 4.52 4.63, 4.84, 5.43, 4.75, 4.89, 4.95, 4.47, and 4.96 centistokes at 40°C. Determine the mean value and the standard deviation (precision index). Assuming no precision

error has been introduced by the viscometer, determine the precession limit of each sample and the mean value using Student's t distribution for a confidence level of 98%. If the viscometer has an accuracy of 1.5% of full-range (2–20 centistokes), determine its corrected mean value and single measurement with 4.45 centistokes.

7. The orifice of an atomizer in a liquid fuel combustor is manufactured in large numbers and whose tolerance in its diameter follows normal distribution. The average diameter of this orifice is measured to be 0.5 mm with a standard deviation of 0.001 mm. What are the probabilities of the following cases: (i) $d \leq 0.512$, (ii) $d \geq 0.503$, and (iii) $0.489 \leq d \leq 0.512$?

References

1. Kline, S. J. and McClintock, F. A., Describing uncertainties in single-sample experiments, *Mechanical Engineering*, 75(1):3, 1953.
2. Holman, J. P., *Experimental Methods for Engineers*, Sixth Edition, McGraw-Hill, Inc., New York, 1994.
3. Draper, N. R. and Smith, H., *Applied Regression Analysis*, Second Edition, John Wiley & Sons, New York, 1981.
4. Coleman, H. W. and Steele, W. G. Jr., *Experimentation and Uncertainty Analysis for Engineers*, John Wiley & Sons, New York, 1989.
5. Doebelin, E. O., *Engineering Experimentation*, McGraw-Hill Book Co., New York, 1995.
6. Day, R. A., *How to Write and Publish a Scientific Paper*, Third Edition, Cambridge University Press, Cambridge, UK, 1989.
7. Stevenson, S. and Whitmore, S., *Strategies for Engineering Communication*, John Wiley & Sons, Inc., New York, 2002.

5

Thermal Flow Measurements

True success of a person should be measured not by name and fame but by virtues possessed by him.

D. P. Mishra

5.1 Introduction

In this chapter, we will introduce the basic principles of flow measurements that can be used in a combustion system. It is assumed that the reader has already been introduced to the basics of intrusive and nonintrusive measurement methods covering both the physical and chemical aspects, particularly with regard to fluid flow and heat transfer. In this chapter, we review the basic measurement pertaining to properties such as gas velocity, pressure, density, temperature, and sound pressure. The problems associated with these measurements in a combustion system will be discussed in detail as they pose interesting problems as compared to the isothermal fluid flow system. Because the combustion process makes the environment hostile due to high temperature, chemical reactions, and a high-pressure environment. Chemical reactions occur locally in combustion systems, so local measurements must be carried out carefully.

5.2 Pressure Measurement

Pressure plays a very important role in any combustion system as most practical combustion systems involve flow in some form or another. Therefore it is important to understand and appreciate how to measure pressure in a combustion system. We know that pressure is a fundamental quantity that is defined as the ratio of force per unit area. In other words, it can be thought of as the force exerted by the bombardment of fluid molecules on a unit area of a container or sensor wall. We know that pressure in a combustion system is manifested due to the exchange of momentum between fluid molecules and the walls of a container. Unlike the fluid in motion, pressure at a point in fluid at rest is same in all directions. Recall that total pressure exerted by the fluid elements on a container or sensor wall is known as absolute pressure. In other words, the absolute pressure can be thought of as the fluid pressure measured above a perfect vacuum with respect to absolute reference, as shown in Figure 5.1. We know that the pressure exerted by the Earth's atmosphere is atmospheric pressure, which is 101.325 kPa under sea-level conditions. In engineering systems, we frequently use pressure measuring systems that indicate gauge pressure instead of absolute pressure. Gauge pressure is basically the difference between absolute and atmospheric pressure in which absolute pressure must be above atmospheric pressure. If absolute pressure is below atmospheric pressure, then the differential pressure is known as vacuum gage pressure (see Figure 5.1). Pressure exerted by fluid can be dependent on several factors: elevation, fluid density, temperature, and flow velocity. In the case of combustion systems, all these variables are likely to change depending on the prevailing situations in the combustor.

The steady state of fluid when flow is not taking place can be measured easily with adequate accuracy because it is purely a measure of the random motion of molecules. But when fluid is flowing, measuring pressure is not that easy, as mentioned above. Static pressure can be measured only when a pressure-sensing

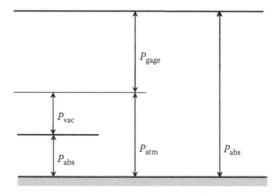

Figure 5.1

Schematic representation of the relationship between absolute pressure, atmospheric pressure, gauge pressure, and vacuum pressure.

5. Thermal Flow Measurements

device will be moved along with the same local velocity in terms of magnitude and direction. According to the Bernoulli equation, static pressure, local fluid velocity, and fluid pressure due to elevation are related for incompressible fluid as given by

$$P + \frac{1}{2}\rho V^2 + \rho gz = \text{constant} \qquad (5.1)$$

where P is the static pressure, V is the local velocity of flow, ρ is the density of fluid, z is the elevation, and g is the acceleration due to gravity. But for the compressible flow, the Bernoulli equation becomes

$$\frac{\gamma}{\gamma-1}\frac{P}{\rho} + \frac{1}{2}V^2 = \text{constant} \qquad (5.2)$$

where γ is the specific heat ratio. Note that for compressible flow, pressure head due to elevation is negligibly small and has been neglected in the above equation. As well, the total pressure can be measured by bringing flowing fluid elements to rest isentropically. If the fluid pressure is transient in nature as in internal combustion (piston engine), we need to use a sensor that has a higher frequency response.

Several measurement methods have been devised over the years for measurement of fluid pressure that can be broadly divided into three categories: (i) gravitational, (ii) direct acting elastic, and (iii) indirect acting elastic. In the case of gravitational types of pressure sensors, the pressure exerted by the static fluid is balanced by the fluid column. Some well known examples of pressure measuring devices are manometers, dead-weight testers, and Mc-lead gauges. We will be discussing certain types of manometers, micromanometers, and dead-weight testers below. In the case of a direct-acting elastic-type of pressure measuring device, the deformation of an elastic sensor due to the pressure force acting on it can be used to measure the acting pressure. Examples of this type of pressure-measuring device are Bourdon tubes, bellows, elastic membranes, and diaphragms. In the case of indirect acting elastic, electrical quantities and concepts such as resistance, conductivity, piezoelectric effects, and ionization, probes can be used for sensing pressure. Examples of pressure-measuring instruments that are commonly used in combustion systems include strain gauges, piezoresistive sensors, piezoelectric sensors, and microphones. For information about other pressure-measuring devises, interested readers may consult [2,3]. We will now discuss various types of manometers.

5.2.1 Manometers

A manometer is the simplest device by which pressure difference in the moderate range can be used to measure. Various kinds of manometers are shown in Figure 5.2. Some of the fluids that are commonly used in manometers include water, alcohol, oil, and mercury. For the measurement of large-pressure differences, mercury is used as the manometric fluid, which is done to reduce to the

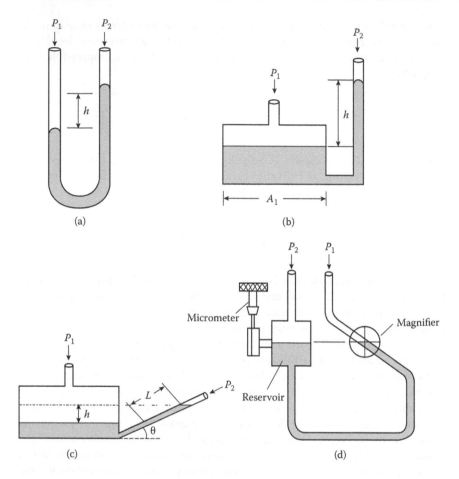

Figure 5.2

Schematic of different types of manometers: (a) U-tube manometer, (b) reservoir-type manometer, (c) inclined manometer, and (d) micromanometer.

height of the manometer. A U-tube (glass) manometer is commonly used for measuring differential pressure, as shown in Figure 5.2a. Let us consider h to be the height of the column of fluid in the U-tube manometer shown in Figure 5.2a that is balanced by the tank (not shown in the figure) pressure, P_1, and atmospheric pressure, P_2, expressed in the following manner:

$$h = \frac{P_1 - P_2}{\rho g} \tag{5.3}$$

where g is the local acceleration due to gravity. It can be noted from the above equation that the sensitivity of a U-tube manometer will be dependent on the density of manometric fluid and the local value of g. The variation of density of manometric fluid with local temperature atmosphere has to be considered properly in order to avoid any error in the measurement of accurate pressure. As a result, the range of the pressure measurements can dictate the choice of manometric fluid. For a small range of pressure measurements, low-density fluid such as alcohol and oil can be used while for a higher range of pressure measurements, a high-density fluid such as mercury can be used. As mentioned earlier, the local value of gravitational acceleration, g, must be used for obtaining an accurate measurement pressure in a manometer. The tube diameter of a manometer should be smaller than 1/10th of the pipe diameter in which the fluid flow occurs so that it does not introduce any disturbances into the flow condition. Therefore, smaller tubes may be used; however, the capillary effect of the tube has to be considered for an accurate measurement, particularly when the tube diameter is less than 10 mm.

There are several types of manometers used by various researchers, but we will restrict our discussion to four types: (a) U-tube manometer, (b) reservoir-type manometer, (c) inclined manometer, and (d) micromanometer, as shown in Figure 5.2. We have already discussed U-tube manometers above. Note that we need to measure the level of manometric fluid in two limbs, which is cumbersome and erroneous to some extent. In order to avoid this problem we can use a reservoir-type manometer (see Figure 5.2b) in which one limb of the U-tube manometer is replaced by a reservoir for manometric fluid. Generally, the cross-sectional area of this reservoir is made larger compared to the manometer tube so that an almost zero or negligible level change in the reservoir will be registered when the pressure differential is applied. This small error can be compensated for during the measurement itself if it is properly designed and fabricated. In order to enhance the sensitivity of a manometer, one limb can be inclined with gravity, and thus an inclined manometer (Figure 5.2c) is used for a small number of different pressure measurements. In this case the actual pressure head is magnified by displacement amplification depending on the inclination angle θ, as

$$h = \frac{P_1 - P_2}{\rho g} = L \sin \theta; \Rightarrow L = \frac{h}{\sin \theta} \tag{5.4}$$

Note that L is the change in length of the manometric fluid changes on the inclined tube due to differential pressure, which is magnified by inclination angle θ. The smaller the inclination angle θ is, the larger the amplification of the pressure head, h, will be; however, the surface tension effects will be higher with a smaller inclination angle. Hence the inclination angle, θ, of an inclined manometer must be greater than 5° to avoid the surface tension effect. As inclination angle θ of an inclined manometer is restricted, it would not be useful for measuring extremely small differential pressure. For this purpose, several types of micromanometers are being designed and developed. We will restrict our discussion to one type of micromanometer, as shown in Figure 5.2d, which consists of an inclined tube and a reservoir with a micrometer. Before starting a measurement, the position of the reservoir is adjusted so that the meniscus of the manometric fluid in the inclined tube is located at the reference point with the help of a magnifier. The manometric liquid displacement in the reservoir is measured using the vernier scale of the micromanometer. On application of an unknown pressure, the meniscus will move away from the hairline of reference point, which can be restored back by slowly varying the position of the reservoir with the help of the micrometer. This change in the position of reservoir measured using the vernier scale of the micromanometer will indicate the small changes in the differential pressure. Low-density fluid such as alcohol can also be used as a manometric fluid for enhancing the sensitivity of a manometer. Besides this, instead of changing the position of the reservoir in Figure 5.2d, if the position of the inclined tube is changed and measured using a micromanometer, then the readability of manometer as low as 0.005 mm of alcohol can be achieved easily.

Example 5.1

A U-tube water manometer is used to measure the pressure of a combustion bomb containing a fuel-air mixture at 35°C. One limb of this manometer is connected to this bomb and the other is exposed to atmospheric pressure at 35°C. If the difference in manometer column height with an error band is 250 ± 0.5 cm, determine the pressure of the combustion bomb and its uncertainty.

Given: $h = 25 \pm 0.1$ cm, $T = 35°C$.

To Find: P.

Solution: The density of atmospheric air can be determined by using the ideal gas law, as given below:

$$\rho = \frac{P_a}{RT} = \frac{101,325}{287 \times 308} = 1.146 \text{ kg/m}^3$$

We can determine the pressure of a combustion bomb as

$$P = P_a + gh(\rho_m - \rho) = 101,325 + 9.81 \times 2.5(996 - 1.146)$$
$$= 101,325 + 24,398.8 = 125,723.8 \text{ Pa} = 125.724 \text{ kPa}$$

Note that although the density of air is quite low compared to manometric fluid ($\rho = 996$ kg/m^3), we have considered it for the sake of completeness. The uncertainty in manometer column height can be estimated as

$$U_h = \frac{5}{2500} \times 100 = 0.2\%$$

Assuming there is no uncertainty in other terms such as manometric fluid density, temperature, and atmospheric pressure, the uncertainty in pressure measurement can be determined as

$$U_P = \frac{2}{100} \times 24,398.8 = 48.8 \text{ Pa}$$

5.2.2 Elastic Pressure Transducers

We learned that manometers can be used for a medium range of pressure. But if the pressure level becomes very high—in the order of megapascals as in certain liquid and gas pipelines for industrial-scale combustion systems—elastic pressure transducers can be used for sensing and measuring pressure. The basic principle of this type of pressure transducer is that the expansion of a flexible metallic material is being sensed and calibrated to measure applied pressure. Based on this principle, as shown in Figure 5.3, several types of pressure transducers are being designed and developed that can be broadly divided into three categories: (i) Bourdon tube, (ii) diaphragm, and (iii) bellows. The Bourdon tube gauge is routinely used in industries and laboratories for indicating the level of pressure. High-precision pressure gauges are used for calibration of pressure transducers. The schematic of a typical pressure gauge is shown in Figure 5.3a. It mainly consists of a curved tube with an elliptical cross section, pinion gear, linkage, and needle, as shown in the figure. One end of this tube is connected to the point at which the pressure is to be measured, while the other end is closed. This closed

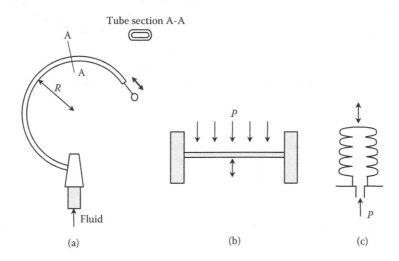

Figure 5.3

Schematic of elastic pressure transducers: (a) Bourdon tube, (b) diaphragm, and (c) bellows.

end is connected to the quadrant gear and pinion arrangement through a link. When a high-pressure fluid is passed through this tube, its elliptical cross section becomes circular and the tube straightens. In other words, there will be an angular deflection of the free end, which indicates the measure of the applied pressure As a result, it creates a motion due to the applied pressure with respect to the atmospheric pressure that is transmitted by a pinion gear mechanism to the needle. The deflection of the needle indicates the pressure in a precalibrated graduate scale; it measures pressure relative to the atmospheric pressure acting on the tube; therefore, it always indicates the gauge pressure, and not absolute pressure. Several other types of Bourdon tubes such as helical, spiral, and twisted can be employed for sensing pressure in a Bourdon tube pressure gauge. Note that the sensitivity of this instrument is dependent on the aspect ratio of the tube cross section, tube length, radius of the curvature of the tube, angle of twist, and the amplification factor of the gear-pinion arrangement. The Bourdon tube pressure gauge is widely used due to its higher accuracy, repeatability, and ruggedness.

The diaphragm-based pressure gauge (see Figure 5.3b) is also used in combustion systems. The diaphragm is a thin circular plate that is fastened tightly with its housing. The structure of this diaphragm can be flat (see Figure 5.3b), corrugated, or capsule-shaped, and is chosen for its high accuracy and better dynamic response. Bellows are basically thin-walled cylindrical shell-like enclosures with deep convolutions, as shown in Figure 5.3c. When measured pressure is applied to its inlet on one end, the other sealed end moves axially, as shown in Figure 5.3c, which is proportional to the applied pressure. In these types of pressure gauges, the deflections of these elastic elements can be measured using various sensors based on several principles including mechanical, resistive, inductive, capacitive, linear variable differential transformer, and potentiometer. The signal from the transducers need to be conditioned and amplified, either mechanically with the help of a gear-pinion, or electrically, and subsequently can be displayed directly using a pointer on a dial or electronically on a digital display, or it can be acquired using an A/D card. In the case of diaphragms, strain gauges are generally used to measure the local strain, which indicates the level of applied pressure.

5.2.3 Dynamic Pressure Transducers

So far we have discussed the measurements of pressure, which remain almost constant over the entire time of measurement. Of course in combustion systems, we do encounter variation in pressures that undergo changes over time. Generally, the frequency of pressure oscillation in combustion systems varies from a few hertz to kilohertz. In a practical combustion system there will be fluctuation of pressure because flow is inherently turbulent. We know that pressure in a turbulent flow can have a wide range of frequency due to the presence of broad-spectrum eddies whose sizes can vary from 0.01 mm to 1.00 cm depending on the level of the Reynolds number of the flow. Pressure fluctuations are also encountered in certain combustion systems due to heat release, flame oscillation, presence of recirculation zones, and so forth, and become more severe particularly when subjected to combustion instabilities. Therefore, it is important to measure pressure fluctuations in order to avoid severe combustion instabilities, which are detrimental to the life of a combustor. If the pressure fluctuations are quite high, then Bourdon-tube pressure gauges will not work. Several dynamic pressure transducers are being designed and developed for measuring pressure fluctuations in combustion systems. Most of these systems use elastic transducers with fast response characteristics. The fluctuations in the pressure of flow creates an oscillatory motion on elastic sensors that is sensed, conditioned, transmitted, and displayed in the form of an electrical signal. As mentioned earlier, several sensors such as variable capacitance, resistance, LVDT, piezoelectric, and piezoresistive can be used to convert mechanical oscillations to an electrical signature. It is expected that an electrical signal will be linearly proportional to the measured pressure. Variable-resistance dynamic pressure transducers use strain gauges that are mounted mainly on elastic membranes with the help of an appropriate adhesive. The range of this kind of dynamic pressure transducer is limited to the range of nonlinear relationships between strain and pressure. It is preferred due to its simplicity, cost-effectiveness, and good dynamic response. However, under high temperature conditions, its characteristics change adversely. For combustion applications, piezoelectric pressure transducers are preferred over other transducers due to their ruggedness and lesser sensitivity to temperature.

5.2.4 Piezoelectric Pressure Transducers

Piezoelectric pressure transducers are an important class of pressure transducer based on the *piezoelectric effect*, which was discovered in 1880 by the brothers Pierre and Jacques Curie. According to this effect, whenever mechanical pressure is exerted on a crystalline substance, an electrical potential/electrostatic charge is developed on it. Interestingly, the reverse effect is observed when an electrical potential/electrostatic charge to a crystal results in a mechanical force. This concept of producing voltage/charge with the application of pressure has been used to design and develop transducer dynamic pressure measurements of flow. Several types of materials with piezoelectric effects are processed that can be used in pressure sensors. The materials for piezoelectric sensors can be broadly divided into two categories: single crystals and ceramic materials. Some of the most popular single crystals are natural quartz, gallium phosphate, Rochelle salt (potassium sodium tartrate), and tourmaline. Synthetic ceramic materials such as barium-titanate and lead-zirconate titanate are preferred over an actual crystal due to their higher sensitivity (piezoelectric constant) and lower cost. The piezoelectric transducer (PZT) is placed onto a thin membrane to which pressure is applied such that this transducer is loaded along a single direction only. Several kinds of configurations, namely disc, plate or etc., can be used for designing and developing pressure sensor from PZT materials. The performance of PZT is dependent on the magnitude of the piezoelectric constant, variation of the piezoelectric constant with temperature, the natural time constant of the material, and the elastic constant. Note that piezoelectric material is quite stiff and can also have a high natural frequency.

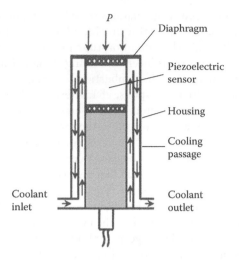

Figure 5.4

Schematic of a typical piezoelectric transducer.

Piezoelectric pressure measuring systems are preferred over other dynamic pressure sensors in combustion systems due to their ruggedness, excellent linearity over a wide amplitude range, and high natural frequency. The schematic of a typical piezoelectric pressure transducer is shown in Figure 5.4, which consists of a diaphragm, piezoelectric element, and housing and cooling systems. Piezoelectric materials such as gallium phosphate and tourmaline remain quite stable even when a temperature sensor goes up to 1000°C. To extend the life of a piezoelectric pressure sensor, a cooling jacket is provided to cool the sensor during the measurement process, particularly when pressure measurement is carried out continuously in a combustion system such as a piston, gas turbine, and rocket engines furnace. Generally this kind of pressure senses pressure force in the transverse direction only.

5.3 Sound Pressure Level Measurements

Combustion systems produce sounds in the audible ranges that are quite intolerable to humans. The road noise produced by IC engine-driven automobiles is of great concern to city dwellers due to the number of vehicles on the road. Noise produced by other combustion systems in industries, rocket engines, gas turbine engines, pulse detonations engines, ramjet and scramjet engines, and so forth are required to be measured and controlled to acceptable levels. Sound is used to suppress the combustion instability in combustion systems, so it is also essential to learn about sound level measurements, which will be discussed briefly in this section.

In its simplest form, sound is a pure tone of a single frequency that is difficult to produce in a real situation. In other words, the actual sound contains pressure fluctuations of several ranges of frequencies due to the nature of the sound source and its interaction with air, other media, and interferences from other sound waves. As a result, the sound wave at the point of observation will be distorted due to reflection, interference, and absorption. In order to avoid this type of distortion, an anechoic chamber is designed for the sound wave to travel from the source outwardly without any distortion. The walls of the anechoic chamber are lined with sound-absorbing materials that under ideal conditions absorb all incident sound waves so that any type of reflection does not occur. This type of environment can be created naturally in an open field near a hill, which is known as a free field sound. In contrast, random sound known as noise is produced from a number of discrete sources that are combined to form a single wave that can be measured by a sound sensor. The sound from a primary source gets distorted during its travel and then it reaches a sensor, which gets influenced by other secondary sources. As a result, randomness both in frequency and amplitude will be incurred that are being sensed by a sound transducer.

Generally, sound pressure in a particular medium is the difference between the instantaneous pressure and average local pressure in the sound wave. A typical sound pressure (arbitrary unit [AU]) with time is shown in Figure 5.5a. As mentioned earlier, sound pressure is the deviation from the equilibrium pressure. Pressure fluctuates randomly and can be described in terms of its rms value as per the following relationship:

$$P_{rms}^2 = \frac{1}{t}\int_0^t P^2 dt \tag{5.5}$$

where P is the instantaneous sound pressure, t is the average measurement time, and P_{rms} is the rms of the sound pressure. For example, the sound pressure of

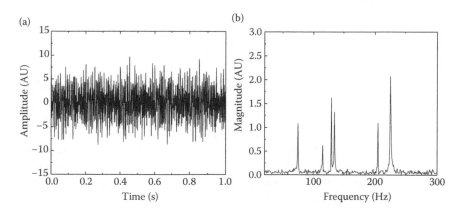

Figure 5.5

Schematic of (a) a typical pressure variation over time and (b) pressure (AU) versus frequency.

1 Pa rms in atmospheric pressure means that it varies between 101,323.6 to 101,326.4 Pa. Note that such a small sound pressure can cause a human to go deaf. Generally, the time interval for a measurement must be large enough so that all the frequencies of interest are sensed by the sound transducers. The sound pressure can be averaged over the enough spatial domain in a similar manner. In order to find out the dominant frequencies in a sound pressure signal, fast Fourier transformation (FFT) analysis can be carried out. The pressure sound signal AU in the frequency domain is shown in Figure 5.5b, which indicates the major dominant frequencies of this sound measurement.

A typical human being can detect the sound of a wide range of amplitudes over a wide range of frequency from 20 to 20,000 Hz. The pressure level produced by a sound wave varies considerably. Therefore, sound pressure level is expressed as a ratio between the greatest sound pressure that a person with normal hearing ability can tolerate without any pain. This is called the reference pressure and it is 20 µPa. The softest discernible pressure level varies over a broad range from 20 µPa to 2 Pa. Sound pressure level (SPL) is expressed in logarithm form as

$$SPL = 10\log\left(\frac{P}{P_{ref}}\right)^2 = 20\log\left(\frac{P}{P_{ref}}\right) = 20\log\left(\frac{P}{0.00002}\right) \text{ (dB)} \tag{5.6}$$

where P is the sound pressure level and P_{ref} is the widely accepted standard reference pressure level that is equal to 20 µPa. The sound pressure level is expressed in decibels. Sound pressure becomes 0 dB when the pressure level is the same as that of the reference pressure level (20 µPa). Although its value looks to be arbitrary, it corresponds to the average threshold of hearing for a person with normal hearing capability. Note that the rms pressure level is used to determine sound pressure level as sounds are random in nature. The sound pressure level can be determined from the power level of a sound in a logarithm scale as

$$SPL = 10\log_{10}\left(\frac{\dot{W}}{\dot{W}_{ref}}\right) \text{ (dB)} \tag{5.7}$$

where $\dot{W}$ is the power of sound expressed in watts (W) and $\dot{W}_{ref}$ is the widely accepted standard reference power of sound. The power of sound is proportional to the square of sound pressure, and therefore the above equation is the same as Equation 5.6 in terms of sound pressure. Let us consider a spherical sound wave, propagating in a radial direction (see Figure 5.6) outwardly from a point source and spreading into larger areas of space. As a result, the sound power per unit cross-sectional area known as sound intensity, I, will decrease in a radial direction. This sound intensity can be expressed as

$$I = \frac{P_{rms}^2}{\rho_o c} = \frac{\dot{W}}{A} = \frac{\dot{W}}{4\pi r^2} \tag{5.8}$$

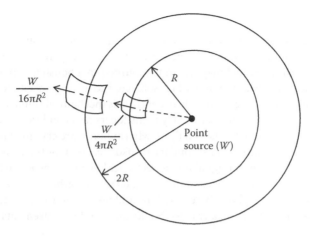

Figure 5.6

Illustration of the relationship between sound intensity and sound power for a spherically propagating sound wave.

where ρ_o is the average density of the medium through which the sound wave propagates, c is the speed of sound, and A is the surface area. The power level varies over a wider range, so the sound power level (SPWL) is expressed in logarithm scale as

$$SPWL(dB) = 10\log_{10}\left(\frac{\dot{W}}{\dot{W}_{ref}}\right) = 10\log_{10}\left(\frac{\dot{W}}{10^{-12}}\right) = 10\log_{10}\left(\frac{I}{I_{ref}}\right) \quad (5.9)$$

Note that the widely accepted standard reference sound power is taken as 10^{-12}. Generally SPWL cannot be measured directly; rather it can be estimated from sound pressure data.

5.3.1 Sound Measurement Devices

Sound measurement devices, like any other measuring instruments, consist of a transducer, a signal conditioning system consisting of filters, amplifiers, and so forth, and a data display unit. The most commonly used transducer for sensing sound waves is the microphone, which will be discussed in the next section. Microphones filter certain unwanted signals and use amplifiers to enhance the strength of the signal to be displayed. If the sound measuring system is designed for getting sound data that may affect the psychological makeup of human beings, then method of non-linear approximation representing the feeling of average human ear is incorporated in the device itself for proper interpretation of sound data. As well, mathematical tools for the analysis of sound signals are included in order to to separate and identify the dominant frequency that may occur during certain physical phenomena. Such a sound measuring device can calibrate the instrument periodically as its characteristics may change from time to time depending on the ambient conditions.

Microphones

A microphone is commonly used for measuring the sound pressure level. It consists of a thin diaphragm, sound damper, housing, secondary transducer, polarizer, and amplifier. The stretched diaphragm is mounted in a housing and is the main mechanical sensor, which vibrates when subjected to acoustic excitation through pressure changes in the medium. Subsequently this mechanical signal is converted into an electrical signal using secondary transducers and conditioned depending on its requirement before further analysis. Microphones are classified into several forms depending on the type of secondary transducer used for the conversion of a mechanical form of a sound signal into an electrical form: (i) capacitance, (ii) electrodynamic, (iii) crystal, (iv) piezoelectric, (v) quartz, (vi) carbon, and (vii) laser. We will discuss these microphone types used in fluid and combustion systems below. Interested readers can refer to advanced books on sound measurements [4–6].

i. Capacitance Microphone

The oldest type of microphone is the capacitance type of microphone, in which a thin diaphragm forms one plate of an air-dielectric capacitor, as shown in Figure 5.7. The sound pressure change makes this diaphragm vibrate and thus changes the distance between the two plates of the capacitor, leading to a change in its capacitance. The signal from this capacitance transducer caused by sound pressure can be sensed with the help of either DC-biased and radio frequency or high frequency sensor. The voltage maintained across the capacitor plates gets changed by the sound pressure, which can be related to a change in the separation distance with time, as given below:

$$V(t) = V_b \frac{ds}{dt} \tag{5.10}$$

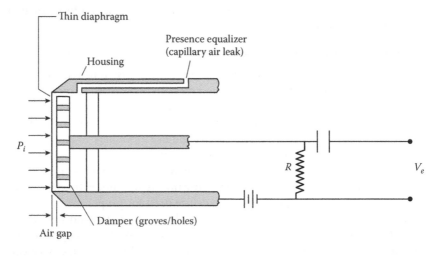

Figure 5.7

Schematic of a typical capacitance microphone.

where V is the voltage across capacitance, V_b is the bias voltage, s is the separation distance between the plates of the capacitor, and t is the time. Note that the value of bias voltage ranges from 100 to 300 V DC. Generally a bias resistor of value higher than 100 MΩ is used, which acts as a highpass filter for a sound signal and a lowpass filter for bias voltage. In most cases, the charge of the capacitance remains almost constant within the time interval of the changes and hence the voltage across the capacitor gets changed either below or above the bias voltage with a slight change in the separation distance between capacitor plates. For recoding a sound signal, the voltage across the resistor is amplified using a suitable electronic circuit. Note that high impendence is generally used in the electronic circuit so it would not contribute any voltage gain. There is another type of condenser known as the electret microphone that is quite popular, especially in laboratory use, due to its low cost and better performance. In this case an external polarizing voltage is replaced by self-polarizing voltage. In an electret, a ferroelectric material is used that is polarized to provide a requisite bias voltage.

ii. Electrodynamic Microphone

The principle of a moving conductor in a magnetic field is used for sensing the sound pressure level in an electrodynamic microphone. When sound pressure acts on the diaphragm, it moves and induces a voltage in the conductor due to the relative motion in the magnetic field provided by a permanent magnet. A typical diagram of an electrodynamic microphone is shown in Figure 5.8, which consists of a diaphragm, a moving induction coil, and a permanent magnet. In some designs both the diaphragm and induction coil are clubbed together, which is known as a ribbon microphone, which senses sound in a bidirectional pattern as it can sense pressure gradient rather than the sound pressure. Because of this, it is not preferred in combustion measurement, but in recent times it has been redesigned to make it unidirectional. Generally a single dynamic element may not respond linearly to all audio frequencies. One way to overcome this problem

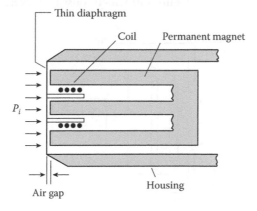

Figure 5.8

Schematic of a typical electrodynamic microphone.

is to use multiple elements for a certain range of a sound spectrum. However, it is not that easy to combine multiple elements in elements that can cover the entire range of a sound spectrum, due to complexities in fabrication.

iii. Piezoelectric Microphone

In a piezoelectric microphone, as shown in Figure 5.9, a piezoelectric element (quartz crystal) is used as a sensor to measure sound pressure. As we know, when a piezoelectric crystal is subjected to a certain pressure (sound), it will produce an electric signal that can be used for measuring sound pressure level, which is why it is also known as a crystal microphone. For higher sensitivity a cantilever element is attached mechanically to the diaphragm. Due to its higher sensitivity and directionality, crystal microphones are used in laboratory combustion experiments.

We have learned about the various types of microphones, so now we need to determine the ideal characteristics one has to look for when choosing a microphone for sound measurements. For this purpose, the characteristics of specific microphones are provided in Table 5.1. As well, one should look for the following general characteristics:

   i. A higher SNR at the lowest sound level
   ii. A flat frequency response over a wide range of sound levels
  iii. Unidirectivity/nondirectivity
  iv. Insensitivity to other environmental properties such as pressure, temperature, humidity, etc.
   v. A higher repeatability over the entire dynamic range
  vi. Lower size, weight, and cost

iv. Sound-Level Meter

A sound-level meter is generally used to measure the sound level from a combustion system for assessing the noise produced by it. The schematic of a typical sound-level meter is shown in Figure 5.10, which consists of several interconnected

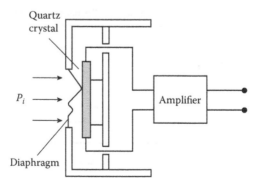

Figure 5.9

Piezoelectric microphone.

Table 5.1 Types of Microphones and Their Characteristics

Type	Frequency Range	Impedance (Ω)	Approximate Open-Circuit Sensitivity (dB)	Advantages	Disadvantages
Capacitance	20 Hz to 20 kHz	10	−45	Less sensitive to vibration • Stable • Inexpensive	Needs high bias voltage • Sensitive to temperature and pressure variation
Electrodynamics	20 Hz to 20 kHz	10	−85	Self-generating,	Large size
Crystal	Audio range and ultrasound	100,000	−50	Self-generating, rugged	Sensitive to vibration
Quartz	Mainly ultrasonic	High	−100	Rugged	Smaller probe
Carbon	Mainly ultrasonic	Medium	−49	High-level signal	Narrow range of frequency response

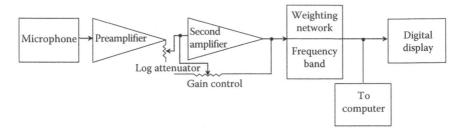

Figure 5.10

Schematic of a typical sound-level meter.

components such as a microphone, preamplifier, weighting networks, amplifier, and digital display. When sound pressure, P_i, is applied on the diaphragm of a microphone, it generates voltage. As discussed earlier, a microphone has a capillary tube to create a slow leak so that the average pressure remains almost at equilibrium with the atmospheric pressure. As a result, due to this slow leak a microphone would not respond to slowly varying pressure. However, this is not of any concern to us with regard to the measurement of audible sound that varies from 20 to 20 kHz. As long as a microphone senses the sound level in this range of frequency, it will serve the purpose of measuring the overall sound level of combustion and flow systems. The output voltage of a microphone is quite small due to the sound pressure, which is in turn due to a high impedance level, and hence an AC preamplifier is used to augment a signal level without distorting the quality of the signal from the microphone. Subsequently, weighting networks (electrical filters) are used so that the frequency response of an instrument can be roughly matched with that of the average human's hearing capability. As a result, this instrument can include a psychoacoustical factor that can provide a measure of a relative sound magnitude as judged by a person known as loudness. This loudness of a sound is dependent on the physical waveform generated from a source and the average sensitivity of the human hearing system and is expressed in terms of a unit known as a phon. The loudness level of a phon is equal to the sound pressure level in decibels at a frequency of 1000 Hz. The value of a phon corresponding to zero indicates the threshold of hearing of a human. The iso-loudness contours obtained from measurements based on pure tones are shown in Figure 5.11, in which sound pressure level is plotted against frequency. For a particular phon, the sound pressure level indicates that a certain pressure amplitude has been applied at a certain frequency so that the human ear can perceive a sensation equal to this loudness. For example, at a loudness level of 40 phon, a sound pressure level of 52 dB at 100 Hz sounds as loud as 40 dB at 1000 Hz. Therefore, it can be inferred that the frequency response of the human ear is not only non-flat but is also nonlinear. The main function of a sound-level meter is to determine loudness perceived by humans and not measurement of sound pressure level. Hence the electrical filters (weighting networks) are used in this instrument for simulating the probable response of an average human to the sound level.

5. Thermal Flow Measurements

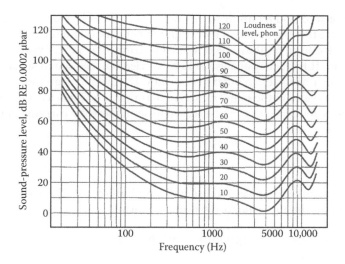

Figure 5.11

Schematic of a free-field iso-loudness contour for pure tones. (From American Standard Measurement for Noise Measurement Z24.2-1942.)

Generally three filters A (40 phon), B (70 phon), and C (100 phon) are provided in a typical sound-level meter. Figure 5.12 indicates the frequency response of these three filters. It can be observed from Figure 5.12 that filter responses selectively between low and high frequencies similar to human ear. Generally filter A is used for a sound-pressure level of 55 dB while filter B is used for a sound-pressure

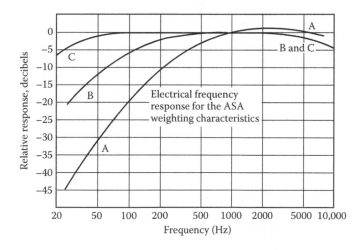

Figure 5.12

Characteristics of three standard weighting networks. (From Broch, J. T., *Acoustic Noise Measurements*, Second Edition, Bruel and Kjaer Instrument Company, New York, 1971.)

level between 55 and 85 dB. If the sound-pressure level is above 85 dB, filter C is used. Note that the output of the weighting network is amplified again, which can be ported to an oscilloscope or virtual instrument panel for visualization of a waveform. If frequency analysis is to be carried out, the sound signals can be sent to an FFT analyzer. In order to determine the overall magnitude of a sound level, the rms value of the sound signal is determined and reported. Of course the rms value of sound signals can be related to sound level by carrying out proper calibrations.

Example 5.2

In an experiment, we need to use two sound sources at 75 dB and 90 dB. Determine the total sound intensity level while using these two sound sources.

Given: $SPWL_1 = 75$ dB, $SPWL_2 = 90$ dB.

To Find: SPWL.

Solution: By invoking a definition of SPWL, we can determine the sound intensity for first source as

$$\log\left(\frac{I_1}{I_{ref}}\right) = \frac{75}{10} = 7.5; \quad I_1 = I_{ref} 10^{7.5}$$

Similarly for the other sound source, we can determine it as

$$\log\left(\frac{I_1}{I_{ref}}\right) = \frac{90}{10} = 9.0; \quad I_1 = I_{ref} 10^{9}$$

The total intensity can be estimated as

$$I_t = I_1 + I_1 = I_{ref} 10^{7.5} + I_{ref} 10^{9} = 1.03 \times 10^{9} I_{ref}$$

The total sound intensity level, SPWL, can be estimated as

$$SPWL = 10\log\left(\frac{I_t}{I_{ref}}\right) = 90.13 \text{ dB}$$

5.4 Gas Velocity Measurements

Gas velocity is essential to characterize the flow in a reacting system as the flow structure can be described in terms of local velocity distribution. Since the reaction occurs within a certain spatial domain in a combustion system, it is important to know whether the local velocity will change due to the local heat release rate. In order to measure the burning velocity, we need to measure the local flow velocity perpendicular to the flame front because the flame front is aligned only when the local flow velocity normal to flame front is equal to its local burning velocity. The mass flow rate across a stream tube can be determined provided that we can measure the local flow velocity. As well, the strength of recirculation eddies can be determined from measurements of the local flow velocity distributions. Keep in mind that recirculation zones are created intentionally in the combustion system for stabilizing the flame.

Both intrusive and nonintrusive methods are available to measure the local velocity in any flow system. Special care must be taken when measuring the local velocity in a combustion system. Intrusive methods were used earlier to measure the local velocity by inserting probes into the flow, which disturbs the local flow itself during the process of measurement. But in recent times, nonintrusive methods using optical techniques are preferred over intrusive methods with the advent of light sources, optics, data acquisition systems, image processing techniques. The oldest and most common optical method is the particle tracking method in which particles such as magnesium oxide, zirconium oxide, and titanium oxide are introduced into the combustion system and whose paths along with the fluid flow are visualized with the help of optical systems. As well, with the advent of laser light sources several optical techniques have been devised and are now used routinely for characterization of the local velocity distribution in a combustion system. Velocity measurement systems for gas can be broadly divided into two categories: (i) intrusive and (ii) nonintrusive methods, as shown in Figure 5.13. In the case of intrusive types, we will be discussing the probe method using pressure-measuring devices and hot-wire anemometers in the following two sections, respectively. In the nonintrusive category, we will learn about laser Doppler velocimetry (LDV) and particle image velocimetry (PIV) when we discuss various combustion diagnostics. In the following section, we discuss three types of probes: (i) pitot-static, (ii) three-hole, and (iii) five-hole.

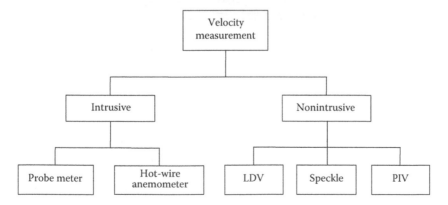

Figure 5.13

Types of velocity measurement systems.

5.4.1 Pitot-Static Probe Method

The local velocity can be measured easily by knowing the total pressure and static pressure at the point of measurement, which can be accomplished using a pitot probe and a static probe together. We can easily determine the local velocity for both incompressible and compressible flow using the respective Bernoulli equation that relates static pressure, total pressure, and local fluid velocity. We know that total pressure P_t is that pressure experienced by an object if a fluid element in motion is brought to rest isentropically. The total pressure can be related to the static pressure and local flow velocity for incompressible and compressible flow, respectively, as

$$P_t = P + \frac{1}{2}\rho V^2 \tag{5.11}$$

$$\frac{\gamma}{\gamma-1}\frac{P_t}{\rho} = \frac{\gamma}{\gamma-1}\frac{P}{\rho} + \frac{1}{2}V^2 \tag{5.12}$$

where P_t is the total pressure, P is the static pressure, V is the local velocity of flow, ρ is the density of fluid, and γ is the specific heat ratio. If there are no losses occurred in the flow, total pressure remains the same. Therefore, a decrease in flow velocity results in an increase in static pressure and vice versa. The measurement of total pressure can be easily accomplished by using a pitot probe, which consists of a simple bent tube with a central opening, as shown in Figure 5.14. This tube is bent around 15 to 20 tube outer diameter from the central opening to avoid the effects of obstruction to the flow on the measurement of total pressure. One end of this probe is aligned with the flow direction while the other end is

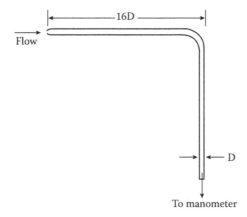

Figure 5.14

Schematic of a pitot probe.

5. Thermal Flow Measurements

connected to a pressure-sensing device such as a manometer or any other strain gauge or piezoelectric sensor. When the incoming stream is entered into the rear opening of the pitot probe, pressure is built up due to the impact of the flow and thus registers an increase in pressure due to the conversion of the dynamic head into pressure. Efforts must be made to align the probe opening properly along with the local direction of the flow or it will measure a lower pressure, incurring errors in the pressure measurement. The extent of the errors is dependent on the extent of the deviation of this yaw angle from the flow. In order to avoid this problem, pressure probe designs such as (i) a channel tube, (ii) a chamfered tube, (iii) a tube with an orifice inside, and (iv) a Keil tube are being used routinely. Among these choices, the Keil tube is the most sensitive to the misalignment of a probe with the flow direction.

We know that the pressure exerted by the fluid element perpendicular to flow direction is known as the static pressure. In other words, static pressure is the pressure experienced by the fluid elements when moving along the local velocity. Hence it is quite difficult to measure the static pressure in a flow as compared to the measurement of total pressure. When we use a probe to measure static pressure, small ports or holes are made around the outer periphery of the tube for sensing the static pressure. These static holes must be located at a sufficient distance away from the downstream of the front portion of the probe so that the measured pressure is almost equal to that of the freestream static pressure. This static pressure is affected by the distance from the front opening, so the pressure gradient created is due to the presence of the probe itself. The measurement of static pressure is very sensitive to the inclination angle of the probe. In order to avoid errors due to misalignments of the probe, four or more mutually perpendicular static pressure holes are used. The schematic of a typical pitot-static probe is shown in Figure 5.15a in which both stagnation and static pressure can be measured simultaneously. We can measure the local velocity by using Equations 5.1 and 5.2 for incompressible and compressible flow, respectively. The measurement of static pressure is very sensitive to the location of its pressure trappings, the location of the probe support, and the alignment of the probe with the flow direction. Let us look at the subsonic flow over a pitot-static probe, as shown in Figure 5.15a, in which streamlines along its centerlines come to rest at the pitot hole. Some of the streamlines turn around the nose of the probe and become accelerated, as shown in Figure 5.15a. For accurate measurements, the static pressure holes should be located away from the stagnation pressure hole. The flow over the pitot-static probe is also decelerated due to the presence of a support stem. The variation of pressure due to the nose and the support stem of a typical pitot probe is shown in Figure 5.16. The increase in pressure due to the presence of the support stem and the decrease in pressure due to the presence of the nose almost cancel each other out around 8D distance from the nose; therefore, the static pressure holes are generally located around 8D for minimizing the error in the measurement of the static pressure. As well, the static pressure measurement is very sensitive to the inclination of the probe with regard to flow direction. The inclination angle between the probe axis and streamline in a plane is known as the yaw

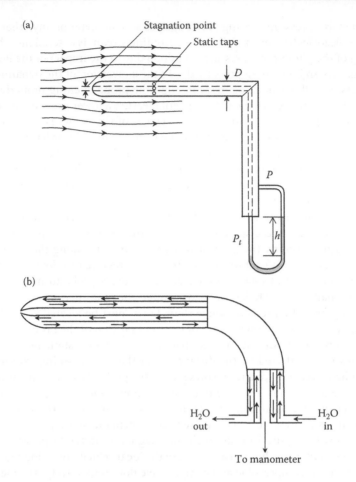

Figure 5.15

Schematic of a pitot-static probe for (a) subsonic cold flow and (b) water-cooled pitot probe.

angle. It has been observed that the effects on static pressure are the opposite of those on stagnation pressure. Hence the effects of yaw angle on a dynamic pressure measurement can be tolerable up to 12°, but beyond this angle, errors in the measurement are intolerable. It is therefore important to use the pitot probe only when the flow direction is known. This probe can be used for a one-dimensional flow, but it should only be used for a two-dimensional flow if it is aligned properly along the streamlines. The probe size must also be much smaller than that of the cross-sectional area of the measuring duct to avoid errors in its measurement due to blockage of flow.

For combustion application, the pitot-static probe can be employed for measuring local velocity. However, for measuring in a flame or small-scale combustion

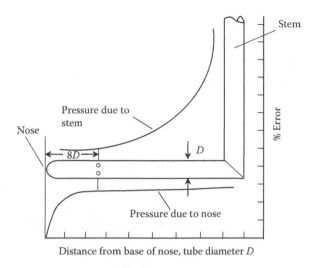

Figure 5.16

Variation of pressure due to nose and support stem of a pitot-static probe.

system, a small-sized probe must be constructed using stainless steel or quartz so that it can withstand high-temperature gas. In most of the bigger combustion systems such as furnaces, boilers, and combustors, a rugged probe with a cooling system is used, which can take care of high-temperature gas laden with dust and liquid droplets. A typical water-cooled pitot-static probe used in a combustion system is shown in Figure 5.15b in which passages are provided to cool the gas. The nose portion of the tube is made of stainless steel, which of course remains uncooled intentionally in order to be heated by the combustion gases so that it can vaporize droplets or burn solid particles that not only block the holes but affect the measurements.

The pitot-static probe is a simple, cheap, and elegant instrument to measure local velocity measurement instruments along with a manometer. The reproducibility of this instrument is quite good as its proportionality constant is almost equal to 1. But it is an intrusive instrument and thus incurs several types of errors, which have been discussed earlier. As well, the measurement of velocity is restricted to 4 m/s if we use a manometer with a least count of 1 mm of water column. However, we can measure up to 0.5 m/s of velocity if we use a micromanometer with a least count of 0.01 mm of water column. Note that the sensitivity of this instrument will be dependent on the sensitivity of the manometer being used for differential pressure measurement.

5.4.2 Three-Hole Probe

We have already discussed how the pitot-static probe can be used for measuring magnitude of local velocity without incurring inordinate errors provided the probe is aligned with the flow direction. Besides this, the flow direction is to be

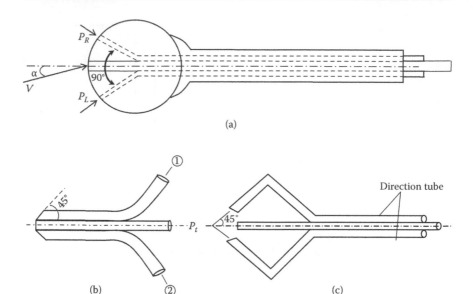

Figure 5.17

Schematics of three-hole probes (yaw): (a) yaw sphere, (b) cobra probe, and (c) claw yaw probe.

measured for several applications. Therefore it is important to know the direction of the flow in a particular plane, for which several types of three-hole probes have been devised for measuring both magnitude and direction of flow. These are also known as yaw probes as they can measure the yaw angle of a flow. Three types of yaw probes are shown in Figure 5.17: (a) yaw sphere, (b) cobra probe, and (c) claw probe. In these probes, three holes are placed in one plane whose top and bottom holes subtends an angle of 45° with respect to the central hole. The front portion of the yaw sphere is a sphere in which three holes are placed in one plane, as shown in Figure 5.17a, in which the top and bottom orifices are placed making 90° while another orifice is placed in the centerline of these two. In the case of a claw probe (Figure 5.17c), orifices are placed in a similar manner except that holes T and B are placed ahead of stagnation hole C. The probe is rotated in a plane until the top and bottom orifice attain the same pressure so that the top flow can be aligned with the bottom flow. By this method, a change in angle can be measured. The actual angle of flow can be measured with respect to a certain reference. If these probes are properly designed and operated correctly, an accuracy of 0.5° can be achieved.

5.4.3 Five-Hole Pitot Probe

The above-mentioned three-hole (yaw) probe cannot be applied for three-dimensional flow. However, most practical flows are three-dimensional in nature. For example, the flow in furnaces, combustors, and boilers will be three-dimensional in nature as it is dominant within a recirculation zone intentionally created for flame stabilization. Several types of flame stabilizers, such as bluff body, swirler, and backward-facing step are being used routinely for stabilizing the flame inside a combustor. As mentioned in Section 5.5.2, we need to measure all three directions and the respective magnitude of velocity at a particular location. Of course one can think of measuring both velocity and direction of a three-dimensional flow by using yaw and pitch probes at a particular location simultaneously, but for this we need to rotate each probe in a plane for measuring its direction, which is quite cumbersome to do in an actual flow system. Therefore, five-hole pitot probes have been devised that are based on the measurement of pressure around the hemispherical nose surface of a probe when it is placed on the flow stream. Several types of five-hole pitot probes are used in practice [1].

Let us discuss a typical hemispherical probe as shown in Figure 5.18 in which four holes are placed around a central hole for measuring pressure. These five holes lead to their respective five tubes for measurement of pressure with the help of pressure-measuring devices. The five holes are numbered according to general convention, as shown in Figure 5.18. This instrument must be calibrated in a subsonic wind tunnel by measuring three differential pressures from which the direction and magnitude of the velocity can be obtained.

In the case of combustion applications, gas must be cooled before being measured in pressure-sensing devices. But in a cooling arrangement, the size of the probe is bigger, which poses problems of errors due to obstruction particularly if used in a small combustor. However, cooling of the probe head can be avoided by using high-temperature resistance materials such as platinum and tungsten, which allows the sensing head to be as small as 3 mm, of course with a hole size of 0.25 mm. But one may face problems of blockage of these holes, particularly when the probe is being used in particle-laden gas.

All the probes described above have certain common limitations. As the sizes of these probes are quite large, they cause disturbances in the flow. Therefore, a distorted flow field occurs with the use of intrusive probes. In combustion systems, the velocity can be quite small with regard to the recirculation region, which calls for measuring less than 1 mm of a water column. For this purpose, a highly sensitive pressure sensor/micromanometer can be used instead of a simple water manometer. The measurement of static pressure is quite cumbersome as it may reach the subatmospheric level in certain locations of the recirculation and separated flow

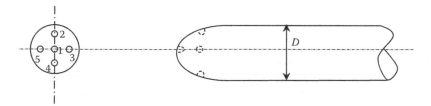

Figure 5.18

Schematic of a typical five-hole probe.

regions. The effects of compressibility seem to be quite small, particularly when flow is below Mach 0.3. As well, the pressure measurements made by these probes are affected considerably by the intensities of turbulence. Errors in pressure measurements using these probes can be tolerable provided the turbulence intensities are less than 10%. Of course, if turbulence intensities are below 20% and the turbulence is isotropic by nature, certain error corrections can be made with the measured data of pressure by using a simple relationship as given below:

$$P_c = P_m - C\rho \overline{V'^2} \tag{5.13}$$

where P_c is the corrected pressure, P_m is the measured pressure, $\overline{V'^2}$ is the fluctuating mean velocity component, and C is the constant that varies from 0 to 1/3. But when the turbulence intensities are beyond 20%, the above simple expression would not be valid. Rather, a detailed procedure should be used for estimating errors in pressure measurements [7].

Example 5.3

A pitot-static probe is placed in a large burner along the flow stream at 35°C to measure velocity for ascertaining uniformity of flow using a water U-tube manometer. If the pressure head across this manometer is 155 mm of water head, determine the local flow velocity.

Given: $T = 35°C$, $h = 155$ mm.

To Find: V.

Solution: The density of atmospheric air can be determined by using the ideal gas law as given below:

$$\rho = \frac{P_a}{RT} = \frac{101,325}{287 \times 308} = 1.146 \text{ kg/m}^3$$

The local flow velocity can be estimated as

$$V = \sqrt{\frac{gh(\rho_m - \rho)}{\rho}} = \sqrt{\frac{9.81 \times 0.155(996 - 1.146)}{1.146}} = 36.33 \text{ m/s}$$

5.4.4 Hot Wire Anemometers

The problems of using a pitot probe in a turbulent flow with a higher turbulence level can be overcome easily by a hot wire anemometer, which is quite sensitive to the fluctuating components of velocity due to its high frequency response. The hot wire probe, as shown in Figure 5.19a, consists of a hot wire that dissipates heat to the flow. In place of wire, a hot film (see Figure 5.19b) can be used as a sensor for measurement of local fluid velocity. The heat transfer from the hot wire is dependent on

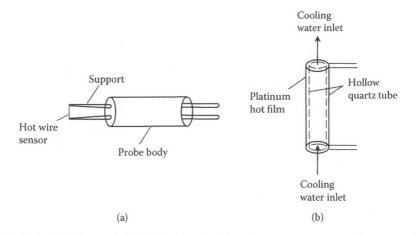

Figure 5.19

Schematic of (a) a typical hot wire probe and (b) hot film.

the local velocity of the flow. In other words, local velocity can be measured quantitatively if this heat transfer can be determined. Generally this heat transfer will be dependent on local velocity only when the air stream is at a constant temperature and pressure, but the hot wire/film sensor is sensitive to temperature, density, and composition of the fluid. For a low-velocity flow, it is not very sensitive to density, but for a high-velocity flow, one has to consider the effects of density while determining heat loss. The heat transfer rate from this hot wire due to electrical heating for isothermal and incompressible flows can be expressed in the following form:

$$q = I^2 R_w = I^2 R_0 \left[1 + \alpha \left(T_w - T_0\right)\right] = h A_w \left(T_w - T_f\right) \tag{5.14}$$

where I is the current in the wire, R_w is the wire resistance, R_0 is the wire resistance at reference temperature T_0, α is the temperature coefficient of resistance, T_w is the temperature of hot wire/film, T_f is the temperature of flowing, h is heat transfer coefficient of hot wire/film, and A_w is the heat transfer area of hot wire. The heat transfer coefficient for a given fluid density is dependent on the local flow velocity as per the following relation as

$$h = C_0 + C_1 \sqrt{V} \tag{5.15}$$

where C_0 and C_1 are constants whose values can be obtained from calibration. This relation is often known as King's law. In order to measure the changes due to flow fluctuation, this hot wire/film is connected to a bridge circuit, as shown in Figure 5.20a, which consists of a hot wire probe, Wheatstone bridge, a potentiometer or voltmeter, an amplifier, and a galvanometer. The current in this bridge can be estimated by measuring voltage across a standard resistance, R_s, as shown in

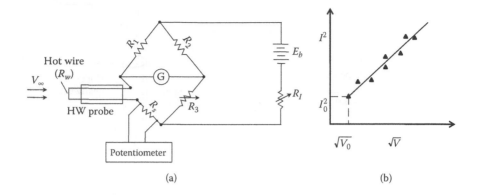

Figure 5.20

Schematic of (a) a typical hot wire flow measurement circuit and (b) a typical calibration curve.

Figure 5.20a, using either a potentiometer or voltmeter. Thus the resistance of hot wire probe, R_w, can be determined easily. By determining current I and the wire resistance, R_w, the local velocity can be determined using Equations 5.14 and 5.15. Of course the calibration of the hot wire probe should be carried out either using a wind tunnel or jet facility with a known fluid velocity. During calibration, the flow velocity is set at a certain known velocity, V_1. Subsequently, the resistance, R_I, is adjusted to have a certain amount of current to flow through the hot wire so that it does not get burned out, but it must be high enough to have ample sensitivity to the measurement of the local flow velocity. As a result, the temperature and the resistance of the hot wire probe attain a certain value. Subsequently, the resistance, R_3, is changed so that the bridge is balanced, indicating a null point in the galvanometer. This balance is needed to ensure that the hot wire remains at a constant temperature for all measured velocities. The measured current I_1 indicates the local velocity V_1 of the flow, which is indicated in a typical calibration plot. If the flow velocity is changed to a new value, there will be a change in temperature and resistance of the hot wire, leading to an unbalanced bridge. Hence, the Wheatstone bridge has to be balanced again by changing the current with the help of variable R_I. The new current that can be measured across the R_s is an indication of the changed velocity. By repeating this procedure, the calibration of the hot wire can be carried out at various local velocity ranges. A typical calibration curve is shown in Figure 5.20b, which indicates that it follows King's relationship from which calibration constants C_0 and C_1 can be determined from the intercept and slope of this curve, respectively.

We will have to use the null/balance mode of the Wheatstone bridge for the measurement of the local velocity under steady flow conditions. But an HWA can also be used for transient flow velocity for which the transient response of the thermal and electrical resistance of the hot wire must be sensed and measured properly. For this purpose, two kinds of electrical compensation—(i) constant temperature (resistance) and (ii) constant current—are used. We have already briefly discussed constant temperature mode and steady flow velocity measurement. But for transient measurement, a feedback control circuit is used to vary the current flow in the hot wire probe so that the wire temperature (resistance) remains almost constant. In contrast to this, in the case of a constant current, a large resistance is added in series with hot wire and a thermal compensating circuit is added to the output AC voltage. In this case, there will be considerable fluctuations in the temperature of the hot wire and support systems. Due to thermal inertia, there will be loss of frequency response and hence this mode is not preferred in a highly fluctuating flow situation. However, the thermal inertia is minimized in the case of a constant temperature mode, as the resistance of the hot wire probe is kept almost constant. The response of the hot wire probe is dependent on the angle of the local flow velocity with the axis of the probe, the diameter of the hot wire probe, and the length-to-diameter ratio. Typical length-to-diameter ratio (L/D) is around 50 for a better response. The upper frequency response of a hot wire probe is restricted to 85 kHz because the diameter of a tungsten hot wire probe size is limited to 5 μm. Of course the hot wire anemometer can be used to measure the direction of the average flow velocity by using the null point method for a three-hole probe [2]. For a two-dimensional flow, X configuration probes can be used with two channel requisite circuits and electronics, but for a three-dimensional flow, three mutually perpendicular probes are used to obtain all three velocities and directions. It has been reported in the literature that apart from local flow velocity and its direction, a hot wire probe can be used to measure fluid temperature, turbulent shear stresses, and concentrations of individual gases in mixture [7]. For more details about hot wire probes, interested readers can refer to [7–12].

The applications of hot wire probes are mostly restricted to low-velocity and dust-free fluid flow as small-diameter hot wires are quite fragile due to inadequate structural rigidity and dust accumulated in the probes affects the measurements adversely. In the high-velocity flow, hot wire probes vibrate and may break due to fluctuating aerodynamic loads. Furthermore, in practical situations it is quite cumbersome to use hot wire probes. In order to overcome these problems, hot film transducers based on the same principles as hot wires are being designed and developed that are quite rugged and insensitive to dust depositions. In this case a thin film of platinum deposited over glass (see Figure 5.19b) is used as a sensor in place of the hot wire probe. For high-temperature applications, hot wires can be cooled easily. The same types of methods and circuits used for hot wires can be employed for hot film sensors. Of course the sensitivity of the hot film will be reduced as compared to a hot wire probe. However, with the advent of advanced electronics and thin film manufacturing methods, its frequency response can be enhanced easily to as high as 50 kHz.

Example 5.4

In a hot wire anemometer, a platinum hot wire sensor is used at 75°C using a bridge circuit in which $R_2 = R_3 = 30\ \Omega$. If the sensor resistance is 2.5 Ω at

25°C and has a thermal coefficient of 0.00425°C^{-1}, determine the setting resistance for this anemometer.

Given: Anemometer circuit.

$T = 75°C$, $R_2 = R_3 = 30\ \Omega$, $R_0 = 2.5\ \Omega$, $\alpha = 0.00425°C^{-1}$.

To Find: R_1 at 75°C.

Solution: We know that the resistance temperature relationship is approximated as

$$R_1 = R_0\ [1 + \alpha(T_s - T_0)] = 2.5\ [1 + 0.00425(75 - 25)] = 3.03\ \Omega$$

When the bridge is balanced, the setting resistance R_1 can be determined as

$$R_1 = R_s\ (R_3/R_2) = 3.03\ \Omega$$

5.5 Temperature Measurements

In a combustion system, temperature measurements of both gas and flame are essential to understand the combustion process. In layman's terms, the temperature of a substance or medium can be expressed in terms of the extent of its hotness or coldness, which can be dependent on the individual's perception of senses. But it can be related to the amount of heat interaction that takes place from the high-temperature to low-temperature zone. As per the kinetic theory of gases, it can be related to the measure of the mean kinetic energy of the molecules of a substance that represents the potential to have a heat flow. Similarly, this concept of temperature can be extended to liquid and solid substances using statistical thermodynamics. Interested readers can refer to other texts for getting a better understanding about the thermodynamic concept of temperature. The temperature of a substance can be determined easily by measuring the change in properties of a substance, such as volume, density, pressure, expansion coefficient, electrical resistance, and radiation, which are influenced by the degree of heat interaction. Based on the change in any of these physical properties, several types of sensors have been designed and developed for the measurement of temperature. Some commonly used temperature-measuring devices are (i) liquid expansion in a tube (e.g., glass thermometer), (ii) thermocouple, (iii) electrical resistance temperature detector (RTD) (e.g., thermistor), (iv) optical pyrometer, (v) radiation thermometer, and (vi) laser scattering methods. We will be restricting our discussions to temperature-measuring devices that can be used in combustion systems. In this chapter, we will be discussing two temperature-measuring devices: (i) thermocouples and (ii) electrical resistance temperature detectors. In Chapter 6, we will be discussing optical and laser scattering temperature measurement systems for combustion systems.

5.5.1 Thermocouples

The most commonly used electrical method of temperature measurement is the thermocouple, which is based on the thermoelectrical property of metals.

If two dissimilar metals are joined together, creating two junctions, as shown in Figure 5.21, and kept in two different temperature baths, an electromotive force (emf) is generated in this circuit. The amount of this emf is dependent on the temperature difference between the two junctions, which can be reproducible and is dependent on the material of the wires in the thermocouple. This phenomenon is known as the Seebeck effect. In contrast, if a current is allowed to pass through the thermocouple, then with two junctions the temperatures will be different from the previous ones as heat is generated or absorbed and thus there will be a heating or cooling effect at each junction. This is known as the Peltier effect, which is caused due to the heating of one junction and cooling of the other. If the voltage across two junctions of a thermocouple is measured using a potentiometer or high-impedance microvoltmeter (1000 MΩ), then the Peltier effect can be neglected as the current flow across the junction becomes zero. As well, if a temperature gradient exists along either or both limbs of the thermocouple, the junction emf will undergo slight variation, which is known

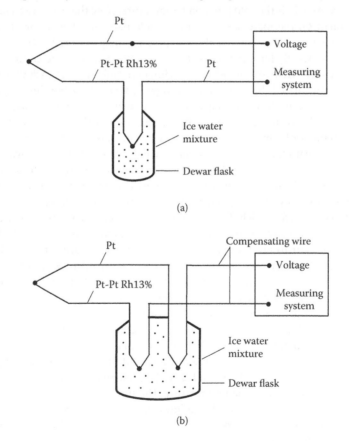

(a)

(b)

Figure 5.21

(a) A simple thermocouple circuit and (b) reference junction compensation.

as the Thomson effect. This effect can be minimized by using either a potentiometer or high-impendance microvoltmeter. Generally, errors in a microvoltmeter circuit are negligibly small but can be determined by knowing the heat transfer condition. It must be kept in mind that for the measurement of temperature, the total voltage across the thermocouple junctions is caused mainly by the Seebeck effect and marginally by the Thomson effect. We know that emf generated due to the Seebeck effect is proportional to the temperature difference between hot and cold junctions, while the Thomson emf is proportional to the square of the temperature difference between hot and cold junctions. Total emf between the two junctions of a thermocouple can be expressed as

$$E = C_S (T_H - T_C) + C_T (T_H - T_C)^2 \qquad (5.16)$$

where E is the total voltage across the thermocouple, C_s is the constant for the Seebeck voltage, C_T is the constant for the Thomson voltage, T_H is the hot junction temperature, and T_C is the cold junction temperature. Note that these two constants are dependent on the materials of thermocouple pairs and hence vary from one thermocouple to another. But the above expression may not predict the measured temperature correctly for a particular thermocouple pair over the entire range of temperature measurement and hence experimental calibration of each thermocouple pair must be carried about over a wide range of its operation. For this purpose, we are only interested in the total voltage generated during temperature measurement rather than the individual contribution of the Seebeck and Thompson effects. Hence the measurement of temperature by using a thermocouple is based on empirical calibration. In real situations, thermocouples will be connected to microvoltmeters, potentiometers, or any other voltage measuring devices through extension wires. Hence the total voltage across the thermocouple during temperature measurement in practical systems will be dependent on the entire circuit and various junctions and their materials, for which the following thermoelectric laws are to be used.

Law of homogeneous metals: If two different metal wires, A and B, form a thermocouple pair to produce emf, when other junctions are introduced in the thermocouple for measurement of voltage (e.g., microvoltmeter), then the temperature changes in the circuit does not affect the net emf provided all wires are made of the same material. This aspect is illustrated in Figure 5.22a.

Law of intermediate metals: A and B form a thermocouple pair with two junctions to produce emf, but when the third metal C is included somewhere in between these two junctions, then the net emf remains unaffected with the addition of metal C. For example, in the case of a Pt and Pt-Rh 13% thermocouple, if a copper metal is inserted as shown in Figure 5.22b with two new junctions, X and Y, which are held at the same temperature, the total emf in the circuit remains unchanged. This thermoelectric law allows the use of a compensation cable or microvoltmeter and also allows the use of joining materials during the fabrication of the thermocouples.

Law of intermediate temperatures: If a simple thermocouple at two different temperatures, T_1 and T_2, produces an emf E_1 and produces emf E_2 when two

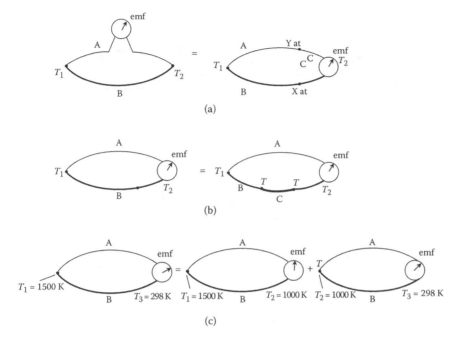

Figure 5.22

Illustration of thermoelectric laws. (a) Homogeneous metal, (b) intermediate metals, and (c) intermediate temperatures.

junction temperatures become T_2 and T_3, then it will produce emf $E_1 + E_2$ when junctions are maintained at T_1 and T_3. Let us consider a Pt and Pt-Rh 13% thermocouple as shown in Figure 5.22c, which operates between two different temperature limits (1500–1000 and 1000–298 K). The emf generated for junctions 1500 and 1000 K becomes 6.54 V while between 1000 and 298 K becomes 6.92 V. If the junctions of this thermocouple are maintained at 1500 and 298 K, according to the law of intermediate temperature, total emf becomes 13.46 V. This law helps us to use thermocouple tables based on a standard reference temperature (273 K) even though reference junction temperature may not be in a standard state during measurement. In addition, direct correction for secondary junctions is possible if its temperature is known.

Voltage to temperature conversion: We use the above thermoelectric laws whenever we are dealing with thermocouple circuits for the measurement of temperature. Let us consider a simple thermocouple system as shown in Figure 5.23a that consists of two junctions—hot and cold—and a potentiometer for the measurement of voltage. The law of intermediate materials ensures that total emf remains the same with the use of a potentiometer or any other compensation wire as long as both connecting junctions remain at the same temperature. By using a potentiometer, one can measure voltage across the thermocouple circuit. But the unknown temperature can be determined provided the temperature of

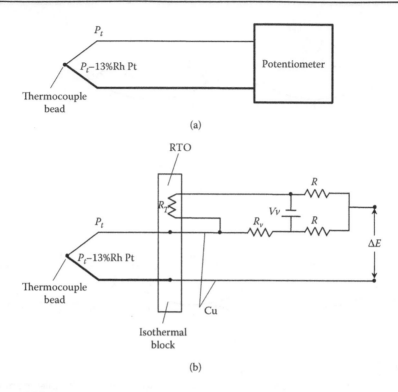

Figure 5.23

(a) A simple thermocouple circuit and (b) reference junction compensation.

the other junction is known. As mentioned earlier, the known temperature junction is labeled as the reference temperature junction. The temperature of the reference junction must be maintained so that it will be accurate, stable, and easily reproducible. Generally, the ice point (273 K) is considered as the reference junction temperature. There are two ways by which the ice point can be accomplished. The simpler one is the ice bath, which can be made by using a mixture of ice and water in a transparent slush form in a vacuum flask. It is not that easy to obtain the actual ice point temperature (0°C), but an ice point of ±0.01°C can be obtained with reasonable effort. Another way of creating a reference junction is by using an electronic circuit as shown in Figure 5.23b, which mainly uses a thermistor/RTD to determine the local ambient temperature. Note that the reference temperature must be in thermal contact with the isothermal block. The voltage across the RTD is adjusted by varying the R_v so that the output voltage across the circuit will be zero at 0°C.

We have already seen that a voltage-balancing potentiometer is traditionally used to measure the actual magnitude of emf generated by a thermocouple so that no current can flow through it from external sources. It is very important to measure the magnitude of voltage carefully as it is in the order of millivolts.

5. Thermal Flow Measurements

However, in recent times high-impendance microvoltmeters are being used to measure millivolts generated by the thermocouple circuit. The range of emf values for certain standard thermocouple pairs over certain temperatures is provided in Table 5.2, which clearly indicates that the voltage range is in the order of millivolts. It can be observed that the reported thermoelectric emf corresponds to the reference junction temperature at 0°C. The standard thermoelectric emf tables for various types of thermocouples have been prepared by the National Institute of Standards and Technology (NIST) [13]. The extended tables that contain emf data with the temperatures for certain thermocouple pairs commonly used in combustion systems are provided in Appendix E. The thermoelectric sensitivity of various thermocouple pairs are given in Table 5.2 along with their temperature limits and applications. It can be noted from this table that Pt-Pt Rh13% (R-type) has lower thermoelectric sensitivity than Cr-Al (K-type).

The emf data, E, in Table 5.2 can be used to determine temperature using the ninth-order polynomial fit as

$$T = A_0 + A_1 E + A_1 E^2 + A_3 E^3 + A_4 E^4 + A_5 E^5 + A_6 E^6 + A_7 E^7 + A_8 E^8 + A_9 E^9$$

where T is the temperature, E is the voltage with respect to ice point (0°C), and A is the polynomial coefficients given in Table 5.3 for certain thermocouple pairs. The corresponding accuracy with which it can mimic the NIST data given in Table E.1 (Appendix E) is indicated in Table 5.3. This equation can be implemented easily in virtual instrument software (LabView) so that temperature measurement can be automated.

As mentioned earlier, most measurement data is acquired by using a data acquisition system (DAS). Special data acquisition cards are being designed and developed for the measurement of thermocouples in which built-in electronic cold junction compensation is provided with the help of a thermostat sensor, as discussed earlier (see Figure 5.23b). This cold junction compensation incurs errors in the range of 0.5°C to 1°C which are clubbed into the acquired data. Hence care must be taken to minimize this bias error. We know that voltage from the thermocouple pairs during the measurement of temperature is in the range of

Table 5.2 Thermocouple Pairs Used in Combustion Systems

Thermocouple	Type	emf (μV/°C)	Temperature Limit (K)	Applications
Chromel-Constantan	E	76	1200	Hot gas
Chromel-alumel	K	41	1200	Hot gas
Platinum, 10% rhodium/ platinum	S	12	2030	Flame
Platinum, 13% rhodium/ platinum	R	11	2030	Flame
Platinum/30% rhodium; platinum/6% rhodium	B	20	2080	Rich flame
Tungsten/5% rhenium; 26% rhenium/tungsten	C	4.8	2800	Rich flame

Table 5.3 Polynomial Coefficients for Certain Standard Thermocouple Combinations Used in Combustion Systems

Type E	Type K	Type R	Type S
Chromel(+) vs. constantan(−)	Chromel(+) vs. alumel	Platinum, 13% Rhodium(+) vs. platinum(−)	Platinum, 10% Rhodium(+) vs. platinum(−)
−100°C to 1000°C	0°C to 1370°C	0°C to 1750°C	0°C to 1750°C
±0.5°C	±0.7°C	±0.5°C	±1°C
9th order	8th order	8th order	9th order
A_0 0.104967248	0.226584602	0.263632917	0.927763167
A_1 17189.45282	24152.109	179075.491	169526.515
A_2 −282639.085	67233.4248	−48840341.37	−31568363.94
A_3 12695339.5	2210340.682	1.90002E + 10	8990730663
A_4 −448703084.6	−860963914.9	−4.82704E + 12	−1.63565E +12
A_5 1.10866E + 10	4.83506E + 10	7.62091E + 14	1.88027E + 14
A_6 −1.76807E + 11	−1.18452E + 12	−7.20026E + 16	−1.37241E + 01
A_7 1.71842E + 12	1.38690E + 13	3.71496E + 18	6.17501E + 17
A_8 −9.19278E + 12	−6.33708E + 13	−8.03104E + 19	−1.56105E + 19
A_9 2.06132E + 13			1.69535E + 20

Source: Holman, J. P., *Experimental Methods for Engineers*, Seventh Edition, McGraw-Hill Publishing Company Ltd, 2007.

millivolts, so the voltage signal from the thermocouple has to be amplified using a suitable signal conditioner (amplifier) for acquiring them through a DAS as it uses A/D converters with ±5 V full scale. The amplification (gain) factor of 100 to 400 is usually employed for temperature measurement using a thermocouple pair. As well, the software associated with this card uses mostly NIST polynomial interpolation for converting voltage to temperature, which is dependent on the materials of the thermocouple and the range of temperature measurements. It must be ensured that the thermocouple used follows the NIST polynomials by carrying out suitable calibration.

5.5.2 Thermocouple Probes

Thermocouple probes can be broadly divided into two categories: (i) bare-wire and (ii) closed. Generally, bare-wire probes are used in the measurement of flames in a combustion system. Two types of bare-wire probes are shown in Figure 5.24a and b. The first probe, shown in Figure 5.24a, consists of a butt type of welded bead; the probe support is mainly meant to measure flame temperature. Generally a fine thermocouple wire of 50 μm or less (R or S type) is used to form a tiny bead that can resolve the thickness of the flame, which happens to be quite small. The supporting stem is placed away from the bead while the thermocouple wire is kept under tension to avoid sagging due to its expansion when used under the high temperature of the flame. Water-cooled probes with bare-wire beads (see Figure 5.24b) housed in metallic tubes are also used in combustion systems. Generally either R- or S-type thermocouples are used in combustion systems. Therefore, it is likely that catalytic reactions may take place on the bead as it is made of platinum and hence may lead to the measurement of erroneous

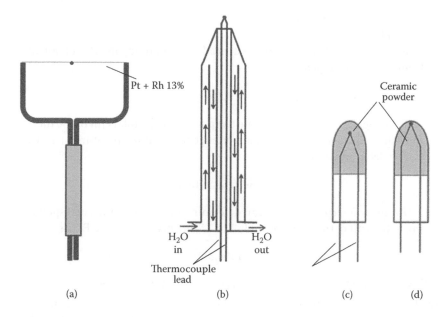

Pt + Rh 13%

H_2O in H_2O out

Thermocouple lead

Ceramic powder

(a)   (b)   (c)   (d)

Figure 5.24

Temperature probes: (a) butt-welded, (b) water-cooled, (c) insulated, and (d) grounded.

temperatures. In order to avoid a reaction to the bead at higher temperatures due to oxidation of the platinum wires, a catalytic coating of alumina-based ceramic (90% Al_2O_3–10% MgO) is used. The coating increases the bead diameter and thereby affects the heat transfer rate by its exposed area. This alumina-based ceramic coating is noncatalytic and the effect of the coating can be neglected by making a very thin coating.

The other category of closed thermocouple probes, which consist of thermocouple wires housed inside a metallic tube known as a sheath (see Figure 5.24c and d), are used in the temperature measurement of hot gases in combustion systems. Common sheath materials include Inconel and stainless steel. Inconel can support higher temperature ranges. The tip of the thermocouple probe is available in two different styles: grounded and ungrounded. In the case of a grounded tip the thermocouple remains in contact with the sheath wall. Note that a grounded junction can provide a fast response time but it is susceptible to electrical ground loops. In contrast, in ungrounded junctions the thermocouple is separated from the sheath wall by a layer of insulation, as shown in Figure 5.24d.

Example 5.5

A Pt-Pt/10%Rh thermocouple is used to measure temperature in a flame with the help of a potentiometer whose reading is 15.58 V. If the potentiometer terminal is at 45°C, determine its measured temperature.

Given: $V = 15.58$ V, $T_a = 45°C$.

To Find: T.

Solution: The thermoelectric potential corresponding to 45°C obtained from Table E.3 (Appendix E) is

$$E_{45°C} = 0.267 \text{ mV}$$

By using the law of intermediate temperatures, the total emf of the thermocouple, E_t, corresponding to a reference junction temperature of 45°C, can be determined as

$$E_t = 15.58 + 0.267 = 15.847 \text{ mV}$$

By linear interpolation, we can determine the temperature corresponding to the emf of the thermocouple $E_t = 15.847$ mV from Table E.3 as

$$T = 1386.36°C$$

5.5.3 Temperature Corrections in a Bare Thermocouple

In the case of a bare thermocouple probe, there will be errors in the measurement of the temperature in a combustion system caused by three modes of heat transfer—convection, conduction, and radiation. The heat is being transferred by conduction from the bead to the thermocouple wire through its lead. A second aspect of affecting the accuracy of thermocouple measurements is dependent on the relative magnitude of radiation heating/cooling in comparison to the convective heating of the bead surface. In the case of flame, significant heat transfer occurs due to radiation heat transfer as the gas temperature is quite high. As a result, there can be large errors in gas temperature measurements made around flames in a combustion system.

Convection: The heat of a gas at temperature T_g is convected to the thermocouple bead, thereby attaining a temperature of T_b. This convected heat is given by

$$\dot{Q}_{conv} = hA_s(T_g - T_b) \tag{5.17}$$

where h is the convective heat transfer coefficient (W/m²K), T_g is the gas temperature (K), T_b is the thermocouple bead temperature (K), and A_s is the surface area of the thermocouple bead (m²).

Radiation: The heat gained by the thermocouple bead is radiated back to the surrounding area, and the measured temperature of the thermocouple is affected by the radiation [14]. The radiated heat transfer rate is given by

$$\dot{Q}_{rad} = \sigma \varepsilon A_s \left(T_b^4 - T_a^4\right) \tag{5.18}$$

where σ is the Stefan–Boltzmann constant = 5.67×10^{-8} W/m²K⁴, ε is the emissivity of the thermocouple surface, and T_a is the surrounding temperature (K).

5. Thermal Flow Measurements

Conduction: The heat from the thermocouple bead is conducted to the lead wire by a heat transfer rate of

$$\dot{Q}_{cond} = kA_s \frac{\Delta T}{\Delta x} \quad (5.19)$$

where k (W/mK) is the thermal conductivity of the thermocouple wire and ΔT (K) is the change in temperature along the length Δx (m).

The convected heat from the gas is conducted, radiated, and some energy is stored in the thermocouple itself. In the steady state, the rate of change of energy stored in the thermocouple is zero. Conduction occurs through a very small diameter wire and, in comparison with convection and radiation, conduction can be neglected. Therefore the energy balance equation becomes

$$hA_s = (T_g - T_b) = \sigma\varepsilon A_s\left(T_b^4 - T_a^4\right) \quad (5.20)$$

Rewriting the above equation

$$T_g = T_b + \frac{\sigma\varepsilon}{h}\left(T_b^4 - T_a^4\right). \quad (5.21)$$

An expression for wire emissivity working with temperatures and thermocouples above 1250 K is used by Hinze [15,16] as

$$\varepsilon = 9.35 \times 10^{-5}\, T_b + 0.06. \quad (5.22)$$

The convective heat transfer coefficient can be estimated from the relation for the Nusselt number as given below [17]:

$$Nu = \frac{hD}{k} = 0.42\, Pr^{0.2} + 0.57\, Pr^{0.33} Re^{0.5}. \quad (5.23)$$

Note that this is valid for $0.01 < Re < 10^5$. D is the diameter of the thermocouple wire, Re is the Reynolds number ($Re = VD/v$), V_g is the average velocity of gas, v is the Kinematic viscosity, Pr is the Prandtl number ($Pr = v/\alpha$), and k is the thermal conductivity coefficient. The relationships for the thermal conductivity coefficient, k, kinematic viscosity, v, and thermal viscosity, α, in terms of thermocouple bead temperature are given below [16]:

$$k = 3.75 \times 10^{-5}\, T_b + 0.04$$

$$v = 2.5 \times 10^{-7}\, T_b - 1.43 \times 10^{-4} \quad (5.24)$$

$$\alpha = 2.89 \times 10^{-7}\, T_b - 1.33 \times 10^{-4}.$$

This procedure can be used to estimate the correction for the measured temperature for radiation.

5.5.4 Suction Pyrometers

We learned that the predominant error during the measurement of temperature using a bare-wire thermocouple in a combustion system is caused by radiation heat losses from the thermocouple bead to its surrounding area. In order to reduce these measurement errors associated with radiation heat losses from a thermocouple bead, a suction pyrometer has been devised that is used extensively in large flames in furnaces and large combustion systems. The schematic of a typical suction pyrometer is shown in Figure 5.25a, which consists of a thermocouple junction (e.g., Pt-PtRh13%), sheath, and radiation shields, and cooling system. The thermocouple is placed on the axis of the refractory (e.g., sintered alumina) sheath, which is surrounded by two refractory layers (e.g., sillimanite) of radiation shields. The type of thermocouple used in this pyrometer will depend on the temperature range of the application. The diameter of the thermocouple wire must be as small as possible but it should not be too fragile that it can break easily due to high-suction velocity. For R- and S-type thermocouples, a wire diameter in the range of 0.25 to 0.5 is preferred depending on the length of thermocouple wire being used in the pyrometer. Generally the thermocouple bead is placed around 4 cm downstream of the suction port, which can be varied

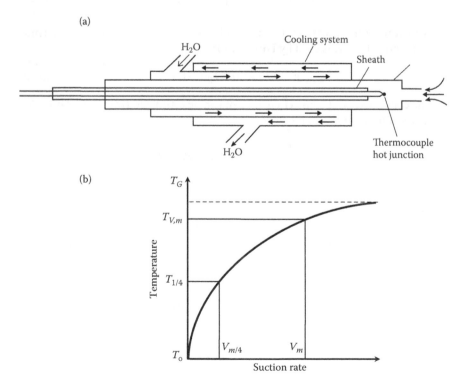

Figure 5.25

(a) Schematic of suction pyrometer and (b) variation of T with suction velocity.

depending on its design consideration. The gas is sucked over the thermocouple bead at a high velocity, which ensures that a higher rate of heat transfer due to convection will take place from the gas to the thermocouple bead. As a result, the temperature measured by the bead will be almost equal to the local gas temperature. Note that the gas is also drawn at a high velocity through the radiation shields. The gas velocity over the thermocouple wire plays an important role and varies from 20 to 200 m/s. The measured temperature T_m must correspond to the actual velocity, which is different than the stagnation temperature T_t corresponding to the stagnation condition ($V = 0$ m/s), as discussed earlier. The measured temperature T_m can be related to stagnation temperature T_0 as per the following relationship:

$$T_m = T_t - (1-\alpha)\frac{V_g^2}{2C_p} \qquad (5.25)$$

where V_g is the gas velocity, C_p is the specific heat of gas, and α is the recovery factor whose values will be around 0.85 ± 0.09 for the thermocouple along the flow direction. There will also be an error incurred due to heat conduction, which can be minimized using a smaller diameter thermocouple wire. The error due to radiation heat transfer can be reduced by using more number of shields and their respective thickness and choosing proper material with low emissivity and conductivity. Hence the accuracy of temperature measurement in a suction pyrometer is dependent on the gas temperature, the suction velocity, the geometry of the pyrometer, the material and construction used, and the intensity of radiation. However, the accuracy of the temperature measurement is predominantly influenced by the magnitude of the suction velocity because a higher heat transfer coefficient is associated with higher velocity. If the suction velocity increases steadily from the zero velocity, the measured temperature increases and reaches an asymptotic value, as shown in Figure 5.25b. The asymptotic value of the temperature coincides with the actual gas temperature, T_g, while the measured temperature, T_m, is less than the gas temperature, T_g. Hence, the effectiveness of the pyrometer, ε, can be expressed as

$$\varepsilon = \frac{T_m - T_0}{T_g - T_0} \qquad (5.26)$$

We can determine the true gas temperature by using several other techniques, such as the sodium line reversal method and spectroradiometers. The effectiveness factor can be obtained by using the proper calibration method for a particular pyrometer and combustion system. If the effectiveness factor, ε, is known, then the gas temperature can be estimated as

$$T_g = T_m - T_0(1-\varepsilon) \qquad (5.27)$$

The error level in a suction pyrometer is found to be around ±5°C at around 1600°C. In a fuel-rich condition, care must be taken to place the pyrometer properly, particularly at high-suction velocity, as the refractory shield may act as a bluff body to stabilize the flame, which can incur a considerable amount of error in the temperature measurement.

5.5.5 Unsteady Temperature Measurements

The measurement of temperature fluctuations in a practical combustion system, particularly in turbulent flow conditions, is quite important but cumbersome to carry out. The simplest way to measure temperature under unsteady conditions is to use fine-wire thermocouples in the combustion system. If bare thermocouple wire is used, the dynamic characteristics of a thermocouple can be analyzed by considering a first-order model, as shown in Figure 5.26, in which the bare bead of a thermocouple is immersed in its surrounding medium. The resistance to heat transfer and heat storage is lumped into the bare bead of the thermocouple sensor. By neglecting the heat losses due to conduction and radiation heat transfer and any catalytic activity on the surface of its bead, conservation of energy can give

$$hA\ (T_g - T)\ dt = mC_p dT \tag{5.28}$$

where h is the local heat transfer coefficient, A is the heat transfer area, m is the mass of the bead, and C_p is the specific heat of the bead. By simplifying the above equation we have

$$T_g = T + \frac{mC_p}{hA}\frac{dT_m}{dt} = T + \tau\frac{dT}{dt} \tag{5.29}$$

where

$$\tau = \frac{mC_p}{hA} = \frac{\rho C_p d^2}{4kNu}$$

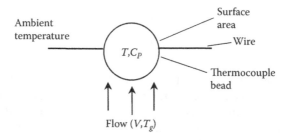

Ambient
temperature

Surface
area

Wire

T, C_p

Thermocouple
bead

Flow (V, T_g)

Figure 5.26

Schematic of a first-order model for a thermocouple temperature sensor.

The solution of the above equation as a function of a time constant would be

$$\frac{T - T_g}{T_0 - T_g} = e^{\left(\frac{-hA}{mC_P}\right)\tau}$$

(5.30)

Note that T_0 is the initial temperature at time zero. τ is the characteristics time constant of the thermocouple bead, d is the diameter of the bead, k is the thermal conductivity, and Nu is the Nusselt number. The response of the thermocouple sensor gets enhanced by decreasing m and C_P or by increasing h and A. Of course, in practice, efforts must be made to enhance the response of temperature measurement by using both criteria. It must be kept in mind that the characteristic time constant, τ, cannot remain constant but will vary depending on the measurement condition as it is in turn dependent on the flow velocity and fluid properties. It is important to determine the characteristic time constant, τ, for a particular thermocouple sensor. Several methods have been devised for evaluating τ of a thermocouple. Generally it can be obtained experimentally by forced convection decay of temperature of the thermocouple bead. For this purpose, the thermocouple bead is heated above its gas temperature either by electrical heating or other means (laser-based or acoustic).

For the purpose of external electrical heating of a thermocouple bead by electrical means, a system as shown in Figure 5.27 can be used that consists of an amplifying unit, a differential active filter to achieve thermal inertia compensation, and a signal processing unit. This signal processing unit can carry out

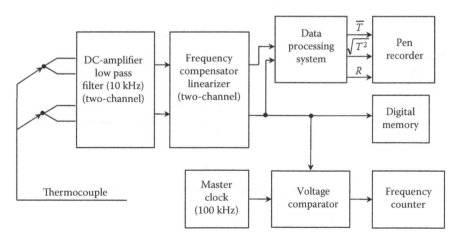

Figure 5.27

A typical compensating circuit for a thermocouple probe. (Reprinted from Symposium (International) on Combustion, 17(1), Yoshida, A. and Tsuji, H., Measurement of fluctuating temperature velocity in a turbulent premixed flame, 945–956, Copyright 1979, with permission from Elsevier.)

several functions: (i) evaluation of the rms value of a probe by a time domain analysis, (ii) measurement of the spectral density of the signal, and (iii) evaluation of the probability density function of the temperature fluctuation.

For a short duration, an AC current (50 kHz) is passed through the thermocouple wire so that its junction temperature gets enhanced suddenly by several hundred degrees Celsius above its surrounding gas temperature. By switching off this current suddenly, the typical emf value of the thermocouple approaches its gas temperature value exponentially, as shown in Figure 5.28, from which a time constant can be determined. The time constant can be determined corresponding to the time taken for the temperature to decay by $1/e$ of its initial value, as described in Figure 5.28a. But this decay cannot be attributed to the temperature difference due to sudden withdrawal external heating alone. Rather, the temperature difference is affected by the supporting wires and the Peltier effect. Therefore, it is suggested in the literature [9] to obtain the appropriate time constant from the plot of the time-resolved time constant with the time of decay, which is known as the plateau method (see Figure 5.28b). The time constant obtained by this method is lower than the time constant obtained by the $1/e$ method. It is also claimed that one can avoid the influence of the Peltier effect due to external heating by following this procedure. Generally, time lag during the measurement of a fluctuating temperature due to thermal inertia is compensated using an electronic compensating circuit, as shown in Figure 5.27. As the emf is related to temperature in a nonlinear manner for a wide range of temperature measurements, a linearizer is employed that uses a polynomial expression for the emf–temperature relationship. The emf signal, being a low value, is amplified using a DC preamplifier. Subsequently, this

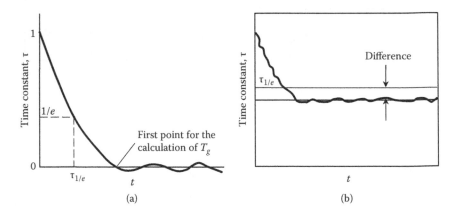

Figure 5.28

Determination of a thermocouple time constant: (a) $1/e$ method and (b) plateau method. (Reprinted from *Progress in Energy Combustion Science*, 19, Heitor, M. V. and Moreira, A. L. N., Thermocouple and sampling probes for combustion studies, 259, Copyright 1993, with permission from Elsevier.)

signal is fed into a frequency compensator and then to a linearizer. Generally a lowpass filter is used to cut off the high-frequency noise and the conditioned emf signal from the thermocouple is recorded in digital form, which can then be processed further to obtain time mean temperature, rms of fluctuating temperature, and so forth. In order to get a true time mean temperature, the signal must be integrated using variable integration time. The time constant can also be estimated from the profile of the probability density function of temperature fluctuations.

Example 5.6

A Pt-Pt/10%Rh thermocouple with a bead diameter of 1.25 mm is used to measure temperature in a combustor in which it is exposed to a high-temperature gas with a heat transfer coefficient of 975 W/m² °C. Assuming that the density of the thermocouple bead is 21,000 kg/m³ and C_p = 133 kJ/kg °C, determine the time constant.

Given: d = 1.25 mm, h = 975 W/m² °C, C_p = 133 kJ/kg °C, ρ = 21,000 kg/m³.
To Find: τ.
Solution: The mass and surface area of the bead can be determined as

$$m = \frac{\rho 4\pi r^3}{3} = \frac{21,000 \times 4\pi (1.25 \times 10^{-3})^3}{3} = 0.17 \times 10^{-3} \, \text{kg/m}^3$$

$$A = 4\pi r^2 = 4\pi \, (1.25 \times 10^{-3})^2 = 0.196 \times 10^{-4} \, \text{m}^2$$

The time constant can be determined as

$$\tau = \frac{mC_P}{hA} = \frac{0.17 \times 10^{-3} \times 133}{975 \times 0.196 \times 10^{-4}} = 1.18 \, \text{s}$$

5.6 Flow Meters

In the last section, we learned how to measure local fluid velocity using intrusive probes. But in most engineering applications, we need to measure flow rate as compared to local velocity. Of course it is possible to determine flow rate by integrating local velocity but it is quite cumbersome and time consuming when the flow rate is measured and reported instantly. Instead of using local velocity for estimating flow rate, it is usually advisable to measure and record the volume/mass that flows through a duct per unit time. A flow meter determines the volume/mass flow rates (liters per minute [LPM]) through ducts or tubes while a velocity meter measures local velocity in the flow. Several types of flow meters have been devised that can be used for combustion systems. We will be discussing only three types of flow meters: (i) obstruction, (ii) variable area (rotameter), and (iii) mass.

5.6.1 Obstruction Flow Meters

Several types of obstruction flow meters—Venturi, orifice, and nozzle—have been devised for measuring the gross flow rate in a conduit. Generally the fluid flow in a conduit is obstructed artificially so that the local velocity is changed, resulting in a change of static pressure between two points. This change in differential pressure can be related to the flow rate in the pipe/conduit. Since this flow rate is measured based on the principle of a changing pressure head across the obstruction, it is also known as a variable head/differential pressure flow meter.

For a deriving general relationship for obstruction flow meters, let us consider a steady one-dimensional incompressible flow in a variable cross-sectional conduit as shown in Figure 5.29. By applying the mass conservation principle between two stations, we have

$$\rho_1 A_1 V_1^2 = \rho_2 A_2 V_2^2 \tag{5.31}$$

where ρ is the density of fluid, A is the cross-sectional area, and V is the local fluid velocity. By assuming the flow to be adiabatic and frictionless, we can apply Bernoulli's equation as given below:

$$\frac{P_1}{\rho_1} + \frac{V_1^2}{2} = \frac{P_2}{\rho_2} + \frac{V_2^2}{2} \tag{5.32}$$

As the flow is incompressible, then $\rho_1 = \rho_2$. By solving Equations 5.31 and 5.32 simultaneously, we have

$$P_1 - P_2 = \frac{\rho V_2^2}{2}\left[1 - \left(\frac{A_2}{A_1}\right)^2\right] \tag{5.33}$$

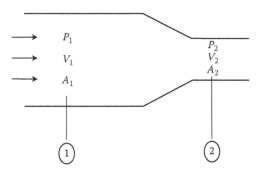

Figure 5.29

Flow through a variable duct.

We can derive an expression for volumetric flow rate Q as

$$Q = A_2 V_2 = \frac{A_2}{\sqrt{1-(A_2/A_1)^2}} \sqrt{\frac{2}{\rho}(P_1 - P_2)} = \frac{A_2}{\sqrt{1-(A_2/A_1)^2}} \sqrt{2g\Delta h} \qquad (5.34)$$

Note that volumetric flow rate Q is proportional to the square root of the pressure difference between two stations and other parameters. This pressure drop across the obstruction flow meter can be measured using differential manometers and expressed in terms of manometric fluid head Δh. In this analysis, fiction is neglected. But the actual flow would incur certain losses due to friction and hence the actual flow will be different than the ideal flow rate. Therefore, a quantity known as discharge coefficient C_D is defined as follows:

$$C_D = \frac{Q_{actual}}{Q_{ideal}} = f(Re_D, \beta = d/D) \qquad (5.35)$$

Discharge coefficient C_D of an obstruction flow meter is obtained by using a suitable calibration method. It is dependent on the Reynolds number and geometric parameter, which will be discussed later. Note that for compressible flow, proper correction must be made in the above expression if the flow in the obstruction flow meter happens to be compressible since the above expressions are derived for incompressible flow.

Let us now discuss three types of widely used obstruction flow meters: (i) orifice, (ii) nozzle, and (iii) Venturi, as shown in Figure 5.30. A typical orifice flow meter is constructed by inserting a flat plate with an orifice between two flanges, as shown in Figure 5.30a. This orifice can have either a square-edged or sharp-edged shape. The pressure ports are located upstream and downstream of this orifice plate across which pressure differential is being measured using a manometer or any other differential pressure sensor. Note that this measured pressure difference is used to determine the flow rate through an orifice. The downstream pressure tap located at the minimum flow is also known as the vena contracta. As per the American Society of Mechanical Engineers (ASME) standard, a downstream pressure tap is located around D/2 (pipe diameter) while an upstream pressure tapping is placed around 1D away from the orifice. The recommended diameter ratio, $\beta = d/D$, for an orifice flow meter is 0.2 to 0.7. The discharge coefficient, C_D, is quite low compared to the other two flow meter types due to higher pressure losses. In this case the discharge coefficient, C_D, decreases with an increase in the Reynolds number, Re_D, due to flow separation.

A nozzle flow meter, which operates on the same principles as an obstruction flow meter (orifice meter), is shown in Figure 5.30b. In the case of a nozzle flow meter a contoured nozzle is used between two flanges in place of the inexpensive orifice plate of an orifice flow meter. The flow pattern is similar to that of an orifice flow meter, of course with a slight vena contracta and less severe flow separation

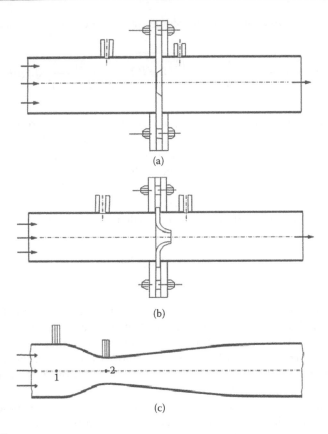

Figure 5.30

Schematic of three types of obstruction flow meters: (a) orifice, (b) nozzle, and (c) Venturi.

leading to relatively higher C_D values as compared to an orifice flow meter for the same Reynolds number. In contrast to the behavior of an orifice flow meter, the discharge coefficient, C_D, decreases with an increase in the Reynolds number, Re_D, due to flow separation.

A Venturi flow meter is the most expensive but most accurate among all three obstruction flow meters. A typical Venturi flow meter is shown in Figure 5.30c, which consists of convergent, throat, and divergent sections. It is designed with a low divergence angle so that it incurs fewer losses as compared to orifice and nozzle flow meters. Note that most of the pressure head loss is incurred due to frictional losses along the wall rather than flow separation losses in the other two flow meters. The discharge coefficient, C_D, is quite high as compared to the other two flow meters due to lower pressure losses. In this case the discharge coefficient, C_D, increases with an increase in the Reynolds number, Re_D, as shown in Figure 5.31. Generally, a Venturi flow meter is preferred where the least pressure loss is desired but this comes with an increased cost of fabrication.

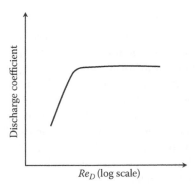

Figure 5.31

Variation of discharge coefficient C_D with the Reynolds number, Re_D, for a typical Venturi flow meter.

Example 5.7

A sharp-edged orifice flow meter is used to measure 200 LPM of kerosene flowing in a pipe of 100-mm diameter. If the discharge coefficient is 0.62, determine the pressure difference across the orifice flow meter if the diameter ratio is 0.65. The density of kerosene is 785 kg/m³.

Given: $Q = 200$ LPM $= 2000/60 = 0.00333$ m³/s, $D = 50$ mm, $d/D = 0.65$, $C_D = 0.62$.

To Find: $P_1 - P_2$.

Solution: The throat diameter would be

$$d = 0.65 \times 100 = 65 \text{ mm} = 0.065 \text{ m}$$

The orfice area becomes

$$A_2 = \pi d^2/4 = \pi(0.065)^2/4 = 0.0033 \text{ m}^3$$

The area ratio of the orifice can be determined as

$$A_2/A_2 = (d/D)^2 = 0.4225$$

By using Equation 5.34 we can determine the pressure difference across the orifice as

$$Q = \frac{C_D A_2}{\sqrt{1-(A_2/A_1)^2}} \sqrt{\frac{2}{\rho}(P_1 - P_2)}$$

$$(P_1 - P_2) = \frac{Q^2\left(1-(A_2/A_1)^2\right)\rho}{2C_D^2 A_2^2} = \frac{0.00333\times(1-0.4225^2)785}{2\times0.62^2\times0.0033^2} = 256{,}493.8 \text{ Pa}$$

$$= 256.493 \text{ kPa}$$

5.6.2 Variable Area Flow Meters

In the previous section, we discussed obstruction flow meters and learned that the flow rate is proportional to the square root of the pressure difference across the obstruction devices (Venturi/orifice/nozzle). As a result, if a wide range of flow is to be measured, then the sensitivity of this measurement gets reduced and thus it may not possible to have a wider range of operation with the same sensitivity. In order to overcome this problem, variable area flow meters, such as rotameters and vane flow meters, have been designed and developed. Since rotameters and turbine flow meters are used extensively in combustion systems for the measurement of volumetric flow rate, we will be restricting our discussion to rotameters and turbine flow meters.

Rotameter

The schematic of a typical rotameter is shown in Figure 5.32, which consists of a conical tapered tube and a float or bob. In this instrument a small weighted object called a float is carried by the fluid flow due to the drag force acting on it. This float can be any shape, but a spherical shape is preferred due to its minimum flow losses. When fluid is flowing over a float, drag force acts on it along with buoyancy force that together balance the weight of the float. As a result, the equilibrium position of this float depends on the amount of flow rate through the tapered tube. When there is no flow in the rotameter, the float or bob rests at the bottom seat of a tapered transparent tube. The position of the float or bob at the equilibrium point indicates the magnitude of the flow rate passing through the rotameter. This type of flow meter is also known as an area meter as the position of the float or bob depends on the annular area between it and the tapered transparent tube.

Let us consider a float or bob at an equilibrium position as shown in Figure 5.32. We have already discussed above that at the equilibrium position of the

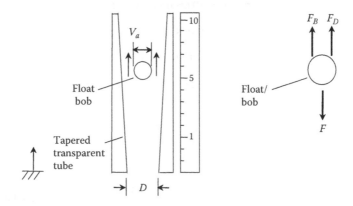

Figure 5.32

Schematic of a typical rotameter.

float, it experiences drag force and buoyancy force due to fluid flow over it which is counterbalanced by its weight. By striking the force balance on the float, we can have

$$F_w = F_D + F_B \tag{5.36}$$

where F_w is the weight of the float, F_D is the drag force due to the upward motion of fluid, and F_B is the buoyancy force acting on the float. The buoyancy force is equal to the weight of the displaced fluid volume, so it can be expressed mathematically as

$$F_B = \rho_f V_b g \tag{5.37}$$

where V_b is the volume of the float or bob, ρ_f is the density of flowing fluid, and g is the acceleration due to gravity. Similarly, the weight of the float or bob can be expressed as

$$F_w = \rho_b V_b g \tag{5.38}$$

where ρ_b is the density of the float or bob. The drag force can be expressed mathematically as

$$F_D = \frac{C_d \rho_f V_a^2 A_b}{2} \tag{5.39}$$

where C_d is the drag coefficient, ρ_f is the density of the fluid, V_a is the velocity of fluid in the annular area between the float and the tapered tube, and A_b is the frontal area of the float/bob. By substituting all the above force terms in Equation 5.36 and solving for the annular fluid velocity, V_a, we get

$$V_a = \sqrt{\frac{2V_b g(\rho_b - \rho_f)}{C_d \rho_f A_b}} \tag{5.40}$$

Generally the float or bob is designed in such a way that the value of C_d is almost independent of the float position. As a result, the annular velocity remains almost constant as other quantities remain constant for a particular rotameter. Then the volumetric flow rate, Q, passing through the rotameter can be expressed as

$$Q = A_a V_a = A_a \sqrt{\frac{2V_b g(\rho_b - \rho_f)}{C_d \rho_f A_b}} \tag{5.41}$$

where A_a is the annular area. As discussed above, the annular velocity remains almost constant, and therefore the volumetric flow rate is dependent on the annular area, A_a. But it is customary to design a rotameter so that the annular area, A_a, is the linear function of the float vertical position y. The annular area, A_a, can be related to the float position as

$$A_a = \frac{\pi}{4}\left[(D+2\alpha y)^2 - d^2\right] \approx \pi\alpha yD; \quad \text{As } D \approx d \text{ when } \alpha \text{ is too small.} \quad (5.42)$$

where D is the diameter of the tube at the base of the rotameter, α is the taper angle, y is the vertical distance, and d is the maximum bob diameter. By combining Equations 5.41 and 5.42, we get

$$Q \approx \pi\alpha yD \sqrt{\frac{2V_b g(\rho_b - \rho_f)}{C_d \rho_f A_b}} \quad (5.43)$$

It can be noted from the above equation that the volumetric flow rate for a particular rotameter is dependent only on the vertical position of the bob. Hence a linear calibration scale is provided on the front side of the rotameter to determine the flow rate by the position of the float. Note from the above equation that the flow rate reading from a rotameter is of course dependent on the composition of the flowing fluid and its properties; namely, pressure, temperature, and viscosity, as the drag coefficient is a function of the Reynolds number and the reading of the rotameter is dependent on the viscosity of the flowing fluid. Generally, a spherical bob is preferred in order to have a constant drag coefficient and thus the reading of the rotameter would not be dependent on the viscosity of the flowing fluid. From Equation 5.43, the mass flow rate can be expressed as

$$\dot{m} \approx \pi\alpha yD\rho_f \sqrt{\frac{2V_b g(\rho_b - \rho_f)}{C_d \rho_f A_b}} = Cy\sqrt{(\rho_b - \rho_f)\rho_f} \quad (5.44)$$

where C is the rotameter constant whose actual value can be obtained by undertaking calibration. The density of the flowing fluid varies due to variation of temperature, and therefore the mass flow reading is likely to change and thus calls for frequent calibration. In order to avoid this problem, the designer has to choose a bob density so that the rotameter reading will change due to variation in the density of a gas. For this purpose let us differentiate the above equation with the density of the fluid and equate it to zero, as given below:

$$\frac{d\dot{m}}{d\rho_f} = 0; \quad \rho_b = 2\rho_f \quad (5.45)$$

5. Thermal Flow Measurements

From the above expression it is clear that if we can choose the density of the bob as twice that of the fluid density, then the mass flow rate would not vary with a variation in the flowing fluid temperature. In that case, the expression for the mass flow rate for a rotameter would be

$$\dot{m} = Cy\rho_f = \frac{Cy\rho_b}{2} \tag{5.46}$$

Therefore, the bob can be designed for a particular fluid to compensate the change in the density of the flowing fluid. A 10% change in fluid density can incur around a 0.5% error in measurement in the mass flow rate for this kind of design. Most of the rotameter readings are obtained from visual observation of the float position. Of course the float position can be sensed remotely without visual observation, but the latter has the advantages of being easier to read, of having a low pressure drop, and of having a linear scale. As well, simple rotameters are used extensively due to their low cost and higher accuracy. Generally the accuracy of a rotameter can be in the range of ±10% while a high-quality rotameter can be in the range of ±2% of full scale, of course with extra cost.

i. Turbine Flow Meter

A turbine flow meter is one of the vane flow meters that is used extensively in combustion systems and hence is discussed briefly in this section. The schematic of a typical turbine flow meter is shown in Figure 5.33, which consists of a turbine wheel, a sensor (magnetic pickup), and a flow straightener. When a fluid is flowing through the turbine, it causes the turbine wheel to turn. Generally a turbine wheel must be made of magnetic material or its blade has a magnetic tip. When the turbine blade is rotated due to the flowing fluid, its blade tip passes through the magnetic pickup and thus generates an electrical pulse for each revolution of the turbine wheel. Note that the output signal is a frequency that is

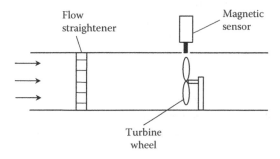

Figure 5.33

Schematic of a turbine flow meter.

proportional to the volumetric flow rate. Therefore, the volumetric flow rate can be related to pulse frequency f as given below

$$Q = \frac{f}{C_f} \qquad (5.47)$$

where C_f is the flow coefficient that is dependent on the flow rate and the kinematic viscosity of the fluid. Generally a linear relationship between Q and f can be obtained in a higher flow rate but becomes nonlinear at a low flow rate. The accuracy range of this turbine flow meter is around ±0.5% of the full scale. The range of this flow meter varies from 1 to 150,000 LPM and from 5 to 100,000 LPM for water and air, respectively, but the accuracy drops down in the lower range of flow. The pressure drop across this turbine flow meter is quite low compared to other flow meters.

5.6.3 Mass Flow Meters

In combustion applications we need to measure mass flow rate rather than volumetric flow rate, particularly while handling the range operation of a gas turbine or rocket engine that is dependent on the mass flow rate of fuel. Several types of mass flow meters are used in combustion systems for the measurement of mass flow rate. The basic principles of mass flow rate measurement are divided broadly into two categories. In the first approach, both density and volumetric flow rate can be measured simultaneously to determine the mass flow rate. In the second approach, parameters such as heat transfer, Coriolis acceleration, gyroscopic action, and shock waves, which are sensitive to changes in mass flow rates, are being used to measure the mass flow meter directly. However, we will be restricting our discussion to the two most widely used mass flow meters: (i) thermal and (ii) Coriolis, which are discussed briefly below.

i. Thermal Mass Flow Meter

The schematic of a typical mass flow meter based on the principle of heat transfer is shown in Figure 5.34, in which the flow is divided into two streams. Note that a small portion of the mass flow rate is passed through the sensing tube. Electric heat is used to transfer a controlled quantity of heat so that there will be rise in temperature across the heating section of the sensing tube. Generally, a resistance temperature sensor is used to measure the temperature between upstream and downstream of the heater, which is influenced by the mass flow rate. In practice, two resistance temperature sensors are used that can also act as heating elements. Note that the flow through the main and sensing tubes must be laminar in nature. As a result, the ratio of the sensing mass flow to the main mass flow rate is independent of the total mass flow rate.

Let us strike an energy balance between the heating zone of the bypass flow-through sensing tube as

$$q = \dot{m}C_p(T_2 - T_1) \qquad (5.48)$$

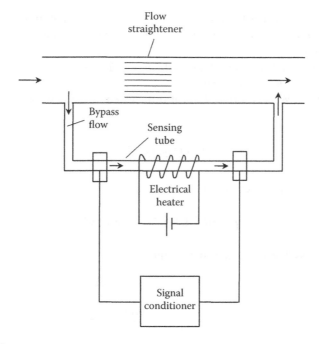

Flow
straightener

Bypass
flow

Sensing
tube

Electrical
heater

Signal
conditioner

Figure 5.34

Schematic of a thermal mass flow meter.

where q is the heat input to the bypass fluid, $\dot{m}$ is the mass flow rate, C_p is the constant specific heat, and T_1 and T_2 are upstream and downstream temperatures across the heating tube. The heater is controlled so that a constant heat input can be maintained. Solving Equation 5.48, we have

$$\dot{m} = \frac{q}{C_p(T_2 - T_1)} \tag{5.49}$$

Heat input to the bypass flow is to be maintained constant over its entire range. As a result, the specific heat is a weak function of temperature and is known to higher accuracy. Therefore, by measuring the temperature difference and knowing the heat input, we can easily determine the mass flow rate passing through this mass flow meter. Generally, a signal conditioner is used to convert temperature sensor data into a linear voltage output that is proportional to the mass flow rate. Note that this type of thermal mass flow meter can be used for both gas and liquid.

Example 5.8

In a thermal mass flow meter, 10% of the total mass flow rate of air is passed through by a sensing tube that is heated by an electric resistor of

5.2 Ω through which 4.3 A of DC current passes through it. If the temperature difference is 6°C, determine the air mass flow rate measured by this instrument.

Given: $I = 4.3$ A, $R = 5.2 \, \Omega$, $\dot{m}_s = 0.1 \dot{m}$, $T_2 - T_1 = 6°C$.

To Find: mass flow rate $= \dot{m}$.

Solution: We know that the heat transferred to the sensing fluid would be

$$q = \dot{m}_s C_P (T_2 - T_1)$$

The resistive heat can be determined as

$$q = I^2 R = 4.3^2 \times 5.2 = 96.15 \text{ W}$$

The mass flow rate can be determined as

$$\dot{m}_s = 0.1 \, \dot{m} = q / C_P (T_2 - T_1)$$

$$\dot{m} = \frac{q}{0.1 \times C_P (T_2 - T_1)} = \frac{96.15}{0.1 \times 1005 \times 6} = 0.16 \text{ kg/s}$$

Note that the accuracy in the temperature measurement would dictate the accuracy of the mass flow rate measurement while keeping the flow laminar.

ii. Coriolis Mass Flow Meter

In the above section, we learned that the thermal mass flow meter can measure mass flow rate directly by measuring temperature difference. However, as it is based on the principle of heat transfer, its calibration depends on the accurate determination of specific heat and heat input. As a result it may incur errors due to the uncertainty of this data. There is another mass flow meter based on the Coriolis principle that can measure mass flow rate directly with high accuracy and fast response. We know that there is a deflection of moving objects when they are viewed in a rotating frame of reference. The French scientist G. Coriolis in 1835 provided a mathematical expression for this, called the Coriolis force, which is proportional to the mass of an object. Later on it was used to determine mass flow, leading to the design and development of the Coriolis mass flow meter, which is discussed briefly below.

A schematic of a typical Coriolis mass flow meter is shown in Figure 5.35, which consists of two U-tubes, an electromagnetic vibrator, and two sensors. Two tubes are attached to a rigidly supported base such as a cantilever. The tube need not only be U-shaped so its shape can vary widely. Generally, an electromagnetic vibrator is used to vibrate the free end of the two tubes at its natural frequency along a vertical direction so that they are vibrated 180° out of phase. Amplitude of this vibration will be zero at the base of the cantilever and will be largest at its

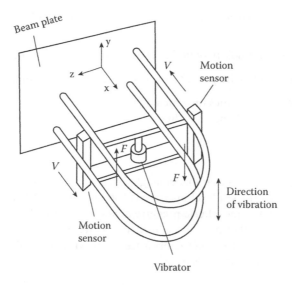

Figure 5.35

Schematic of a Coriolis mass flow meter.

end. When the fluid flows through the tube, the Coriolis force can cause a variation in the phase shift, which can be detected by the sensor. This resulting output signal from the senor due to the relative twisting motion of the two tubes is proportional to the mass flow rate. It can also measure the density of the fluid as the tubes are vibrated at a natural frequency because natural frequency vibration depends not only on the mass flow rate but also on the density of the fluid passing through the tube. Generally, Coriolis mass flow measurement is fast, accurate, and inherently bidirectional. The accuracy of this mass flow meter in the order of ±0.5% of a full-scale reading for a gas flow rate can be achieved with proper calibration. The Coriolis mass flow meter can accurately measure the flow rate of various types of fluids such as gases, liquids, slurries, and two-phase mixtures.

Review Questions

1. What is a pitot probe? What kinds of errors are incurred during its uses? Can it be used for velocity measurement? Explain with the help of a requisite expression.

2. What is a pitot-static probe? Can it be used for local flow velocity measurement? Explain with the help of a requisite expression.

3. You want to design a pitot-static probe for the measurement of local velocity in the range of 4 m/s to 40 m/s. What are the design precautions you will be considering for minimizing errors due to flow angularity?

4. What is a three-hole pitot probe? How is it different from a simple pitot probe?

5. What is a five-hole pitot probe? How is it different from a three-hole probe?

6. What is a hot wire probe? How is it different from a hot film probe?

7. What are the modes on which a hot wire probe can be operated? Which is the preferred one and why?

Problems

1. The pressure of a chamber at height of 2000 m from sea level is measured to 65 kPa. The sea-level pressure is 760 mm of Hg. What is the absolute pressure? If atmospheric pressure at sea level instead of the pressure at a height of 2000 m is not considered, determine the error in absolute pressure. The atmospheric pressure is varied with height as $P = P_0 \exp$ (0.000119 h), where h is in m.

2. A liquid propellant tank contents are compressed with a high-pressure nitrogen gas along with Jet A1 fuel with a density of 805 kg/m³. A U-tube mercury manometer is connected to the side bottom of this tank as shown in Figure P5.2 to measure its pressure. The mercury column heights of this U-tube manometer are $h_1 = 2.3$ m, $h_2 = 0.1$ m and $h_3 = 1.0$ m, respectively. Determine the pressure of this tank.

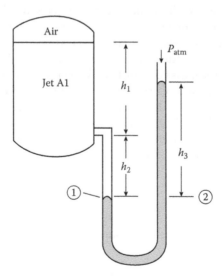

Figure P5.2

3. A U-tube mercury manometer is used to measure the pressure of a combustion chamber containing a fuel-air mixture at 25°C. One limb of this

manometer is connected to this chamber and the other is exposed to atmospheric pressure at 25°C. If the difference in the manometer column height is 25 cm, determine the pressure of the combustion chamber.

4. A differential U-tube manometer is connected to an orifice meter as shown in Figure P5.4 for measurement of air. This manometer with a water column has the following heights $h_1 = 0.8$ m and $h_2 = 0.5$ m. Determine the pressure differential between point A and B.

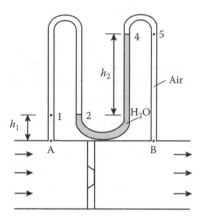

Figure P5.4

5. A U-tube mercury manometer is connected to measure the pressure of a combustion chamber containing a fuel-air mixture at 25°C. One limb of this manometer is connected to this chamber and the other is exposed to atmospheric pressure at 25°C. If the difference in the manometer column height is 25 cm, determine the pressure of the combustion chamber.

6. Determine the sound pressure level in decibels if the pressure of a sound measured at 1.5 m from a combustor is found to be 5.5 N/m².

7. In a combustion lab, one engine produces a sound pressure level (SPL) of 100 dB while a blower produces SPL at 65 dB. Determine the total sound intensity level while using these two sound sources.

8. Two combustors with individual sound pressure levels of 85 dB and 90 dB are operated simultaneously in a lab. The background noise from a blower is around 65 dB. Determine the total sound pressure level.

9. A gas turbine engine produces a sound level of 100 dB at 10 m. Determine the sound pressure level at 10 m and 50 m.

10. A pitot-static probe is mounted on the fuselage of an aircraft flying at a 15-km altitude and a velocity of 250 km/hr. Determine (1) static pressure, (2) density, and (3) pressure difference across this pistol-static probe.

11. A pitot-static probe is used to measure velocity in a combustor along the flow stream at 65°C for ascertaining uniformity of flow using a water U-tube manometer. If the pressure head across this manometer is 15 mm of Hg head, determine the local flow velocity.

12. A three-hole yaw probe is used to measure angle by using the null method. The angle measured in an air stream at 90 kPa and 35°C with a free stream velocity of 60 m/s is found to be 10°. Determine the pressure measured by this probe if the static pressure error $\dfrac{P_{t1,Y} - P_t}{\frac{1}{2}\rho V_1^2}$ is equal to –2.

13. In a hot wire anemometer, a platinum hot wire sensor is used at 55°C using a bridge circuit in which $R_2 = R_3 = 34\ \Omega$. If the sensor resistance is 1.5 Ω at 25°C and has a thermal coefficient of 0.00345°C^{-1}, determine the setting resistance for this anemometer.

14. A hot wire anemometer and a tungsten wire of 50-mm diameter and a length of 3 mm with a resistance of 3.5 Ω is used to measure air velocity at 25°C. When 500 mA of current is passed through the hot wire, it attains a temperature of 95°C. Note that it follows King's law with a calibration constant $C_o = 37.6$, $C_1 = 0.95$, determine the air flow velocity.

15. A chromel-alumel thermocouple exhibits an approximate linear relationship between 600°C and 1000°C. If emf at 600°C and 700°C are 24.905 and 29.129 mV, respectively, determine the sensitivity of this thermocouple. Determine the corrections to be made if the cold junction temperature is 35°C. If the indicated emf is 35.95 mV, with respect to an ambient temperature of 25°C, determine the temperature of the hot junction.

16. An R-type thermocouple probe with a bead diameter of 0.54×10^{-3} m is used to measure the temperature of hot gases at the exit of a combustor. The total mass flow rate at the exit of the combustor for this case is around 0.005 kg/s. The density of the product gases at the exit of the combustor based on equilibrium calculations is around 0.148 kg/m³. The area at the exit of the combustion chamber is 1.2×10^{-3} m². If the measured temperature is 1632 K, determine the actual gas temperature considering the radiation correction.

17. A Venturi flow meter is to be designed using an inlet mild steel pipe diameter of 50 cm for maximum air flow rate 3.5 m³/s at $P = 2$ bar, $T = 35°C$. If the maximum allowable pressure drop is 200 Pa, determine the throat diameter. Consider $C_D = 0.98$ when $0.3 \le d/D \le 0.75$.

18. A rotameter with a bob or float ($\rho_b = 2750$ kg/m³) is used for water. However, it will be used for Jet A1 fuel with a density of 805 kg/m³. Determine the conversion factor.

19. A rotameter is used for N_2 gas at 101.325 kPa and 25°C. However, it will be used for propane gas at 2 bar and 35°C. Determine the conversion factor.

20. One designer has proposed to design a thermal mass flow meter for a flow rate of 0.084 kg/s without using any bypass system. An electric DC current of 5.4 A is passed through a resister of 6.5 Ω, which is used to heat the entire flow. Determine the temperature difference across this instrument. Is it prudent to avoid a bypass system in this thermal mass flow meter?

References

1. Dominy, R. G. and Hudson, H. P., An Investigation of factors influencing the calibration of five-hole probes for three-dimensional flow measurements, *Journal of Turbomachinery*, 115:513–519, 1993.
2. Kettle, D. J. and Benedict, R., *Fundamentals of Temperature, Pressure, and Flow Measurement*, Third Edition, Wiley India Pvt. Ltd., 2011.
3. Figliola, R. S. and Beasley, D. E., *Theory and Design for Mechanical Measurements*, Third Edition, John Wiley & Sons, Inc., Hoboken, NJ, 2000.
4. Doebelin, E. O., *Measurement Systems Application and Design*, Fourth Edition, McGraw-Hill Publishing Co., New York, 1990.
5. Broch, J. T., *Acoustic Noise Measurements*, Second Edition, Bruel and Kjaer Instrument Company, New York, 1971.
6. Peterson, A. P. G. and Gross, E. E., *Handbook of Noise Measurements*, Eighth Edition, GenRad, Inc., Concord, MA, 1980.
7. Beer, J. M. and Chigier, N. A., *Combustion Aerodynamics*, John Wiley & Sons, Inc., New York, 1972.
8. Lowell, H. H., Design and applications of hot wire anemometers for steady state measurements at transonic and supersonic speeds, NACA Technical Note 2117, 1950.
9. Perry, A. E., *Hot Wire Anemometry*, Oxford University Press, 1982.
10. Lowell, H. H., Design and applications of hot wire anemometers for steady state measurements at transonic and supersonic speeds, NACA Tech Note 2117, 1950.
11. Holman, J. P., *Experimental Methods for Engineers*, Seventh Edition, McGraw-Hill Publishing Company Ltd., Boston, 2001.
12. NIST Thermocouple Reference Table, Monograph 175, revised to ITS-90.
14. Carvalho J. A. and Dos Santos, W. F. N., Radiation errors in temperature measurements with thermocouples in a cylindrical combustor, *International Journal of Heat and the Mass Transfer*, 17:663–673, 1990.
15. Hinze, J. O., *Turbulence*, McGraw-Hill, New York, 1959.
16. Bradley, D. and Matthews, K. J., Measurements of high gas temperatures with fine wire thermocouples, *Journal of Mechanical Engineering Science*, 10:299–305, 1968.

17. Martins, C. A., Pimenta, A. P., Carvalho, J. A., Ferreira, M. A., and Calderia-Pires, A. A., CH and C2 radicals characterization in natural gas turbulent diffusion flames, *Journal of The Brazilian Society of Mechanical Sciences and Engineering*, 27:110–118, 2005.
18. Yoshida, A. and Tsuji, H., Measurement of fluctuating temperature velocity in a turbulent premixed flame, in Symposium (International) on Combustion, 17(1):945–956, 1979.
19. Heitor, M. V. and Moreira, A. L. N., Thermocouple and sampling probes for combustion studies, *Progress in Energy Combustion Science*, 19:259, 1993.

6

Gas Composition Measurements

I consider nature a vast chemical laboratory in which all kinds of composition and decompositions are formed. Vegetation is the basic instrument, the creator uses to set all nature in motion.

Antoine Lavoisier

6.1 Introduction

In order to understand the complex processes involved during combustion, it is very important to measure the concentration of various species as they are used and generated during this process. But what do we mean by "species"? The species in a combustion system consists of atoms, molecules, radicals, and ions. We know that molecules are the uncharged species with paired electron spins. Examples of molecules are monoatomic (He, Ar), diatomic (CO, H_2), and poly-atomic species (CH_4, C_2H_6, C_3H_8, NH_3, etc.). Species with unpaired electron spins are called radicals. For example, species such as O, H, OH, CH, and CH_3 are radicals. Monatomic molecules that are paired can be labeled as atoms. These radicals are more reactive in nature than stable molecules due to the presence of unpaired electrons that can cause the chemical reactions to be faster. As a result, the lifetimes of radicals are quite short due to higher reactivity. This is mostly confined to the high-temperature zones of combustion. If molecules are charged, they are called ions. For example, H^+ and O^- are ions. Generally ions will be more reactive than radicals but their concentrations are too negligibly small to contribute to the chemical conversion of fuel and oxidizer into its product. The determination of species, both stable molecules and radicals, are quite important in the reaction scheme used in the simulation of a combustion system. The utilities of species in a reaction scheme can be divided into three categories: reactants, intermediate, and products. We know that any atom, radical, and molecule can be termed as an intermediate species if its concentration attains a peak value across the flame. The products can include atoms, radicals, and molecules although molecules are the more desirable product in a combustion system.

6.2 Quantification of Composition

We know that the flame is the genesis of a combustion process in most conventional combustion systems. Hence it is essential to understand the structure of the flame, which can be characterized and depicted with variations in compositions of both stable and unstable species along the distance of the flame from burned to unburned zones. The composition of species across the flame can be expressed in terms of a scalar quantity known as concentration, which represents the amount of species per unit volume (kmol/m³). The concentration can be expressed in terms of mass, partial pressure, and number of molecules per unit volume apart from the number of moles per unit volume. It is important to ascertain the reaction rate of the fuel and other species in terms of the rate of change of concentration with time (kmol/m³ s or kg/m³ s). In certain situations, species flux, expressed as amount of species per unit area in a unit time (kmol/m² s or kg/m² s), is used for analyzing the flame [1].

It is essential to determine the emission/composition levels of various pollutants such as CO_x, NO_x, SO_x, and UHC from the exhaust of a combustion system because it is now mandatory due to the growing concern about environment pollution problems. This emission level can be expressed in various ways, and the quantity of species can be expressed either on a volume or mass basis. The unit for mass would be grams of ith species per unit volume of all species. The volumetric unit of species is parts per million (ppm), which is defined as

$$1 \, \text{ppm} = \frac{\text{Volume of } i\text{th species}}{10^6 \, \text{Volume of mixture (all species)}} \tag{6.1}$$

For example, the emission level of a particular species from a furnace can be reported in terms of ppm at 3% oxygen level while in a gas turbine engine it is customary to report ppm at 15% oxygen level. The species concentration in terms of ppm at a given temperature and pressure can be converted to mass by invoking the ideal gas law as

$$\frac{m_i}{V} = \frac{M_i P}{R_u T} 10^{-6} \text{ (ppm)} \qquad (6.2)$$

where $\frac{m_i}{V}$ is the mass concentration of the ith species, P is the pressure, T is the temperature, R_u is the universal gas constant, and M_i is the molecular weight of the ith species. In SI units, it is expressed as kg/m³, but for convenience, it is mostly expressed as µg/m³, and for automobile engines, it is reported in terms of grams of pollutant species per km. For other combustion systems such as boilers, dryers, and burners, the emission level is expressed in terms of gm per kW. In recent times, an elegant way of normalizing the emission level with fuel flow rate known as the emission index is being used for expressing the emission level. The emission index of an ith species is defined as the ratio of mass flow rate of the ith species to the mass flow rate of fuel burned during combustion, which can be expressed as (Section 7.4.1 of ref. [1]);

$$EI_i = \frac{\dot{m}_i}{\dot{m}_F} \qquad (6.3)$$

where $\dot{m}_i$ is the mass flow rate of the ith pollutant species from the exhaust and $\dot{m}_F$ is the mass flow rate of fuel. The emission index is a nondimensional quantity, but it is customarily expressed in the unit of g/kg. It is used extensively because it can provide a direct measure of the amount of emitted pollutant of a particular species per unit mass of fuel during combustion that is independent of the dilution level of the exhaust gas and combustion efficiency.

The measurement of species in a combustion system can be accomplished in two ways: (i) intrusive and (ii) nonintrusive [2,3]. In the intrusive method, a sampling probe is introduced into the combustion system that sucks the gas samples for analysis. The walls of the probe are cooled to prevent further chemical reactions in the probe itself. This frozen sample is analyzed using various analytical instruments that will be discussed in the next section for determination of the chemical composition. On the other hand, in the nonintrusive method, the concentration of species can be measured in the combustion system without inserting a probe into it for taking out the gas from the measurement zone. We will be discussing some of these laser-based methods in Chapter 7 as they have assumed importance in recent times due to the increasing demand for validating numerical models. The major problems with the nonintrusive method are proper calibration, which is very costly, and higher level of complexities. Besides these issues, it calls for several types of methods that would be required for complete analysis. In contrast, the sampling method is quite simple and elegant even though it physically disturbs the combustion system. Of course, if proper sampling and its storage can be ensured, then various types of analytical methods can be used for measurement of species at leisure. Besides this, the nonintrusive method has sound theoretical basis and is hence noncontroversial, unlike the *in situ* method. As the sampling of species plays an important role in the measurement of species, we now discuss it in detail.

Example 6.1

An IC engine uses CNG gas. The measured compositions of exhaust gas at dry conditions are

$$O_2: 3.5\%$$

$$CO: 0.25\%$$

$$CO_2: 15.1\%$$

$$NO_x: 25 \text{ ppm}$$

Determine the emission index of NO_x.

Given: The measured compositions of exhaust gas are given under dry conditions.

To Find: The emission index of NO_x.

Solution: Assume that CNG gas contains only methane.

The emission index of NO_x, EI_{NO_x} can be estimated as

$$EI_{NO_x} = \left(\frac{X_{NO_x}}{X_{CO} + X_{CO_2}} \right) \left(\frac{n \, MW_{NO_x}}{MW_F} \right) = \left(\frac{25 \times 10^{-6}}{0.0025 + 0.151} \right) \left(\frac{1 \times 30}{16} \right)$$

$$= 305.37 \times 10^{-6} \, kg/kg = 0.305 \, g/kg$$

Note that the emission index for NO_x is independent of the dilution of air. As a result, it is preferred over other forms of expressing emission levels.

6.3 Sampling Systems

Gas sampling and its subsequent analysis is performed by the simple and elegant method of measuring the composition of species in a combustion system. But an adequate quantity of gas that represents the local condition must be sampled by using the proper sampling system. In other words, the sampling system must be designed properly so that errors during sampling can be minimized to a large extent. Sampling systems can be broadly divided into two categories: (i) online and (ii) offline (batch) [2,3]. In the online sampling system, sample gas withdrawn from the combustion system is fed directly into the analytical instrument. Figure 6.1 depicts various features of these two sampling systems. Both sampling systems consist of a sampling probe, heated filter, heated sample line, moisture

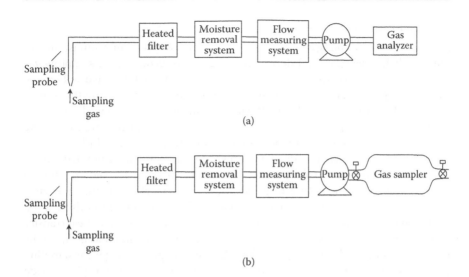

Figure 6.1

Sampling systems: (a) online and (b) offline.

removal system, and flow control valves. But in the case of the batch/offline sampling system (see Figure 6.1b), sample gas is withdrawn from the combustion system and is conditioned and stored in sampling bulbs or bottles made of glass, metal, or polymer for subsequent analysis. In the case of the online sampling system, a gas analyzer is connected to the system (see Figure 6.1a). We will be discussing more about sampling probes since the accuracy of species measurement is dependent mostly on the probe design and its operating practices. For meaningful measurements of species composition in combustion systems, it is essential to be aware of the types of error and their sources during the sampling processes. Major errors incurred during the sampling of gas in combustion systems are

i. Aerodynamic and thermal disturbances due to insertion of the probe
ii. Change in local compositions around the capture area due to change of flow by insertion of the probe
iii. Change in concentration due to inadequate quenching of the chemical reaction
iv. Enhancement or abatement of the reaction rate due to catalytic reactions occurring inside the probe
v. Averaging errors due to nonhomogeneous flow in the capture volume of the sampling. particularly in turbulent flow conditions

Before focusing more on these errors associated with sampling probes, let us look at types of sampling probes that are routinely being used.

6.3.1 Sampling Probes

Several types of sampling probes for measurements of species in combustion systems are being designed and developed. Since major errors in the sampling gas to be measured are dependent strongly on the rates of reactions involved during the sampling of species, the gas must be cooled properly in the sampling probe. The quenching of reactions during the measurement of species must be achieved inside the sampling probe. However, it is quite difficult to quench highly reactive species such as free radicals and ions in the sampling probe itself. Either convective cooling or aerodynamic quenching methods are being incorporated during the design of sampling probes. Sample gas can be mixed with cold and diluted gas to reduce its temperature to avoid any further reactions within the sampling probe. The quenching of a gas sample can be accomplished in three ways: (i) dilution of the sample by using cold and inert gas, (ii) convective heat transfer from the sample using a suitable coolant, and (iii) aerodynamic cooling due to the rapid expansion of the gas sample. The schematic diagrams of typical sampling probes are provided in Figure 6.2. In the first diagram, a pitot-type probe made

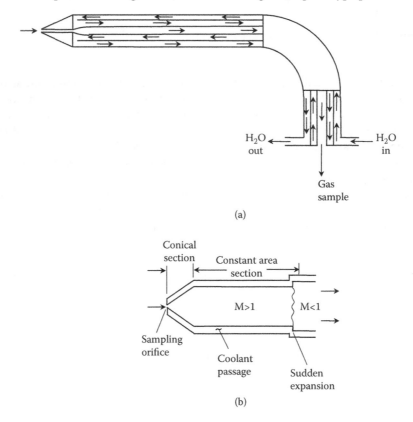

Figure 6.2

Types of sampling probes: (a) isokinetic and (b) aeroquench.

6. Gas Composition Measurements

of mostly stainless steel (see Figure 6.2a) is used, which must be cooled using water or any other liquid. The sampling gas must be sucked at a particular rate so that the sampling gas flow velocity will match with the local stream velocity. This sampling process is known as isokinetic sampling. If the gas sample velocity is more than the local flow velocity, then suction streamlines get converged toward the opening of the probe (see Figure 6.3). If the local flow velocity is more than the gas sample velocity, then the flow gets diverged from the probe opening. As a result, measured concentration C_m of the sampling gas will be different from the actual concentration C if gas is not sampled isokinetically, leading to the erroneous measurement of the concentration of the sampling gas. Observe that change in concentration is proportional to change in velocity for the same probe's cross-section area, which can be expressed mathematically as

$$\frac{C_m - C}{C} = \alpha \frac{V - V_m}{V}; \quad \Rightarrow \frac{C_m}{C} = 1 - \alpha + \alpha \frac{V}{V_m} \tag{6.4}$$

where V_m is the sampling velocity, V is the freestream fluid velocity, and α is the isokinetic factor indicating extent of isokinetic sampling. The isokinetic factor must be zero for an ideal case. Unfortunately α cannot be exactly zero in practice. One should try to minimize the isokinetic factor, α, close to zero. Generally,

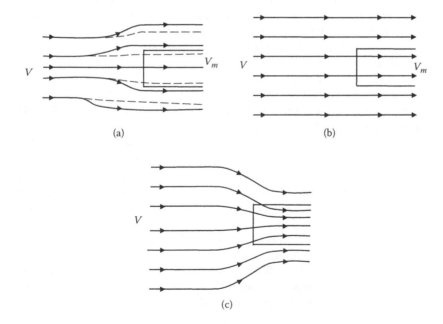

(a) (b)

(c)

Figure 6.3

Aerodynamics around a sampling probe: (a) low sampling velocity, (b) isokinetic sampling, and (c) high sampling velocity.

an internal static pressure tap is provided in the gas sampling probe that allows checking the level of isokinetic sampling by comparing the external and internal static pressure. Accordingly, the suction rate is adjusted to ensure the isokinetic sampling of gas. In this type of probe, convective cooling is employed to quench the sampling gas. One can achieve a quenching rate of around 10^5 K/s, which may be sufficient for quenching the concentration of stable species, particularly those sampled from the postflame regions. But such type of probe cannot be employed if the gas sample from the flame zone is to be measured because the cooling rate in the order of 10^8 K/s is required for quenching the sampled gas. Note that quenching of the sampled gas occurs almost at a constant pressure. Therefore, quenching of low-activation energy reactions ($E < 20$ kJ/mol) cannot be achieved because their reaction rates decrease with lowering temperatures [3], and so the bimolecular reaction rate varies inversely with temperature at a constant pressure. In order to overcome this problem, it is prudent to decrease both the temperature and pressure of the sampling gas by expanding it rapidly to ensure adequate quenching of the measured gas. Aerodynamic sampling probes are designed and developed based on this principle, [5,6]. The schematic of a typical marco aeroquench sampling probe is shown in Figure 6.2b, in which a sample is accelerated quickly to a supersonic flow (Mach number, $M = V/a > 1$, where a is the speed of sound) by reducing its static pressure and temperature rapidly. This supersonic flow is maintained in the probe until the stagnation temperature of the sample is reduced to a low value so that flow can return back to a subsonic flow while undergoing shock waves. Subsequently, convective cooling is used to reduce the static pressure and temperature to such an extent that most chemical reactions get quenched. This type of probe must be designed carefully so that quenching effectiveness is not reduced due to short probe distance and shock recovery reduced due to friction.

For this probe to work effectively, the back pressure in the downstream of the probe must be lower than 0.08 bar. Generally this is preferred when the gas temperature exceeds 1800 K. For more details, interested readers can refer to [4,5]. It must be emphasized that measurements of concentrations must be carried out as quickly as possible without recompression of the sampling gas, because three body chemical reactions can occur in the sampling itself. Although the aeroquench probe ensures a higher cooling rate at a lower pressure [6], radical reactions are likely to occur in the sampling probe itself. For example, some important reactions, such as $H_2 + OH \rightarrow H_2O + H$, even at moderate temperature occur at a much faster rate than can be quenched using an aeroquench sampling probe. In order to sample highly reactive reactions involving radicals, gas from the flame must be sampled in collision-free conditions that can be achieved using a high vacuum system. This technique is being used routinely for mass spectroscopy. A brief account of molecular beam sampling is given below.

6.3.2 Molecular Beam Sampling Probe

The schematic of a typical molecular beam sampling probe for measurement of concentrations from flames is shown in Figure 6.4, which consists of two

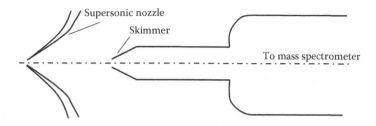

Figure 6.4

Schematic of a molecular beam sampling probe.

nozzles—the sonic nozzle and the skimmer. The sampling gas is expanded in the supersonic nozzle to subatmospheric pressure. The central portion of this supersonic jet does not collide with the walls because a wide-angle sampling cone in the range of 50° to 100° is being used. This collision-free gas jet along the centerline is separated by a skimmer and expanded in another orifice (skimmer) into the collimating chamber of the mass spectrometer. The design of this molecular beam-sampling probe is quite challenging as sampling composition can be distorted by several physical mechanisms. The radial pressure gradient within the supersonic expanding jet can increase the concentration of heavier species along the centerline. The barrel shock and attached shocks formed in front of the skimmer nozzle can cause species separation, amd condensation of certain species may occur in the supersonic nozzle due to the occurrence of lower temperatures caused by excessive expansion. These problems can be avoided by keeping a higher Reynolds number ($Re > 10^3$) in the first orifice so that the pressure diffusion can be reduced and distortion across the barrel shock can be minimized. The skimmer nozzle is designed so that its Knundsen number is almost equal to the local Mach number. Beyond the skimmer nozzle, the gas flow of the sample will be almost in the free molecular regime. As a result, lighter components of the sample gas even beyond the skimmer spread more rapidly along the radial direction due to the difference in the thermal velocity of molecules with different masses. Note that even the micron-order orifice distorts the sampling space due to higher concentration gradient as pressure gradient across the sampling probe is quite enormous. Hence proper design and fabrication with close tolerance are essential for proper measurement of concentrations in a flame system.

6.3.3 Langmuir Probe

We know that the reaction zone of a flame, particularly hydrocarbon, contain considerable concentrations of ions in the range of 10^9 to 10^{12} ions/cm^3. Several types of ions that are mostly positive (around 99%) with an atomic mass in the range of 2 to 70 are present in the hydrocarbon flame. Generally, the concentration of positive ions will be dependent on the type of fuel and fuel-air mixture ratio. The positive ions are formed due to electron collisions prevailing in chemical reactions in the flame itself. The concentration of ions is dependent on the number of carbon atoms involved during combustion in the case of a hydrocarbon flame.

The simplest and earliest device designed and developed for measurement of ion concentration in a flame is the Langmuir probe. The schematic of a typical Langmuir probe is shown in Figure 6.5, which consists of large-area and small-area electrodes flanked by insulation. Generally, a small-area electrode is made of noncorrosive and heat-resistant material (e.g., Pt-Rh13%), which is enclosed in an insulating material (e.g., quartz). Around this insulation, an outer casing is provided whose surface area will be much higher than the small-area electrode. The probe can be cooled using water jackets around the electrode. Generally an electric potential of –30 V is used to ensure saturation current in this probe. The ion concentration in a flame can be determined by measuring the current flow between two electrodes with different electric potentials. The actual value of the current will be dependent on the applied voltage and electrode area. The current is limited by the arrival of ions to the surface area of a small-diameter electrode, because the ratio between a large and small electrode must be very large for higher sensitivity as electrons have a higher mobility rate in the average velocity range of 10^5–10^6 m/s in a typical hydrocarbon flame. Besides this, the current is limited depending on whether a small electrode is positive or negative. For example, if it is negative, the current is proportional to and limited by a positive ion concentration; if it is positive, the current is proportional to and limited by a negative ion concentration. The measurement of this current is proportional to the reaction rate of the fuel concentration as the magnitude of ions is proportional to the quantity of carbon atoms being consumed in the flame. It can be used to measure the reaction zone thickness in the flame. As its sensitivity is quite high, it can be used even in a turbulent flame. It is also used for measurement of velocity for both detonation and deflagration waves and as a sensor, particularly as an ionization detector in several analytical instruments such as gas chromatography (GC) and unburned hydrocarbon gas analyzers.

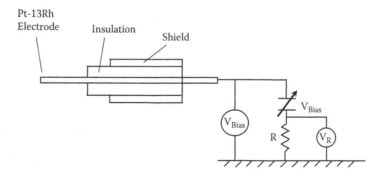

Figure 6.5

Schematic of Langmuir probe for the measurement of ions in a flame.

6.4 Analytical Methods for Combustion Product Measurement

We know that the gas sampled from a combustion system is a multicomponent mixture (e.g., exhaust gas from an engine or burner). Generally, there are two basic approaches that can be adopted. One way is to separate the mixture into individual components and then measure these individual components by a suitable analytical method. In this method, one must ensure that the separation of individual gases is complete. In the other approach, one can choose an analytical method that is only sensitive to one component among the mixture of compounds. In this approach, one must be sure that the analytical instrument is not influenced by other constituents in the mixture.

Several methods are available for analyzing batch samples taken from combustion zones. Some suitable analytical techniques can be adopted for the analysis of species. However, a combination of several techniques can be adopted depending on the requirement of accuracy and availability of resources, particularly when a batch-sampling method is used. Of course, the choice of a particular method is very often dictated by local availability of instruments, prior experience, and personal conviction. However, it is very important to understand the basic principles of the method used in a particular instrument in order to know the pros and cons of the common analytical methods available for measuring compositions within combustion systems. The range of analytical methods is quite large, so we will restrict our discussion to a few widely used methods for combustion systems. Interested readers can refer to specialized books and monographs for other analytical methods [3,7,8]. The applications of these techniques for the analysis of species are discussed briefly in the following sections.

6.4.1 Orsat Apparatus

The Orsat apparatus is one of the oldest methods of analyzing a component in a mixture of exhaust gases from a combustion system. This method is based on the principle of measuring the partial pressure of a gaseous species after eliminating the other components. In this method, species are first separated out by using the methods of adsorption, freezing out constituents, or reacting them with reagents, and then the partial pressure of a particular gaseous constituent is measured by employing a pressure-measuring device.

The schematic of a typical Orsat apparatus is shown in Figure 6.6, which consists of a measuring burette and three reagent pipettes. These three pipettes are used to absorb CO_2, O_2, and CO from the gas mixture using chemical reagents. Generally, potassium hydroxide (KOH) is employed in the first reagent pipette from the gas inlet to absorb carbon dioxide gas, but for absorption of oxygen gas, a mixture of pyrogallic acid and a solution of potassium hydroxide is used in a second pipette. In the third pipette, cuprous chloride is used to absorb carbon monoxide (CO). Initially, the sample of exhaust gas from the combustion system is supplied into the measuring burette. Subsequently, the sampling manifold is

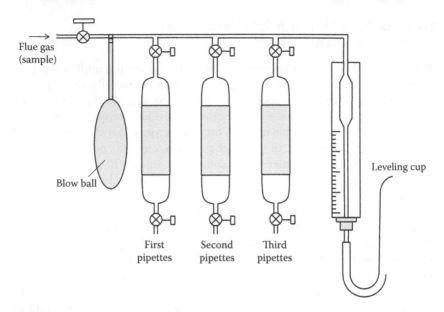

Figure 6.6

Schematic of Orsat Apparatus.

closed so that the sample will be forced to enter into the first pipette in which CO_2 will be absorbed. The sample is then allowed to flow into the burette for measuring the decrease in volume due to absorption of CO_2 gas. Then the sample should be brought into contact with the reagents in the second pipette. The above procedure is repeated to measure O_2 and CO successively. Care must be taken to use fresh reagents to minimize errors. Sufficient time must be provided so that the regents are reacted completely with a specific species. This method is very simple in operation and has a low cost. However, it is not currently being used because it is very time consuming and labor intensive. However, it is quite suitable for measuring combustion products in percentages, particularly when it will not be used continuously.

6.4.2 Flame Ionization Detector

The flame ionization detector (FID) is one of the most widely used detectors, particularly in gas chromatography and process analysis. It is based on the principle of ionization of gases in the presence of a flame. The basic design of a typical ionization detector is shown in Figure 6.7, which consists of hydrogen-fueled burner, ionization detector, and its associated electronics.

The sample and hydrogen are mixed beforehand and fed into the burner tip (see Figure 6.7) where a stable flame is established with the help of an ignition coil. The air is supplied to this chamber to form a hydrogen diffusion flame. A certain amount of DC potential (voltage 100–300 V) is applied between the

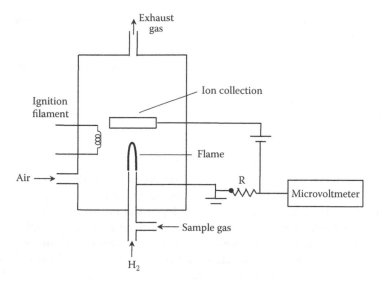

Figure 6.7

Schematic of a flame ionization detector.

burner tip and the ion collector. When a certain amount of organic compound is introduced into the hydrogen diffusion flames, a measurable amount of ions are produced in the flames. With the potential present across the burner tip and the ion collector, the ions will form a flow of current whose value is proportional to the concentration of the organic compound present in the hydrogen. This current can be measured easily by a sensitive ammeter. The response of the FID is approximately proportional to the number of carbon atoms in the sample and to the amount of the gas sample in the stream. In other words, the signal output is directly proportional to the molecular weight of the sample, which is linear over a wide range. For example, a given quantity of ethane produces a signal two times that of methane because it has two carbon atoms whereas the methane has only one. The ratio of the mixture of hydrogen and air as the sample gas for establishing a flame is very critical as this determines the signal output of the detector. Interestingly, the FID does not respond to carbon dioxide (CO_2) or water (H_2O) molecules but it does respond mildly to a carbon compound containing sulfur, oxygen, and halogen. Of course, a higher flow rate of the sample will introduce more carbon atoms, leading to a higher signal strength. However, the stability of the flame is very sensitive to the flow rate and therefore the hydrogen gas must be regulated properly. As well, a small-sized flame is preferred to reduce the consumption of hydrogen. The major advantages of the FID are that it is very reliable and is free from operational hazards.

The sensitivity of the FID is quite high, which can be used even to detect very low concentrations of hydrocarbon in the order of a 1-ppb (parts per billion) level. However, special measures must be taken for this type of measurement, as given below:

i. The hydrogen and air must be ultrapure to avoid spurious signals due to contamination of the gases
ii. The supply lines must be perfectly clean
iii. The flow rate must be properly controlled as per calibration requirements

6.4.3 Thermal Conductivity Detectors

A thermal conductivity detector is based on the principle that a hot body will lose heat by heat conduction, which is dependent on the heat capacity of the gas medium. The rate of heat loss is also dependent on the quantity of the gas medium, and hence can be used to measure the quantity of a gas sample.

A typical thermal conductivity detector (TCD) is shown in Figure 6.8, which consists of two tungsten heated filaments that act as a resistance element for sensing changes in thermal conductivity. One filament is placed in the sample chamber while the other filament is placed in the reference chamber. When a measured quantity of sample gas is passed through the sample chamber, heat loss from the heated filament takes place in the sample gas. The amount of heat loss is dependent on the flow rate and thermal conductivity of the sample gas. Since the flow rate is controlled, this heat loss is dependent on the thermal conductivity of the sample. As a result, the electrical resistance of the filaments gets changed. In the reference chamber, the heated element is cooled by pure carrier gas. Generally, helium is used as a carrier gas since its thermal conductivity is very high. As a result, the resistance of the reference element is changed considerably by the change in temperature due to heat loss. This difference in resistance between the detector element and reference element is dependent on the extent of the instantaneous heat conduction of the gas in the sample chamber. A Wheatstone bridge

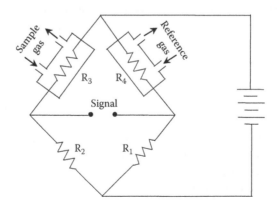

Figure 6.8

Schematic of a typical conductivity detector.

is used to measure the change in resistance by which the concentration of gas can be measured. This type of sensor can detect most compounds as its ability to distinguish is related to the difference in the thermal conductivity between the sample gas and carrier gas. Hence, it is very often known as a universal detector, routinely used in gas chromatography, and which we discuss briefly in the next section.

6.4.4 Gas Chromatography

Gas chromatography is one of the most popular and commonly used methods in both laboratories and industrial process applications. Gas chromatography is the process of separating the individual constituents of a mixture by permitting the gas mixture to flow through the column of adsorbent materials in which different substances are selectively separated and measured quantitatively by a detector individually. A schematic of a typical simple gas chromatograph is shown in Figure 6.9, which consists of columns, a carrier gas line, flow splitter, sample injection system, detector, and oven.

A gas chromatograph column is a long tube with a small internal diameter that is coated with adsorbent or inert materials. These materials will retain or slow down the flow of gas molecules moving with different speeds depending on the individual gas molecule's physical or chemical properties. The column retains some components longer than others, resulting in each component emerging from the column exit at its own specific time. Hence, the main purpose of the column is to separate the individual components in a gas mixture. When each component gas exits the column individually, gas detectors are used to measure it. Several different types of gas detectors can be used in gas chromatography. Of course, the type of detectors used in GC is dependent on its applications. The most popular are FID and TCD, whose basic principles have been discussed in the previous sections. TCD is typically used for high concentrations while FID is generally employed for low ppm ranges.

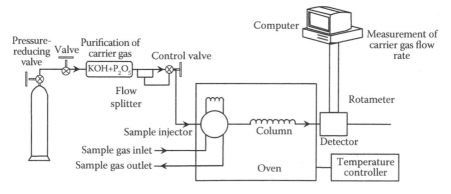

Figure 6.9

Schematic of a gas chromatography.

Generally, the detector is placed in an oven that can be at a constant temperature. Sometimes, the temperature of the oven can be programmed as per a specific function set by the user. The main objective of placing the column at a certain temperature is to achieve complete separation of each of the components and produce a narrow peak to be recorded. Since the column is quite long with a small internal diameter (2-mm inner diameter stainless steel or capillary column 0.1–0.5-mm ID glass tube) packed with adsorbent materials, the sample injected cannot be moved across the column without external pressure. Hence, a pressurized gas known as carrier gas is used whose main purpose is to push the sample through the column. The most common carrier gases used in GC are air, nitrogen, helium, and hydrogen. The injection part is basically a series of specialized valve assemblies with interconnecting fittings and stainless steel tubes. In actual operation, a carrier gas stream is constantly flowing through the column. The sample to be analyzed is injected into the carrier gas and gets separated depending on the affinity of the specific gas to the packing materials of the column. By choosing the column materials and operating temperature judiciously, one can manage to separate the components of a mixture completely. When this separated gas passes through the detector, it produces a signal that can be recorded as a function of time. Let us look at a typical chromatogram as shown in Figure 6.10. Each component of the gas mixture can be identified based on the fact that each component has a characteristic retention time for a particular

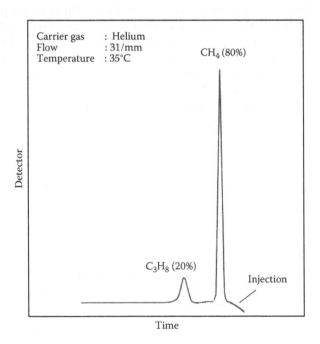

Figure 6.10

A typical chromatogram.

operating condition (column, oven, temperature, etc.). This retention time of each component is measured along the baseline from the injection point to the apex of the component (see Figure 6.10). By comparing with previously recorded chromatograms of known mixtures, one can identify each component. After identifying the component, we must analyze each component quantitatively, which is generally based on the relationship between the peak parameters of area under a peak and the concentration of each component in the sample mixture. Such quantitative analysis can only be accomplished by using the calibration curve, which can be constructed by using a known component as a standard. Of course, the estimation of the peak area under the chromatogram for each component must be carried out accurately to analyze the mixture quantity more precisely. A gas chromatograph is one of the most versatile instruments for the analysis of exhaust gases. However, it is used more effectively for unburned hydrocarbon. It is relatively simple and inexpensive, requiring only average operator training. But the disadvantage with this instrument is that it can be operated only in batches, and therefore it would not be used in a real-time gas analyzer routinely employed online in a practical combustor to control emissions.

6.4.5 Mass Spectroscopy

Mass spectroscopy is one of the analytical methods that is used in measuring species in combustion systems. It is being used extensively in flame studies as it can provide adequate sensitivity to detect most minor species in a typical flame. In this case, the molecules in the gas sample are bombarded by electron beam energy that produces positive ions due to either removal of electrons or the breaking of a chemical bond. Of course, during this process of bombardment, a few negative ions are likely to be formed as most molecules may not have stable levels for electron capture, but the positive ions formed during the bombardment of the molecules can be detected by mass spectroscopy. Mass spectroscopy basically measures the mass-to-charge ratio of charged particles (positive ions). These positive ions are separated based on their mass-to-charge ratio by applying an electromagnetic field and are detected by quantitative methods such as time of flight, quadrupole, and magnetic sector. The ion signal can be processed into mass spectra. Hence a typical mass spectrometer has three distinct components: (i) ion generator, (ii) mass analyzer, and (iii) detector. As discussed above, the function of an ion generator is to convert molecules in gas samples into ions. The mass analyzer is meant to sort out the ions based on their mass-to-charge ratio, and a detector is used to measure the quantity of positive ions present in the sample. Based on the method of ionization and detection, several types of mass spectrometers are being designed and developed. We will be restricting our discussion to two of the most widely used mass spectrometers: (i) time-of-flight and (ii) quadrupole.

i. Time-of-Flight Mass Spectrometer

The schematic of a time-of-flight spectrometer is shown in Figure 6.11, which consists of three basic components: (i) ion generator, (ii) mass analyzer, and

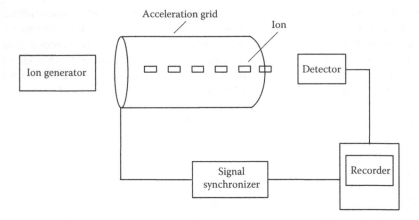

Figure 6.11

A typical time-of-flight mass spectrometer.

(iii) detector. In this case, a pulsed electron beam is used to ionize the molecules of a sample gas. Generally, a short ionization pulse in the order of a few hundred nanoseconds is employed that produces a bunch of electrons and subsequently accelerated using an acceleration grid. These electrons are allowed to pass through a flight tube, as shown in Figure 6.11, so that they can arrive at the detector head. Note that an electric field is applied to impart the kinetic energy proportional to the charge of an ion. As a result, the terminal velocity that an ion can have is dependent on the mass of the ion. In other words, each ion has a characteristic terminal velocity due to its own unique mass. As the accelerated ions are allowed to pass through the flight tube and eventually impinge into the detector head, a time-varying signal can be generated by a single pulse of electron beam. Note that the time taken by the ions to arrive at the detector is proportional to the square root of their respective masses. The signal from the detector can be passed through the signal processing units. The signals from this detector can be digitized and stored easily for further analysis. It is quite easy to obtain the complete spectra in the order of a millisecond. The real-time data can be easily obtained from this instrument as compared to other mass spectrometers.

Generally, the pulse rate of ionization in the order of 5 to 100 kHz is used in commercial instruments. A higher pulse rate is preferred when higher sensitivity and time resolution are required. Hence, instead of a pulsed electron beam, a laser pulse can be used to ionize the molecules of sample, which can enhance its sensitivity. But for molecules with ultrahigh mass, a lower pulse ionization is preferred.

ii. Quadrupole Mass Spectrometer

Quadrupole mass spectrometers are one of the most widely used spectrometers in combustion systems due to their higher sensitivity. The schematic of a typical quadrupole mass spectrometer is shown in Figure 6.12, which consists of

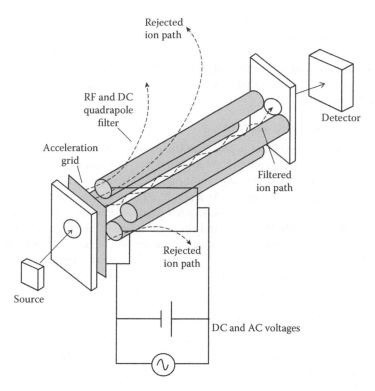

Figure 6.12

A typical quadruple mass spectrometer.

four metal rods, an acceleration grid, quadrupole filter, and detectors. In this case, four metal rods, as shown in Figure 6.12, are placed parallel to each other. Generally, circular rods with a specific ratio of diameter to spacing that can be approximated as hyperbolas are used for making manufacturing easier. It is very critical to maintain the proper ratio of diameter to spacing as even a small variation in this ratio would affect the resolution and peak shape drastically. Hence, different manufacturers use their own standard to maintain proper operating characteristics. Note that one opposing rod pair is connected electrically while one pair of rods is subjected to a radio frequency. Of course, a DC field is applied over this RF voltage. The AC and DC fields are applied to these metallic rods in the quadrature so that only a single open mass-to-charge trajectory is passed through the axis of this quadrature. In other words, the trajectory of ion can be controlled by the size of the rod, oscillating electric fields, and their frequencies. Only ions of a certain mass or charge ratio can arrive at the detector for a particular ratio of voltage while other ions follow unstable trajectories and can collide with the rods. As a result, one can select an ion with a particular mass-to-charge ratio continuously by varying the applied voltage. The frequency of

RF voltage can be varied with a mass number if a wide range of mass of various species as in a flame system is to be measured. As a result, higher sensitivity and good mass resolution can be achieved in this spectrometer as compared to other instruments. Interestingly, this instrument offers almost the same mass resolution even when the mass number varies drastically, which is quite common in combustion systems.

6.4.6 Spectroscopic Technique

The spectroscopic technique is based on the principle that each compound can absorb a particular spectrum of light as a function of frequency. The amount of light absorbed by a gas contained in a sample cell is dependent on the concentration of that gas in the sample. As a result, a particular gas can be measured both qualitatively and quantitatively. Such analysis can be broadly classified into two categories: (i) dispersive and (ii) nondispersive. In the dispersive type of spectrometer, a narrow band of spectrum is employed for analyzing the gas. In contrast, a wide range of frequency is used in the case of a nondispersive type of spectrometer. The dispersive type is preferred in laboratories whereas the nondispersive type is used in industries for analysis of air pollutants.

i. Dispersive Spectroscopy

A schematic of a dispersive spectrometer is shown in Figure 6.13 depicting its basic features. This dispersive spectrometer consists of a light source, monochromator, and detector. The function of the monochromator is to disperse the light radiated from the source into narrow bands. Either a prism or a diffraction grating is used to provide a narrow frequency band. The diffraction grating will produce a constant band over the entire frequency range, but the bandwidth of a prism monochromator varies strongly with the frequency. The frequency ranges employed in chemical analysis cover ultraviolet, visible, and infrared. Most spectroscopic analysis is carried out in the midinfrared range, particularly in the range of 2.5 to 15 μm. The detector is meant to measure the intensity of

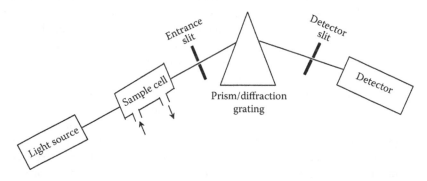

Figure 6.13

Basic principle of dispersive spectroscopy.

light. A number of lens, mirrors, and slits are used to focus the radiation and provide a constant amount of radiation over the whole range of the spectrometer. Generally, mirrors are used over lens, as they produce less aberration at a lower cost. When the absorbing species are present at different concentrations, the intensity of radiation reached at the surface of the detector can be related to the incident radiation by the Beer–Lambert law as

$$\frac{I}{I_0} = \exp(-\alpha_i C_i L) \tag{6.5}$$

where I_0 is the incident radiation intensity, I is the radiation intensity reaching the detector, α_i is the molar absorption coefficient that is independent of concentration, L is thickness of the sample cell, and C_i is the concentration of the ith species that absorbs radiation.

In order to relate the concentration to absorbance, one has to obtain a calibration curve. This is generally carried out by noting the change in absorbance with a concentration of a convenient peak on the spectrum. Of course, a strong peak is preferred with minimum overlaps with other species present in the sample to ensure accurate and precise measurement. For multicomponent mixtures, it is essential to have some idea of compounds with their respective spectra. For analysis of components, we must measure absorbance peaks. Generally, the peaks are chosen so that overlap between other peaks in the mixture must be avoided. However, it would not be possible to avoid overlaps in a real situation. In such a situation, a system of algebraic equations can be solved to find out individual concentrations. This type of spectroscopic analysis is carried out in laboratories to measure concentration of species.

ii. Nondispersive Internal Spectrometry

Nondispersive internal spectrometry is one of the standard instruments used for measuring exhaust emissions. In this method, nondispersive infrared (NDIR) radiation is used to detect and measure the unknown compound. A typical NDIR analyzer is shown in Figure 6.14, which consists of two infrared sources, a chopper, a small cell, filter cell, reference cell, and detector. The reference cell is filled with a nonabsorbing gas such as nitrogen, He, or Ar, whereas the sample cell is continuously flushed with carrier gas along with the sample. The detector consists of two compartments separated by a flexible diaphragm as shown in Figure 6.14, which actuates the capacitance or any other sensor to create a signal that is proportional to the concentration level in the sample gas. Both compartments are filled with the same gas to be measured. For example, if an NDIR gas analyzer is meant to measure CO, these two compartments of the detector are filled with CO, which absorbs radiation in the wavelength range from 4.5 to 5 μm. For CO_2 gas measurement, the absorbing wavelength ranges from 4 to 4.4 μm. Similarly, the gases that have a distinct absorbing range can be measured using an NDIR gas analyzer. However, it incurs an error due to interference when the exhaust

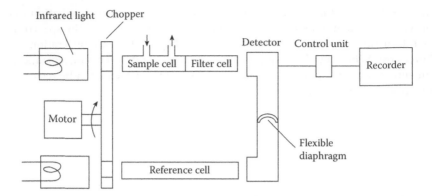

Figure 6.14

NDIR analyzer.

samples containing certain species absorb the radiation in the same range as that of the measured species. In order to reduce the effect of any interfering gas, the filter cell is filled with a large concentration of interfering gas.

This analyzer operates in the following manner. The radiant energy from the infrared source, after passing through a chopper, passes through both the reference cell and sample cell. A chopper is used so that the radiant energy on each side of the detector will be the same. It also produces an optical signal that can be easily handled by its detectors and associated electronics. Filters are sometimes used to ensure the narrow absorption wavelength of a particular gas to be measured. The reference cell passes through almost the entire infrared radiation without any absorption to one compartment of the detector chamber. In contrast, the infrared radiation is being absorbed by the gas to be measured when it passes through the sample cell. As a result, the sample cell side detector compartment will receive less radiation. These two chambers are sealed and insulated properly so that the temperature of the gas in this detector chamber increases due to the absorption of radiation.

This differential in the intensity of the radiation between two compartments of the detector will cause a pressure differential across the flexible diaphragm to move back and forth and thus sets up a signal that can be amplified and subsequently recorded. The magnitude of the diaphragm displacement is proportional to the amount of concentration of gas that absorbs the radiation in the sample cell. The commercial NDIR gas sensor for CO can measure up to a level of 1 ppm. Nondispersive gas analyzers are very convenient instruments, particularly for the analysis of a single compound such as CO or CO_2, but one has to avoid interference from other components in the mixture. For example, water vapors interfere while measuring the concentration of CO gas, and so water vapor must be removed beforehand. Note that NDIR gas analyzers have a rapid response and good sensitivity over a wide range of concentrations.

6. Gas Composition Measurements

6.5 Luminescence-Based Analyzer

We know that gas molecules emit light when they are excited with a sufficient amount of energy. Such phenomena of luminescence can be observed in a flame. Generally, two types of luminescence are used in transducer devices to measure species: (i) photoluminescence and (ii) chemiluminescence. Photoluminescence is basically the light emitted by gas molecules when they are at a high temperature; for example, light emitted from a flame or from ultraviolet light. In contrast, chemiluminescence is emitted during certain chemical reactions. The analyzers based on these luminescent effects are used for analyzing the pollutants from combustion systems since they have higher sensitivity. We will restrict our discussion to three types of analyzers: (i) fluorescence sulfur dioxide (SO_2), (ii) flame photometric, and (iii) chemiluminescence NO_x analyzers.

i. Fluorescence Sulfur Dioxide Analyzers

Fluorescence sulfur dioxide analyzers are based on the principle that when ultraviolet light interacts with sulfur dioxide molecules, it produces fluorescence with a different wavelength. The photomultiplier tube measures this fluorescence light. A schematic of a typical fluorescence sulfur dioxide analyzer is shown in Figure 6.15, which consists of an ultraviolet light source, two bandpass filter sample cells, and a photomultiplier tube (PMT). The ultraviolet radiation emitted from a light source is passed through a bandpass filter to allow a specific wavelength by which the sulfur dioxide molecules in the sample are excited to fluorescence with a different wavelength. With the help of another bandpass filter, the fluorescence light emitted from the SO_2 is detected and measured by

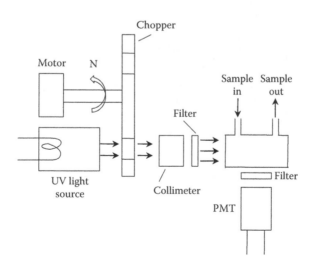

Figure 6.15

Schematic of a typical fluorescence sulfur dioxide analyzer.

the PMT. The fluorescence sulfur dioxide analyzer can be used for measuring SO_2 from 0.5 ppm up to a few thousand ppm. It has a very good sensitivity and it is a user-friendly instrument that can be used continuously in stake emission monitoring systems.

ii. Flame Photometric Analyzers

The flame photometric analyzer is based on the photometric principles of a sulfur compound. When a sulfur compound passes through a hydrogen air flame, it is excited to luminescence and emits radiation around 394 nm. The luminescence signal at this wavelength is measured using a narrow bandpass filter and PMT. The main function of a PMT is to amplify a small luminescence signal to produce an electrical signal that can be measured easily. A schematic of a typical flame photometric analyzer is shown in Figure 6.16, which consists of a hydrogen air flame burner, bandpass filter, and PMT. As discussed earlier, the sample containing the sulfur compounds is allowed to pass through the fuel line of the burner. When a flame is established, the sulfur compounds are excited to luminescence and they emit radiation at around 394 nm, which is measured by the PMT. The amount of light emitted is proportional to the concentration of sulfur compounds present in the flame. This instrument measures the total sulfur content. Several sulfur compounds such as hydrogen sulfide, sulfurdioxide, and mercaptans can be produced during combustion of sulfur-containing fuel. This analyzer is very sensitive even up to 0.01 ppm of total sulfur content, which finds applications in stack monitoring systems.

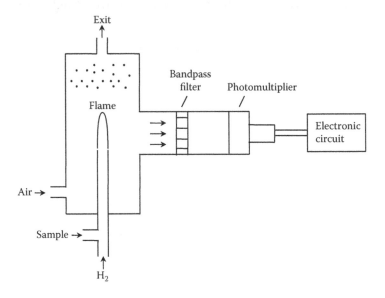

Figure 6.16

Schematic of a typical photometric analyzer.

iii. Chemiluminescence Analyzers

In the chemiluminescence analyzer method, the luminescence is caused by a chemical reaction. For example, when nitric oxide (NO) reacts with ozone, it produces infrared radiation at around 0.5 to 3 μm. This process of chemiluminescence can be described by following chemical reactions:

$$NO + \dot{O}_3 \rightarrow NO_2^* + O_2 \tag{6.6}$$

$$NO_2^* \rightarrow NO_2 + h\nu \tag{6.7}$$

First, NO is reacted with ozone to produce nitrogen dioxide, NO_2^* at an excited state that subsequently reverts back to its ground state with the emission of radiant energy. The intensity of the radiation is proportional to the concentration of nitric oxide present in the sample, which can be measured by using PMT.

A schematic of a typical chemiluminescence analyzer is shown in Figure 6.17, which consists of an ozone generator, reaction chamber, NO converter, bandpass filter, and PMT. A controlled amount of sample and ozone are fed into the reaction chamber where a luminescence signal is produced that is measured by a PMT through a filter. Ozone can be generated either by ultraviolet radiation or high voltage by breaking up the dry O_2 molecules flowing in a quartz tube. Note that the chemiluminescence reaction with ozone is specific to the presence of NO. It is very important to convert the NO_2 present in the sample to NO in a NO converter. In this converter, catalysts are generally used to convert NO_2 to NO, which can be fed into the reaction chamber. The chemiluminescence analyzer is a continuous analyzer that is used for the measurement of NO and NO_2

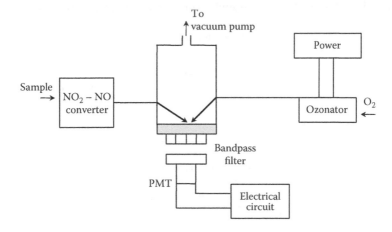

Figure 6.17

Schematic of a typical chemiluminescence NO_x analyzer.

from the exhaust gases of combustion systems. It is a very accurate and sensitive instrument that can measure even as low as 4 ppb. It is also a low-cost analyzer compared to the spectrophotometric devices, and hence, it is very popular in industry. This technique is a standard method for measuring NO_x in both ambient air and combustion devices.

6.6 Electrochemical Gas Analyzers

The electrochemical sensor is based on the principle that when a gas to be measured is reacted with an electrode, it produces an electrical signal that is proportional to the gas concentration. This current flowing through the anode and the cathode can be measured by a microammeter to determine the concentration of the gas to be measured in the exhaust gas. This type of electrochemical sensor is often termed a microfuel cell type gas analyzer.

A typical electrochemical sensor is shown in Figure 6.18, which consists of a sensing electrode and a counterelectrode separated by a thin layer of electrolyte. Besides these elements, a hydrophobic membrane and reference electrode are also used.

The gas to be measured first passes through a small capillary type opening and then diffuses through a hydrophobic membrane. The function of this membrane is to control the amount of gas reaching the electrode surface. Such a barrier is made of a low-porosity Teflon membrane that not only provides mechanical protection to the sensing electrode but also filters out unwanted particulates. It is important to select the proper amount of gas molecules that can reach the sensing electrode. This membrane also prevents the leakage of liquid electrolyte and thus prevents the drying out of the sensor too quickly. The selection of electrode material is very important while designing this type of analyzer. Generally, a noble metal such as platinum or gold, which can be catalyzed for an effective reaction with the gas molecules, is used as an electrode. Different types of materials for the electrodes can be selected depending on the application and design considerations.

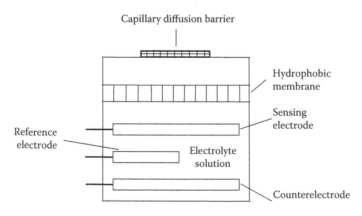

Figure 6.18

Schematic of an electrochemical gas analyzer.

6. Gas Composition Measurements

The main function of the electrolyte is to facilitate the cell reaction and carry the ionic charges across the electrodes efficiently. Besides this, it must also form a stable reference potential with the reference electrode. The electrolyte must not evaporate quickly as the signal of the sensor may deteriorate its performance. In order to improve the performance of the sensor, a reference electrode is placed closer to the sensing electrode within the electrolyte. The main function of this reference electrode is to maintain a fixed voltage at this sensing electrode. Keep in mind that no current flows to or from the reference electrode. The gas reacts at the surface of the sensing electrode after getting diffused through the hydrophobic membrane. The reactions at the sensing electrode (anode) for certain gases such as CO, H_2S, NO, NO, and H_2 are given below:

$$CO + H_2O \rightarrow CO_2 + 2H + 2e^- \tag{6.8}$$

$$H_2S + 4H_2O \rightarrow H_2SO_4 + 8H^+ + 8e^- \tag{6.9}$$

$$NO + 2H_2O \rightarrow HNO_3 + 3H^+ + 3e^- \tag{6.10}$$

$$H_2 \rightarrow 2H^+ + 2H^+ + 2e^- \tag{6.11}$$

During the oxidation of the anode, electrons are released that then travel via an external circuit to the cathode where oxygen molecules consume the electrodes involving oxygen and H^+, as given below:

$$O_2 + 4H^+ + 4e^- \rightarrow 2H_2O \tag{6.12}$$

Hence, an adequate supply of oxygen must be provided for better performance of this type of electrochemical cell.

Electrochemical analyzers are not very sensitive to pressure changes but they are quite sensitive to temperature and hence provisions should be made to maintain the temperature within the sensor to be as stable as possible. The temperature effect is typically 0.5% to 1.0% per °C depending on the type of sensor. Electrochemical sensors are generally quite selective to the measured gas for which they are designed. Different types of filters are used to minimize the effects of interference of other gases on the sensor. The life of expectancy of an electrochemical sensor depends on several factors such as the amount of gas exposed, environmental conditions, types of electrolytes, and filters used in the sensor. Generally, the life of this type of sensor ranges from 1 to 2 years. Since electrochemical sensors are very cheap and require little training to operate, they are used extensively in portable instruments for combustion systems.

6.7 Paramagnetic Oxygen Analyzers

We need to measure oxygen gas level in the exhaust from the combustion system because it is important to know the extent of combustion. It will also be useful to interpret the emission data to identify the causes of high emission levels in combustion systems. The oxygen level in exhaust gas can be measured using an electrochemical sensor as discussed in the section above. Another method used widely for the measurement of oxygen level is the paramagnetic oxygen analyzer, which is based on the paramagnetic property of oxygen gas. Note that oxygen exhibits a strong paramagnetic property as compared to other common gases.

The schematic of a typical paramagnetic gas analyzer is shown in Figure 6.19, which consists of two diamagnetic dumbbell-shaped hollow spheres, a light source, mirror, detector, magnets, amplifier, indicator, casing (measuring cell), and recorder. These two quartz spheres filled with nitrogen gas are suspended by a fiber in a magnetic field, as shown in Figure 6.19, so that it can be rotated out of this field. In the beginning, the two spheres are kept in a balance position in a homogeneous magnetic field with the help of an optoelectric compensation circuit. When the sample gas containing oxygen flows through the measuring cell, the paramagnetic characteristics of the oxygen molecules will change the magnetic fields. As a result, torque is produced on the dumbbell spheres, whose magnitude is proportional to the concentration of oxygen in the sample gas. The deviation of the spheres can be detected with the light source, reflecting mirror, and optical detector (photodiode). Hence, a current is passed through the feedback loop so that the spheres can return to the initial balanced state. Note that

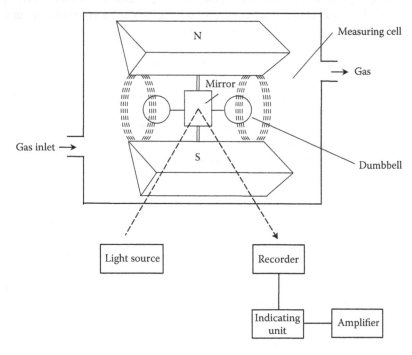

Figure 6.19

Paramagnetic oxygen analyzer.

the current passing through the feedback loop is proportional to the oxygen concentration. Thus, the oxygen concentration is converted into an electric signal. As the principle of measurement is dependent on the strong magnetic property of oxygen molecules, measurement of oxygen gas is not affected by other molecules or other common gases. Hence, interference of other exhaust gases such as CO, CO_2, and NO during measurement of the oxygen gas is quite negligible as these gases do not exhibit minimal sensitivity to a paramagnetic property compared to oxygen gas. Water is considered to have an interference effect on oxygen gas measurement, so the sample is passed through a silica-gel-packed container to remove water vapor from the sample before being fed to the paramagnetic gas analyzer. Generally this instrument is calibrated using pure nitrogen gas at a zero setting and span gas of oxygen at a specific setting.

6.8 Smoke Meters

The smoke meter is used routinely to measure smoke concentration in the exhaust gas of piston engines and other combustion systems. Generally, the exhaust gas from a combustion system contains black smoke due to high-temperature fuel combustion and white smoke due to unburned oil, unburned fuel, and water. In earlier days, a certain quantity of exhaust gas from the engine and combustion systems was allowed to pass through a filter paper of a specified quality. Depending on the smoke density, the smoke was deposited on the filter paper and filter paper would become darkened with smoke, which could be quantified by a light reflectance method.

But in recent times, smoke meters based on the light absorption principle governed by the Beer–Lambert law are used to measure smoke concentration. The schematic of a typical absorption light smoke meter is shown in Figure 6.20, which consists of a sample cell, exhaust fan, light source, lens, slit, and optical

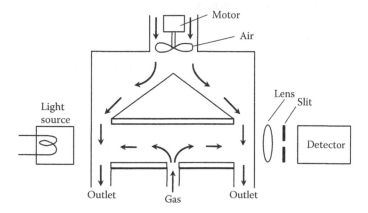

Figure 6.20

Light absorption smoke meter.

detector. An incandescent lamp or a light-emitting diode with a wavelength in the spectral range of 550–570 nm can be used as the light source. The visible light from a source is passed through a sample cell of a known length, which reaches the optical sensor placed at the opposite end. When the exhaust gas is passed through this sample cell, light intensity transmitted through the sample cell to the optical sensor gets reduced due to absorption of light by smoke particles. In other words, the smoke opacity is the fraction of light that is not being allowed to reach the photodetector due to the absorption of light by smoke particles. As a result, the light intensity is attenuated depending on the level of smoke in the exhaust gas. In other words, the transmitted light that falls on the photodetector is dependent on the smoke concentration. That means the strength of the absorbed light indicates the smoke concentration of the exhaust gas. The smoke opacity is being measured with this instrument and hence it is often known as a smoke opacity meter. Generally, the smoke concentration in this instrument can be expressed as smoke density, which is defined as the ratio of the voltage output of the photodetector with an exhaust sample and clean air. As per the Beer–Lambert law, the absolute smoke density can be characterized by the light absorption coefficient, which can be expressed as

$$I = I_0 e^{-kL} = I_0 \left(1 - \frac{N}{100} \right) \tag{6.13}$$

$$N = (1 - I/I_0)100 \tag{6.14}$$

where I is the light intensity with smoke, I_0 is the light intensity without any smoke, L is the sample cell length, k is the absorption constant, and N is the opacity/smoke number. By measuring the light intensity for a particular instrument, we can determine the smoke opacity/smoke number due to presence of smoke in the exhaust gas by using Equation 6.14.

Review Questions

1. What is an emission index? Why is it preferred over emission levels in ppm/percentage?

2. What are the methods of sampling gas for measurement of emission levels from a combustion system? Explain them.

3. What is isokinetic sampling? Why is it essential for accurate measurement of emission levels?

4. Draw the stream lines around a typical sampling probe with sampling velocity for (a) greater than isokinetic sampling velocity, (b) equal to isokinetic sampling velocity, and (c) less than isokinetic sampling velocity, and explain its importance for measurement.

5. Describe the working principle of a simple gas chromatography with the help of a schematic diagram.

6. Can gas chromatography be combined with mass spectroscopy? Explain.

7. What is an FID sensor? Can it be used in gas chromatography?

8. What is the basic principle based on which an NDIR gas analyzer is designed and developed? Draw the schematic of a typical NDIR gas analyzer and explain its working principle.

9. What are the methods to measure the NO_x concentration of exhaust gas? Can the NDIR principle be used for its measurement? Explain.

10. What are the ways of measuring the oxygen level in exhaust gas from a combustion system? Explain the working principles of each instrument by using schematic diagrams.

11. How can the smoke level in a piston engine be measured? Which is the best method and why?

Problems

1. The measured compositions of exhaust gas at dry condition and atmospheric pressure from a combustor are given below:

$$CO: 0.025\%$$

$$CO_2: 12.5\%$$

$$NO_x: 160 \text{ ppm}$$

Determine the partial pressure of each of these gases in this dry mixture.

2. A gas turbine engine uses propane gas. The measured compositions of exhaust gas at dry condition are given below:

$$O_2: 15.5\%$$

$$CO: 0.35\%$$

$$CO_2: 18.2\%$$

$$NO_x: 60 \text{ ppm}$$

Determine the emission index of NO_x.

3. An orsat apparatus is used to measure gas composition from a combustor operated on propane gas. The following volumes are measured as follows:

   Initial volume = 98 cm³; after CO_2 absorption = 85 cm³; after O_2 absorption = 82 cm³; after CO absorption = 80 cm³. Determine the air-fuel ratio and equivalence ratio.

4. A sampling probe is designed to ensure isokinetic sampling. An error in the measured concentration of 3% is acceptable. If the isokinetic factor is 0.9 and local velocity is 5.4 m/s, determine the sampling velocity.

5. A smoke meter uses an initial light intensity flux of 1.5 W/cm² that has a path length of 30 cm. If the absorption coefficient is $1.25 \dfrac{1}{m}$, determine the smoke number and light intensity measured by its detector.

References

1. Mishra, D. P., *Fundamentals of Combustion*, PHI Ltd., New Delhi, India, 2010.
2. Chedaille, J. and Braud, Y., *Measurements in Flames*, Vol. 1, Edward Arnold, London, 1972.
3. Fristrom, R. M., *Flame Structure and Process*, Oxford University Press, New York, 1995.
4. Gayden, A. G. and Wolfhard, H. G., *Flames: Their Structure, Radiation, and Temperature*, Fourth Edition, Chapman & Hall, London, 1979.
5. Heitor, M. V. and Moreira, A. L. N., Thermocouple and sampling probes for combustion studies, *Progress in Energy Combustion Science*, 19:259, 1993.
6. Colket, M. B., Chiappetta, L., Guile, R. N., Zabielski, M. F. and Seery, D. J., Internal aerodynamics of gas sampling, *Combustion and Flame*, 44:3–14, 1982.
7. Skoog, D. A., Holler, F. J. and Crouch, S. R., *Principles of Instrumental Analysis*, 6th Edition, Cengage Learning Pvt. Ltd., Stamford, CT, 2007.
8. Settle, F. A. (ed.), *Handbook of Instrumental Techniques for Analytical Chemistry*, Prentice Hall PTR (ECS Professional), Upper Saddle River, NJ, 1997.

7
Optical Combustion Diagnostics

Optical diagnostic tools cannot be an end but the humble beginning of understanding the intricacies of a combustion process.

D. P. Mishra

7.1 Introduction

In modern times, optical combustion diagnostics are being used extensively to unravel underlying complex phenomena that occurring during combustion that are essential to enhance not only the performance of the combustion system but also for the reduction of harmful pollutant emissions. Combustion diagnostics are being used to generate experimental data that are required to validate the modeling work. Several types of combustion diagnostic tools have been developed and applied in designing new combustion systems to meet the demand of energy and stringent emission regulations. The field of combustion diagnostics is evolving at a rapid rate, and hence some of the important optical-based measurement methods are discussed in this chapter.

In this chapter we will learn about the types of light sources that are important for flow visualizations and both qualitative and quantitative measurements in combustion systems. In this section, apart from other light sources, several types of laser sources are discussed briefly. Subsequently, the basic principles of scattering are discussed briefly as they are the backbone of most modern combustion diagnostic tools. We will be learning about how to carry out direct flame photography for both qualitative and quantitative measurements. Both Schlieren and shadowgraph flame visualization techniques are covered in detail and the interferometer technique for flame visualization and other measurements is discussed briefly. Several laser-based advanced techniques for quantitative measurements, such as laser Doppler velocimetry (LDV), particle image velocimetry (PIV), planar-laser-induced fluorescence (PLIF), and Rayleigh scattering thermometry, are also discussed briefly in this chapter. For more advanced treatment of these topics, interested readers may refer to specific advanced books and review papers [1–7,12].

7.2 Types of Light Sources

Most combustion diagnostic methods, such as Schlieren, shadowgraph, and interferometer, rely on certain types of electrical light sources that in recent times are being substituted by coherent and solid state light sources. Some of the light sources—incandescent lamp, compact arc lamp, xenon flash tube, and LED—shown in Figure 7.1 being used for flow and flame visualizations are discussed briefly below.

The incandescent lamp is a continuous wave light source that is suitable for Schlieren and interferometer systems. The tungsten halogen lamp (see Figure 7.1a) is one of the inexpensive, stable, and dependable light sources for the Schlieren system. This type of lamp has a tightly wound coil element that produces a white light with a yellowish cast. It has a high luminous radiant exitance/emittance that is an order of magnitude less than the arc lamp, but has a reduced lifetime even though it produces a higher emittance at around 15–20 V, particularly for automotive applications.

Another popular light source is the compact arc lamp, which can provide a higher luminous radiant emittance (10–100 times) than the tungsten filament lamp. In a compact arc lamp, a small spark is produced between the electrodes

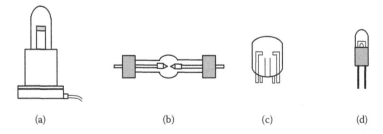

(a)   (b)   (c)   (d)

Figure 7.1

Schematics of typical nonlaser light sources: (a) tungsten-filament, (b) compact arc lamp, (c) xenon flash tube, and (d) LED.

confined in a glass bulb and light is emitted from the plasma produced during spark. This type of light source can produce either continuous wave (CW) or pulse mode. Two light sources—high-pressure mercury arc lamps (Figure 7.1b) and xenon lamps (Figure 7.1c)—are used extensively for flow visualization. A xenon lamp has a better white-light emittance than a mercury light source.

In recent times, LEDs (Figure 7.1d) are being used for flow visualization as they are available easily in a monochromatic wavelength covering white-light emissions. LEDs have a relatively low light emittance compared to arc light sources. However, by using a better design of light source with the help of multiple LEDs and proper methods of enhancing emittance, higher power levels can be achieved that can replace incandescent lamps. LEDs can be used as a pulsed light source even in the range of high frame rate, which can be used for high-speed flow visualization.

7.3 Laser Lights

Most combustion diagnostic methods, such as LDV, PIV, PLIF, Raman spectroscopy, and Rayleigh scattering—with the exception of certain flame visualization techniques such as Schlieren, shadowgraph, and interferometer—rely mainly on laser light as it has several optical properties (coherent, monochromatic, uniform polarization, etc.). A laser light has a high degree of both temporal and spatial coherence. Recall that the word "laser" stands for "light amplification by stimulated emission of radiation." Several types of lasers have been designed and developed over the past few decades that find applications in modern times. Most of these lasers are comprised of three crucial elements: (i) a laser material that gains and amplifies radiation, (ii) a pumping source that excites the laser materials, and (iii) a mirror arrangement (resonator) that allows oscillation within the laser material, as shown in Figure 7.2. The basic principle on which a laser works is

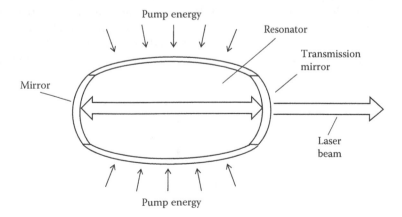

Figure 7.2

Schematic of a typical laser.

described briefly below. Interested readers may refer to [7], which is an advanced book on laser physics.

We know that an atom of laser material accepts energy from an external source and becomes elevated to a higher energy state due to the pumping action of incident light. In other words, an electron in the atom jumps from an inner orbit to an outer orbit. This process is known as absorption, as shown in Figure 7.3a, during which hv amount of energy is absorbed by an atom. After a short time, this excited atom at higher energy level E_2, as shown in Figure 7.3b, being at metastable state, returns back to its original energy level E_1 while undergoing spontaneous emission of the same amount of incident photon energy without any external influence. This is known as *spontaneous emission of light energy*. On the other hand, if an atom interacts with light energy with an appropriate frequency then it may either get elevated to a higher energy level by absorbing a photon or get stimulated to emit a certain photon energy nonspontaneously. The latter part of this process involving amplification of light energy is known as stimulated emission, as shown in Figure 7.3c, which is the crux of producing laser light. The physical principle of this stimulated emission was discovered by Albert Einstein in 1916. In the case of a laser, a large number of atoms are subjected to absorption of light energy, stimulated radiation of light, or both simultaneously. The stimulated emission of light will be predominant only when a higher number of atoms convieniently expressed in terms of population density N_2 (atoms/m^3) at higher energy level E_2 are subjected to stimulated emission. That means a lower number of atoms (population density N_1) at lower energy level E_1 can absorb pumped light energy. In other words, lasing of light can be possible only when the population inversion ($N_2 > N_1$) is caused during pumping of external energy. Note that it would not be possible to have population inversion using two energy systems, as described in Figure 7.4a. The number of transitions from upper level E_2 to lower level E_1 and vice versa remains almost the same; therefore, more than two levels of energy medium can be used for population inversion for producing a laser medium. Generally, four levels of energy medium are preferred over a third level (see Figure 7.4b) as it is easier for population inversion because in a four-level laser, lower laser level E_2 is higher than the basic energy level, as shown in Figure 7.4c. For example, if energy level E_4 is reached by the atom due to light

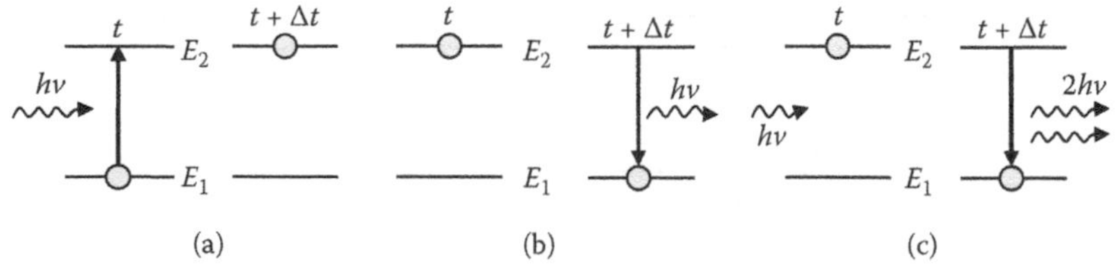

Figure 7.3

Processes involved between energy interactions during the lasing of an atom.

7. Optical Combustion Diagnostics

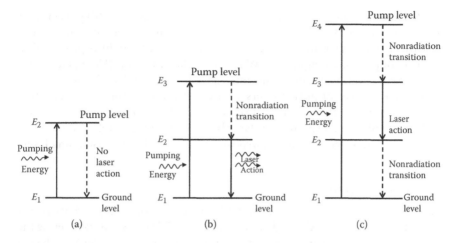

Figure 7.4

Energy level diagram: (a) two-level, (b) three-level, and (c) four-level.

pumping at a certain frequency, it may jump back to upper energy level E_3 undergoing a nonradiative transition. The atom remains in a metastable state at E_3 for a longer duration of time before it jumps back to E_2 due to the lasing action. As a result, population inversion can take place efficiently and thus the spontaneous emission of light can occur in all directions. Hence, the adjacent atoms get excited, leading to the rapid increase of stimulated radiation like in a chain reaction.

Note that pumping energy for a laser can be provided by various means depending on the types of laser materials being used. Sources of pumping energy such as electrical discharge, flash lamp, chemical reactions, or another laser beam can be used for producing a laser. For example, electromagnetic radiation is used for pumping in the case of a solid laser material while semiconductor laser materials are pumped by an electronic current. The collision of atoms/molecules using electrons and ions are used in gas lasers, where acceleration of free electrons by the electric field excites the gaseous medium for lasing. But in a solid state laser, a rod-shaped flash lamp is used to pump light energy on the laser material by a cylindrical mirror with an elliptical cross section. Similarly, several types of resonators based on mirror arrangements have been devised for producing laser light. The confocal resonator shown in Figure 7.2 is quite stable and easy to control and therefore finds a wide variety of applications.

7.3.1 Types of Lasers

Several types of lasers have been designed and developed over the years for a wide variety of applications. Based on mode of light emission, these can be divided into two categories: (i) continuous wave (CW) laser and (ii) pulsed laser. In the case of a CW laser, the power of the laser beam remains continuous over time. In contrast, in the case of a pulsed laser, laser light is produced in the form of pulses. Examples of CW lasers are helium-neon (He-Ne) and CO_2 lasers. Neodymium–yttrium-aluminum-garnet (Nd-YAG) and copper-vapor lasers are examples of pulsed lasers whose pulse varies from a few hundred hertz to kilohertz. The power level of lasers varies from microwatts to a few watts. In combustion experiments, laser lights in both the visible and ultraviolet range are used for diagnostic purposes. We will be briefly discussing lasers used in combustion diagnostics.

7.3.2 He-Ne Lasers

The He-Ne laser is the most efficient gas laser in the visible range whose power ranges from 1 to 10 mW. Although a wide range of visible He-Ne lasers covering the red, orange, yellow, and green parts of the visible spectrum can be devised, the red part at a wavelength of 632.8 nm is widely used due to its higher gain in population inversion. In this laser, a mixture of helium and neon gases in the ratio of 1:10 at low pressure is used as a lasing medium that is housed in an optical cavity consisting of two concave mirrors. One mirror with less reflectance (99%) allows the laser light to pass through it. A high voltage of electrical discharge is used to pump the requisite energy for lasing He-Ne gas. During this process of electrical discharge, helium atoms get excited due to the inelastic collision of energetic electrons from the ground state to a higher energy state. Subsequently the collisions between the He atoms in the metastable state and ground state neon atoms result in an efficient transfer of energy to the neon atoms, thus causing amplification of a stimulated emission of radiation. This process occurs in the optical oscillator that reflects back and forth and thus the power of the radiation becomes enhanced rapidly and results in a stable continuous laser beam. As mentioned earlier, the visible He-Ne Laser (632.8 nm) finds applications in several measuring systems namely interferometer, LDA, etc. due to coherent, excellent spatial distribution of beam intensity.

7.3.3 Argon-Ion Lasers

The argon-ion laser is a gas laser similar to the He-Ne laser described above that uses Argon as the active medium. In this case, an electric discharge is used to pump energy to the laser medium. But a very high current must be used for achieving ionization and excitation of argon (Ar) gas. The plasma of an Ar-ion laser consists of a high-current-density glow discharge in the presence of a magnetic field. It is capable of emitting a continuous wave output of several watts to tens of watts in several wavelengths covering both the visible and ultraviolet range. Although light emission with several wavelengths can be achieved using a broadband mirror system, individual wavelengths can be produced using a Brewster prism in the resonator. The Ar-ion laser that produces wavelengths of 514.5 and 488 nm is used extensively in measurement systems. It finds applications in LDV and PIV systems, discussed later in this chapter.

7.3.4 Nd-YAG Lasers

The Nd-YAG laser is one of the most important and widely used solid state lasers particularly for combustion diagnostics due to its high amplification and good mechanical, thermal, and optical properties. In this case, a crystal made of neodymium-doped yttrium-aluminum-garnet ($Nd:Y_3Al_5O_{12}$) is employed as a lasing medium. The laser beam is produced by triply ionized neodymium (Nd^{+3}) due to its lasing activity in the Nd:YAG crystal when excitation is achieved with a flash lamp or laser diodes over a broad energy band and nonradiative transition to the upper energy level in the lattice of the crystal. The oscillator causes periodic excitation of the crystal that leads to energy bands formed by the upper energy. As a result, the energy band levels of this system are continuous in nature. It has a low laser threshold and hence population inversion takes place as soon as the threshold is reached. Of course this threshold value is dependent on the design of the laser cavity. In this manner, several laser pulses can be produced during the pumping pulse of the flash lamp. This laser can be operated on a trigger mode by using a quality switch (Q-switch) that is placed in the cavity of the laser. This optical switch gets opened as soon as maximum population inversion takes place in the neodymium ions. The light wave can then pass through the cavity and depopulate the excited laser medium at maximum population inversion. During this process, the cavity gets resonated at the most energetic point of the flash lamp cycle. As a result, a very powerful laser pulse (e.g., 250 MW with pulse duration of 10 ns) can be achieved easily. A Q-switch is comprised of a polarizer and a Pockel cell, by which the quality of the optical resonator can be changed by adjusting the voltage across this cell. Conventionally, an Nd:YAG laser emits light with a wavelength of 1064 nm. The high-intensity laser that pulses at a wavelength of 1064 nm can be efficiently frequency-doubled using deuterated potassium dihydrogen phosphate (DKDP) crystal to generate light at 532 nm, used routinely for PIV. The Nd-YAG laser is also used for the pumping purposes of a dye laser used in PLIF systems for measurement of species in the flame. Nd-YAG laser may also find application in modern laser ignition systems in advanced combustion systems due to higher power per pulse.

7.3.5 Dye Lasers

A dye laser is used when a tunable laser source is required. Basically, a dye laser consists of an organic dye mixed with a solvent that is circulated through a dye cell. A fast flash lamp or an external laser (Nd-YAG laser) is used to pump the photons into dyed liquid so that it can attain its lasing threshold point. As in other lasers, it also uses mirrors to oscillate the produced laser light so that it can be amplified with each pass. Most of the dyes contain large organic molecules and thus fluoresce over a wide range of wavelengths when excited with a pumping light. Note that the dye is transparent to the lasing wavelength, but within a few seconds, the dye molecules attain a triplet state and the molecules begin to phosphoresce and start absorbing the lasing wavelength. As a result, the dye becomes opaque to the lasing wavelength. Hence, a flash lamp or laser used for pumping must have a short duration capability with a higher power level as most liquid dyes have a higher lasing threshold. As well, the dye is being recirculated at a high speed so that the triplet molecules will be out of the path of the laser beam. By using the proper lasing medium in dye lasers, light pulses in the range of a few femtoseconds are being developed for combustion diagnostics. In this chapter, we will discuss the microsecond dye lasers used routinely in PLIF systems.

7.4 Basic Principles of Light Scattering

We need to understand the basic principles of light scattering as both visualization of flow and flame result due to the interaction of light with matter. Most optical diagnostic tools are based on light scattering principles. The physical process during the interaction between light and matter is often referred to as light scattering because when light rays fall on a group of particles (atoms, molecules, crystals, etc.), they can be either deflected partially from their earlier straight path due to diffraction, reflection, and so forth, or get absorbed partially. As a result, both the direction and intensity of incident light changes during the scattering process. Apart from changing its direction and intensity, the frequency of an incident light wave may be changed depending on the nature of the scattering process.

Let us try to understand how light interacts with matter. For this we will be considering a simplified model as shown in Figure 7.5 that can depict the physical processes during the interaction of matter and light. As mentioned above, both properties of matter and light are altered during the scattering process. Therefore, we need to look at various properties of light such as wavelength, frequency, photon energy, light intensity, and polarization of light, depending on the type of matter, its structure, size, and shape. Various optical phenomena such as reflection, refraction, diffraction, absorption, and remission of light energy must be considered during the scattering of light. For analysis of the energy of light and its interaction with matter, we will be using the concept of a photon that indicates the quantum of light energy. Recall that the energy and momentum

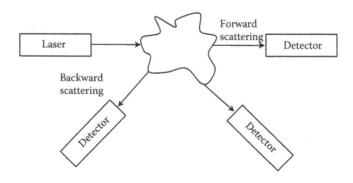

Figure 7.5

Basic process of light scattering.

associated with a photon as per Planck's law can be related to its frequency f and wavelength λ as

$$E = \frac{hc}{\lambda} = hf \qquad (7.1)$$

where c is the speed of light with which photons move and h is the Planck constant $= 6.626 \times 10^{-34}$ J-s. We will be discussing energy interaction between incident light and matter below.

Let us consider a typical experimental arrangement for light scattering consisting of incident laser light, scattering volume, condensing lens, aperture, and detector. It can be observed that an incident beam of light (I_0)—preferably a laser—for measurements is concentrated by a condensing lens in the range of a few hundred microns with a width of around 1 mm and thus creates a very small scattering volume as shown in Figure 7.6. The incident light will be scattered from the particles, aerosol, molecule, atoms, or ions along all directions. However, the light intensity I from the scattering volume that subtends a solid angle of Ω can be measured using a requisite detector with the help of the condensing lens and aperture, as shown in Figure 7.6. The scattering light intensity, I, can be related to the incident light identity, I_0 (W/cm^2), number density of scattering species, N, distance between incident beam and detector, L, differential scattering cross section, $d\sigma/d\Omega$ (cm^2/sr), solid angle between detected (collector) and scattering volume, Ω_c (sr), as

$$I = \beta I_0 N L \Omega_c \left(\frac{d\sigma}{d\Omega} \right) \qquad (7.2)$$

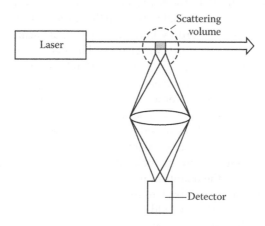

Figure 7.6

Schematic of a typical experimental setup for light scattering.

Table 7.1 Typical Data of Differential Scattering Cross Sections of Various Scattering Processes

Scattering Process	Scatterer	Cross-Section (m²/Steradian)	Frequency Shift
Mie scattering ($d/\lambda \gg 1$)	$D = 0.1$–$10\ \mu m$ (soot, dust, seeding particles)	10^{-9} to 10^{-3}	0
Rayleigh scattering ($d/\lambda \ll 1$)	$N2$ (molecules, soot, dust, seeding particles)	9×10^{-24}	0
Bragg scattering	Structure of crystal	—	$\pm(10^{-4} - 10^{-2})$ (with Brillouin scattering)
Brillouin scattering	Crystals and liquid	—	$\pm(10^{-6} - 10^{-5})$
Raman scattering	N_2 Molecules/atoms	5×10^{-26} (rotational) 5×10^{-27} (vibrational)	$\pm(10^{-5} - 10^{-1})$
Fluorescence	Atoms/molecules	$10^{-12} - 10^{-9}$(atom) $10^{-18} - 10^{-15}$(molecule)	$\pm(0 - 10^{-2})$
Absorption	Atoms/molecules		0

where β is the detection efficiency. Note that differential scattering cross sections given in Table 7.1 indicate the relative magnitude of various scattering processes that will be discussed below.

7.4.1 Rayleigh Scattering

As mentioned earlier, the strength of scattering can be measured by the loss in energy in light incident on matter. It should not be confused with absorption of light alone. During absorption, light energy is converted into the internal energy of the medium whereas during scattering, light energy is radiated in all directions. The strength of the scattering light will be dependent on the wavelength of light apart from the size of the scattering particle, and its shape. When the wavelength of light is greater than the scattering particles ($d/\lambda \ll 1$), the strength of scattering is inversely proportional to the fourth power of the wavelength ($1/\lambda^4$) of incident light as put forward by Lord Rayleigh. That means the intensity of scattering in the case of a shorter wavelength (blue) is much higher than longer wavelength light toward the red end of the visible spectrum. This is known as Rayleigh scattering, which is responsible for the blue appearance of sky during daytime. Note that the size of a particle and its structure would affect the strength of scattering. As the particle size is significantly less compared to the wavelength of the particle, laws of reflection, refraction, diffraction would not be applicable to the Rayleigh scattering phenomenon. When light falls on a molecule, it absorbs a photon from the incident radiation and thus raises its energy to a virtual state as per the quantum mechanical model. As this virtual state is unstable, the molecule returns back to its original state while emitting one photon of energy. Therefore, there would not be any permanent exchange of energy between light and the particle because the amount of absorbed energy is equal to the amount of emitted energy (see Figure 7.7). Hence, Rayleigh scattering comes under the category of elastic scattering. In spite of a restriction of energy

7. Optical Combustion Diagnostics

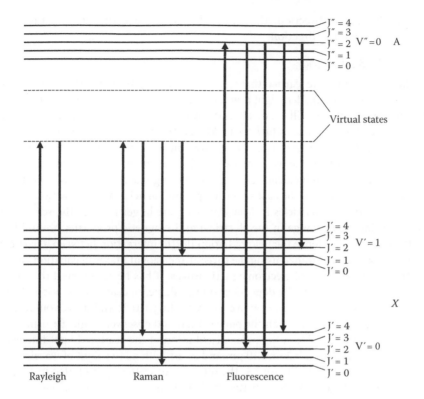

J" = 4
J" = 3
J" = 2 V" = 0 A
J" = 1
J" = 0

Virtual states

J' = 4
J' = 3 V' = 1
J' = 2
J' = 1
J' = 0

X

J' = 4
J' = 3
J' = 2 V' = 0
J' = 1
J' = 0

Rayleigh   Raman   Fluorescence

Figure 7.7

Energy diagrams: Rayleigh, Raman, and fluorescence. (From Mayinger, F. and Feldmann, O. [eds.], *Optical Measurements: Techniques and Applications,* Second Edition, Springer-Verlag, Berlin, 2001.)

exchange, there is a change in direction of incident light. However, there would not be any change in the frequency of incident light, as per Equation 7.1. The scattered light signal would be specific to any particular species in the spectral sense as frequency remains constant during scattering. As a result, it cannot be used for concentration measurements of any species; rather, it can be employed for measurement of total density and temperature in a combustion system. We will be discussing the Rayleigh temperature measurement system in this chapter.

Rayleigh scattering describes the elastic scattering of light by molecules much smaller than the wavelength of light. The Rayleigh scattering model breaks down when the particle size becomes larger than 10% of the wavelength of the incident radiation. In the case of particles with dimensions greater than this limit, the Mie scattering model can be used to find the intensity of the scattered radiation. The intensity of the Mie scattered radiation is given by the summation of an infinite series of terms rather than by a simple mathematical expression. It can be shown, however, that Mie scattering differs from Rayleigh scattering in several aspects; it is roughly independent of wavelength and it is larger in the forward direction than in the reverse direction. The greater the particle size, the more light is scattered in the forward direction.

7.4.2 Mie Scattering

In the above section we learned that Rayleigh scattering from a particle will occur if the wavelength of incident light is greater than the scattering particles ($d/\lambda \ll 1$) but the Mie scattering will occur when the wavelength of incident light is smaller than the scattering particles ($d/\lambda \gg 1$). Mie scattering will occur from dust particles whose diameter varies from 1 to 10 μm whenever visible light falls on them. Note that there will be change in phase, amplitude, and polarization of scattering light compared to the incident light. Apart from light scattering from a particle, reflection, refraction (in the case of a transparent particle), and diffraction are to be considered when particles in Mie scattering are larger in size. The scattering processes induce a change in the electrical and magnetic properties around the scattering particle. Hence the field radiation due to interaction between a particle and light can be determined using the three-dimensional Maxwell equation based on the classical theory of electromagnetic waves. It has been observed that light scattered by particles will be dependent on the shape, size, and orientation of the particle and the ratio of the refractive index of the particle and its surroundings. Scattered light also depends on the polarization and observed angle. By using the Mie theory, one can obtain the polar distribution of the scattered light intensity of an oil droplet with $d = 1$ and 10 μm and in air with a laser light of wavelength $\lambda = 532$ nm, as shown in Figure 7.8. The intensity of scattering is plotted in a logarithm scale so that it can have better readability as a neighboring circle can be depicted even with a variation factor of 100. It can be observed that a particle does not obstruct light; rather, it scatters it in all directions. In other words, the intensity of scattered light varies along the angle of observation. The intensity of scattered light along the forward direction is higher compared to the backward direction. The overall intensity of scattered light increases with increasing particle diameter but there will be rapid fluctuations in the light intensity, as is evident in Figure 7.8. Therefore it is important to choose the right particle size to achieve a higher SNR for a meaningful measurement. The particles need not be spherical all the time. In a practical system, nonspherical particles do exist, and the light scattering from them is quite complex compared to spherical particles. In a real measurement system, the relative scattering intensity is used, which depends on the incident wavelength, diameter of the particle, and direction of observation, as discussed earlier. In most measurement systems, the size and shape of scattering particles vary to the minimum possible level that does not pose any serious problems for measurement.

The Mie scattering is used in several measurement systems apart from flow and flame visualizations. Based on the Mie scattering of laser light, several non-intrusive flow velocity measurements such as LDV and PIV are being developed and used for both flow and combustion problems; these will be discussed in this chapter. The droplet size of a spray and its distribution can be measured based on the Mie scattering technique and soot particles and their distribution in the flame can be determined using the Mie scattering principle.

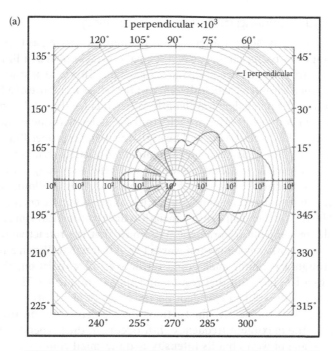

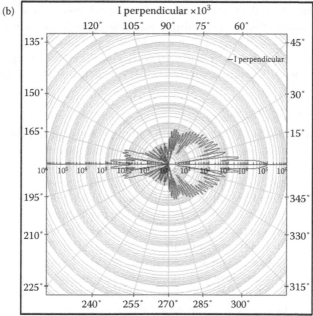

Figure 7.8

Polar diagram of light scattering of an oil droplet in air: (a) diameter = 1 μm and (b) diameter = 10 μm.

7.4.3 Bragg Scattering and Brillouin Scattering

Let us discuss briefly Bragg scattering and Brillouin scattering as they are used in LDV. Bragg scattering can occur in a special crystal known as a Bragg cell in which atoms are arranged in equidistance in a plane. This type of scattering will occur only when the inter-particle distance in the lattice structure of a Bragg cell is of the same order of incident wavelength. When incident light falls on an atom of a crystal, the light emitted from each atom interferes with the incident light. The intensity of Bragg scattering will be high only when a proper Bragg angle is maintained. The typical distance between atoms in a lattice structure is around 1 angstrom (A°), so the scattering light will be in the x-ray range. This scattering method is used in the analysis of solid state materials. Brillouin scattering occurs when light interacts in a medium such as air, water, or crystal in which there will be time-dependent density variation due to acoustical vibration, magnetic mode, or temperature gradient. We know that the refractive index of a medium changes when compression waves travel through it. When a light wave travels through a medium undergoing refractive index variation, the light is subjected to a Doppler shift that results in a shift in frequency. The amount of the frequency shift will be dependent on the angle of observation because the relative velocity of the density fluctuation wave is dependent on the angle of observation. The maximum frequency shift occurs in the perpendicular direction to the plane of waves but its intensity is quite small compared to inelastic scattering. However, the highest intensity can be achieved only when the angle between the incident light and the direction of the acoustical wave is equal to the Bragg angle. Bragg scattering in such a situation is also considered to be inelastic as there is a frequency shift in Brillouin scattering. Therefore, Bragg scattering is used along with Brillouin scattering in LDV, which will also be discussed in this chapter.

7.4.4 Raman Scattering

We have discussed Rayleigh scattering in which photons are scattered elastically from atoms or molecules with the same frequency and wavelength as that of incident light. Apart from Rayleigh scattering of incident light, a small portion of scattering photons are inelastic in nature due to differences in frequency as compared to the incident light. This shift in frequency is known as the Raman shift and this scattering is known as Raman scattering (see Figure 7.8), which causes a change in the energy of the molecules other than that of ground energy. It occurs instantaneously in less than around 10^{-2} s. This type of light scattering was first observed by C. V. Raman and K. S. Krishnan in 1928. Raman scattering is due to the absorption of a photon and its subsequent emission via an intermediate quantum state of molecule. During this interaction of light and molecules, there will be change in energy levels in rotational or vibration or electronic excitation depending on the nature of the energy exchange between the molecule and incident light. The energy diagram depicting rotational and vibrational Raman scattering is shown in Figure 7.9. The rotational quantum number is denoted by J and the vibrational quantum number is denoted by V. The difference between

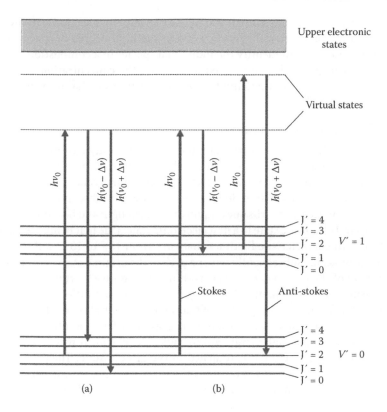

Figure 7.9

Energy diagram for (a) rotational and (b) vibrational Raman scattering.

the two adjacent rotational levels is much lower than that of the vibrational levels, and hence the Raman line due to rotational transition is much closer to the excited line compared to the vibrational line, as shown in Figure 7.9. This Raman shift in frequency will be dependent on the energy difference between the ground and excited levels of the molecules. If the molecule emits a lower energy than the absorbed photon, they will be downshifted in the frequency of scattered light. This type of scattering is called Stokes Raman scattering. On the other hand, if a photon emitted by a molecule is higher than an absorbed photon, there will be upshifted frequency, which is called anti-Stokes Raman scattering. Anti-Stokes Raman scattering can occur only when the energy exchange between the molecule and incident photon can cause sufficient increase in the excited state particularly at an elevated temperature level. Note that the energy difference between the absorbed and emitted photon by the molecule will be dependent on the two resonant states of the molecule. Hence both the Stokes and anti-Stokes spectra form a symmetric pattern above and below the absorbed photon energy. Of course, the intensity of the scattering light will be dependent on the population of the initial state of the molecule. Raman scattering has a low intensity of scattered light that is caused due to a small Raman scattering cross section. However, it is preferred in combustion diagnostics due to the high degree of specificity of Raman spectra, as it can identify the individual species in a combustion system. As a result, this principle can be used to measure the concentration of several species in a measuring volume simultaneously as each species has its own specific Raman frequency shift.

7.4.5 Fluorescence Emission

We know that fluorescence is basically the emission of light from an atom or molecule at an excited state, which can be achieved due to various means such as electron bombardment, chemical reaction (chemiluminescence), and photoabsorption. We will be discussing fluorescence emission by absorption of photon. In this case, a molecule absorbs one photon of incident light and becomes elevated from a ground electronic state to a first electronic state. Due to it being metastable in nature with characteristic lifetimes that vary from 10^{-5} to 10^{-10} s, the excited state of the molecule subsequently returns back to the original state (ground electronic state) of the lower energy level, as shown in Figure 7.9, while undergoing spontaneous emission of light radiations (fluorescence) with different energy and frequency levels. During this process of fluorescence, several types of energy loss due to occurrences such as dissociation, energy transfer to another molecule, energy transfer to internal energy within molecule, chemical reactions, and so forth can be possible before it is returned to the ground state. The process that competes with fluorescence is known as the quenching process, which reduces the signal intensity from fluorescence emission. The emitted fluorescence signal is dependent on the concentration and temperature of a molecule. As the energy difference between excitation and its emission will be specific to each molecule, induced light with proper frequency and wavelength will be chosen depending on the molecule to be excited for the fluorescence signal. Therefore, simultaneous measurement of various species as in Raman scattering cannot be possible using the fluorescence emission method. but it is preferred for combustion diagnostics as it is species-specific with a higher SNR compared to Raman scattering. We will be discussing more about this while learning about the PLIF system for species measurement.

7.5 Flow Visualization

In any combustion (reacting flow) system, we need to be able to visualize the fluid motion in order to gain an in-depth understanding of aerodynamic motion, which is crucial for better understanding the intricate involved phenomena. The basic principle of flow visualization is to make fluid elements visible using various direct and indirect techniques. In some cases, the motion of fluid elements can be observed by using foreign materials such as smoke, small tufts, wood particles, ceramic powders, and chemical coatings for visualization of the flow using illuminating light and a camera. In other indirect methods, optical patterns due to the variation of optical properties such as density, temperature, and refractive index are used to visualize the flow. Some widely used flow visualization techniques are (i) smoke flow visualization, (ii) chemical coating, (iii) Schlieren, (iv) shadowgraph, and (v) interferometer. Some of these methods are also used in reacting flow systems, which are discussed below.

7.6 Flame Visualization

The visualization of flame is a very important experimental tool for understanding the intricate processes in combustion systems. As most combustion processes occur in high temperatures and with higher pressure than in a single/multiphase flow, flame visualization is quite cumbersome compared to the simple flow visualization techniques mentioned above. Some of the features discussed with regard to isothermal flow visualization can be also used for flame visualization. Several methods have evolved over the years for the visualization of flames in a combustion system that can be used for both qualitative and quantitative analysis. Some of them, such as (i) luminous photography, (ii) Schlieren imaging, (iii) the shadowgraph method, (iv) interferometers, and (v) chemiluminescence photography, are discussed briefly in the following sections.

7.6.1 Luminous Photography

Most flames are inherently luminous in nature and radiate visible light originating mainly from the reaction zone. In the case of hydrocarbon flame, species such as CH, HCO, CO, and C_2 radiate visible light [8,9]. It has been reported that the combination reactions between CO and O in the flame region produce a low level of radiation in the range of a wavelength of blue color [9]. For example, let us consider luminious photographs of three laminar CNG-air premixed Bunsen flames at equivalence ratio $\phi = 0.9$ (lean), $\phi = 1.0$ (stoichiometric), and $\phi = 1.3$ (rich), respectively, as shown in Figure 7.10. It can be observed that a lean flame has one luminous

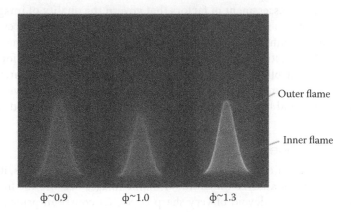

$\phi \sim 0.9$ $\phi \sim 1.0$ $\phi \sim 1.3$

Outer flame

Inner flame

Figure 7.10

Luminous photographs of a CNG-air Bunsen flame.

surface while a rich flame has one inner cone that is enveloped by an outer cone. The inner cone of hydrocarbon flame has a higher intensity to be captured by the camera with less exposure time in the order of seconds. The color of the flame cone surface is blue (shown as white in Figure 7.10) due to the presence of higher concentrations of excited CH* radicals. For a rich fuel-air mixture, the flame cone surface becomes green, which is attributed to the presence of a large concentration of excited C_2 molecules. When the fuel-air mixture becomes rich, the flame also emits yellow color due to the radiation of carbon particles (soot), particularly at high temperatures. In the case of hydrocarbon diffusion flames, the color of the flame is mostly yellow due to the generation of a large amount of soot. As a flame provides most of the radiation in the visible range due to bimolecular combination reaction of radicals, the luminous zone can be used as a signature for proper identification of the flame surface. The intensity of the luminosity is strongly dependent on the concentration of radicals that occur in the reaction zone; therefore a visible flame photograph corresponding to peak luminosity can provide qualitative information about the peak reaction zone. From luminous photographs of flames, the peak radical concentration can be measured qualitatively in a relative sense. The flame thickness can be measured crudely from the images of a luminous flame.

There are several complications that arise due to direct photography of visible flames. The visible plane images will be biased as the flame surfaces do not lie in the single focal plane. Flame image also suffers optical distortion due to the presence of stronger refractive index gradients as the temperature gradient prevailing across the flame is quite stiff. The refractive index is inversely proportional to the pressure and becomes negligibly small at a lower pressure (20 kPa); this effect can be minimized by using a sharply focused lens while blocking out the central cone of light. It is important to design a photographic system for flame studies that consider the requisite resolution of flame images, allowable distortion, exposure time limitation, flame brightness, and so forth. With this objective assessment of factors that govern the quality of images and analysis, one can choose appropriate optical systems (lens, mirrors, etc.), and shutter speed, magnification ratio, and type of camera, to achieve the desired objectives of the investigation. As discussed earlier, the exposure time for taking photographs is controlled by the shutter speed of the camera. For research work, we prefer a simple camera that allows manual adjustment and has a wide range of shutter speeds and higher pixel levels rather than a fully automated modern camera.

7.6.2 Chemiluminescence Photography

In order to take certain quantitative measurements from flame images, it would not be that simple to analyze the luminous flame images because the flame has finite thickness and is devoid of very sharp edges. Of course in principle, it is possible to measure flame thickness from luminous flame images by using the intensity of light, but the proper criteria must be chosen. Another way to identify flame surface is to use chemiluminescence photography [8]. The intensity of CH*

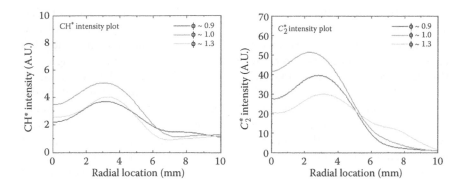

Figure 7.11

Variation of intensity from C_2^* and CH* along the radial direction.

and C_2^* from the chemiluminescence photographs (not shown) is plotted along the radial direction for CNG-air premixed flame (Figure 7.11). Note that both the peak values of CH* and C_2^* radicals occur at the flame surface itself and also varies with equivalence ratios.

7.6.3 Optical Flame Visualization

In the above section, we discussed direct flame photography, which has the limitation of not being able to obtain a sharp image. In order to overcome this problem, several optical flame visualization techniques have been designed and developed. We will be discussing three major flame visualization methods: (i) Schlieren, (ii) shadowgraph, and (iii) interferometry, which are used routinely by researchers for the analysis of flame. These methods are mainly based on the principle of the variation of a refractive index of light when it is passed through a flow field with a density gradient. The density gradient in the flow field can be caused by compressibility of flow, turbulent mixing, temperature gradient, concentration gradient, and so forth. But in the case of flame, a higher density gradient exists due to the large temperature gradient across the flame. Apart from temperature gradient due to the heat release in the flame, turbulent mixing can also cause density nonhomogeneity in the flame, which can be visualized by using optical flame visualization methods. In order to visualize these changes in density of the flow field, three techniques are employed: (i) Schlieren, (ii) shadowgraph, and (iii) interferometer, as mentioned above. The basic principles of these three methods are briefly discussed below.

7.7 Light Refraction in an Inhomogeneous Medium

Before venturing into the study of Schlieren and shadowgraph visualization techniques, let us understand how the bending of a light beam takes place due to a change in density. We know that the light beam gets deflected by the change in the refractive index of the medium along a certain direction. The amount of deviation of the light beam can be determined both by wave and geometric theory. We will resort to the geometric theory as it is simple to enumerate the basic principles and it is good enough to allow determination of the deviation angle. For this, let us consider a parallel beam of light travelling along the z-direction as shown in Figure 7.12, through a medium whose refractive index varies along the y-direction. Let us consider the light beam in position (z_1) at time (t), which has moved through a distance Δz at time $t + \Delta t$ (see Figure 7.12). During this time, Δt light beam, traveling through an inhomogeneous at the speed c_0/n, has turned by angle $\Delta\alpha$. Considering the deviation to be small, the distance traveled Δz during time Δt can be related to the value of the local speed of light (c_0/n) as given by

$$\Delta z = \frac{c_0}{n}\Delta t \tag{7.3}$$

where c_0 is the speed of light in the vacuum and n is the refractive index. From Figure 7.12, the deflection angle of the light beam can be expressed as

$$\Delta\alpha = \frac{\left(\dfrac{c_0}{n_2} - \dfrac{c_0}{n_1}\right)}{\Delta y}\Delta t \tag{7.4}$$

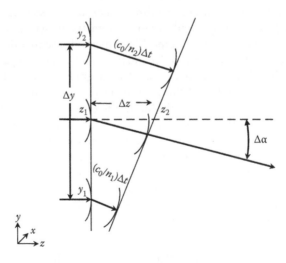

Figure 7.12

Schematic of elemental light refraction by refractive index gradient.

7. Optical Combustion Diagnostics

By combining Equations 7.3 and 7.4, we get,

$$\Delta\alpha = \frac{n\left(\dfrac{c_0}{n_2} - \dfrac{c_0}{n_1}\right)}{c_0 \quad \Delta y}\Delta z \tag{7.5}$$

By simplifying the above equation, we have

$$\Delta\alpha = \frac{n}{n_1 n_2}\frac{(n_1 - n_2)}{\Delta y}\Delta z \tag{7.6}$$

Note that the term $\dfrac{n}{n_1 n_2}$ can be approximately equal to $\dfrac{1}{n}$, when Δy is tending toward zero. With this limiting condition, the above equation becomes

$$\frac{d\alpha}{dz} = \frac{1}{n}\frac{(n_1 - n_2)}{\Delta y} = \frac{1}{n}\frac{dn}{dy} \tag{7.7}$$

As the deflection angle α is small, this angle can be approximated as the slope of dy/dz, and then the above equation can be expressed as

$$\frac{d^2 y}{dz^2} = \frac{1}{n}\frac{dn}{dy} \tag{7.8}$$

For a two-dimensional gradient field, there will be a change in the refractive index along the x and y directions, and then the above equation can be expressed in partial derivative form as

$$\frac{\partial^2 y}{\partial z^2} = \frac{1}{n}\frac{\partial n}{\partial y} \tag{7.9}$$

For the other component along the y-direction, we have

$$\frac{\partial^2 y}{\partial z^2} = \frac{1}{n}\frac{\partial n}{\partial x} \tag{7.10}$$

For a small angle, Equation 7.9 is valid for the entire light path through the test (disturbed) section (region). If the entering angle is zero, the angle at the exit of the test section can be expressed as

$$\alpha = \int \frac{1}{n} \frac{\partial n}{\partial y} dz \qquad (7.11)$$

7.8 Schlieren Methods

The Schlieren method of flow visualization used routinely for flame visualization derives its name from the German word "Schliere" (plural Schlieren), which means "inhomogeneous regions/medium." The inhomogeneous medium may be caused by hot air rises in the flame due to change in its density, leading to a change in the refractive index. Similar effects resulting in changes in the refractive index of light can also occur due to concentration gradient (e.g., dissolving of salt in water, a change in thickness in a glass plate). In other words, this method is based on the basic principle of visualizing the density inhomogeneity (gradient) in a flow. This method of flow visualization was first introduced by German physicist August Toepler in 1864 while studying supersonic flow and it will be discussed briefly. This method of flame visualization is preferred over the luminous photography method as it has several advantages such as a sharp flame surface, the ability to identify the structure of a flame and turbulence structure, and the least exposure time. In the case of flame visualization by the Schlieren method, the requisite exposure time is in the order of 10^{-3} s during which a direct flame photograph cannot be captured due to insufficient light emanated from the flame surface. The Schlieren method is also routinely used in turbulent flow for capturing flow structure due to turbulence, onset of turbulence, transition of flow, aerodynamics for location of shock fronts, determination of detonation velocity, and so forth. Several methods of different optical arrangements have been designed and developed for Schlieren visualization of flow and flame [7–9]. We will restrict our discussion to two widely used systems: (i) two-lens Schlieren arrangement and (ii) two-mirror Schlieren arrangement, which will be discussed next.

7.8.1 Two-lens Schlieren Arrangement

The schematic of a typical two-lens Schlieren system is shown in Figure 7.13, which consists of a light source, condensing lens, slit, collimating lens (L_1) and Schlieren head (L_2) [10], knife edge, and camera. The light from its source is passed through a condensing lens so that it can pass through a slit to produce a well-defined bright source with sharp boundaries at the focal plane of the first Schlieren lens (L_1) (see Figure 7.13). Then the collimated beam of light passes through the flame in the test section, which is subsequently focused

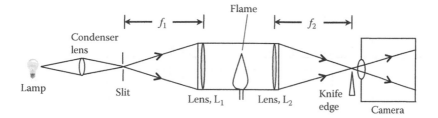

Figure 7.13

Two-lens Schlieren optical system.

with the second Schlieren lens (L_2 in Figure 7.13) onto a knife edge (opaque mask). The function of the knife edge is to cut off light from the test section so that the density gradient due to a change in the refractive index can be captured completely. The camera is placed just behind the knife edge, which can capture Schlieren images of flame using its focal lens for obtaining suitable flame images. As mentioned earlier, the variation of the refractive index in the flame will cause the light beam to deviate so that it can pass through the knife edge and subsequently can be captured by the camera. We should recall that the light beam is deviated in the direction of the increasing refractive index. In other words, light is tuned toward the region of higher density. For example, light passing through a flame will be bent toward the unburned side of the flame. Generally, the typical turning angle of a light beam in a Schlieren system that can be captured on a screen (x-y plane) is quite small, in the order of 10^{-3} to 10^{-6} radians. In order to understand how a knife edge helps in enhancing the sharpness of Schlieren images, we can consider an optical ray diagram in a typical Schlieren system as shown in Figure 7.14a. Let us assume the light is a cross section with a dimension of a_s and b_s and becoming a parallel beam of light while passing through lens L_1. If no disturbance is present in the test section, the light beam forms a cross section with dimension a and b at the focus of lens L_2, as shown in Figure 7.15a, which can be related to the initial light source dimensions as

$$\frac{a}{a_s} = \frac{b}{b_s} = \frac{f_2}{f_1}$$

(7.12)

where f_1 and f_2 are the focal length of two lenses L_1 and L_2, respectively (see Figure 7.13). The light intensity, I, arriving at any location (x, y) of the screen remains almost constant when there is no disturbance in the test section. This intensity of light can be related to the initial intensity of light, I_s, at the source, as discussed earlier. When the light beam at any position in the test section is deflected by angle α (see Figure 7.14) due to presence of a flame, then the light

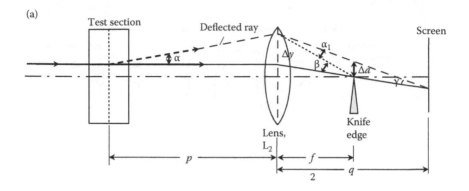

(a)

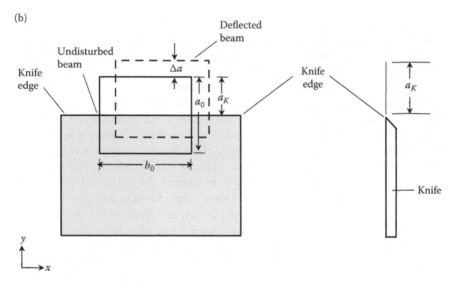

Figure 7.14

(a) The deviation of an optical ray at a knife edge and (b) deflected beams at the knife edge.

beam on the plane of the knife edge is shifted by Δa, which can be related to deflection angle α and f_2, as

$$\Delta a = \alpha_1 f_2 = (\beta - \gamma) f_2 = \Delta y \left(\frac{1}{f_2} - \frac{1}{q} \right) f_2 = \frac{\Delta y}{P_s} f_2 = \alpha f_2 \tag{7.13}$$

Note that this change in the light beam can be positive or negative depending on the orientation of the knife edge. This change in light beam Δa does not

7. Optical Combustion Diagnostics

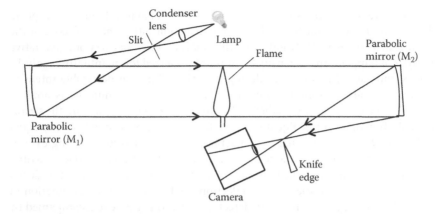

Figure 7.15

Twin-mirror Schlieren optical system.

depend on the distance between the test section and lens L_2. The change in light intensity on the screen can be expressed as

$$\Delta I = I_d - I_K = \frac{\Delta a}{a_K} I_K \qquad (7.14)$$

Then the relative intensity with respect to the light source known as contrast can be expressed as

$$\frac{\Delta I}{I_K} = \frac{\Delta a}{a_K} = \frac{\alpha f_2}{a_K} \qquad (7.15)$$

Assuming a two-dimensional field, a local change in light intensity in the Schlieren image can be related to the gradient of the refractive index by using Equation 7.11 as

$$\frac{\Delta I}{I_K} = \frac{f_2}{a_K} \int \frac{1}{n} \frac{\partial n}{\partial y} dz \qquad (7.16)$$

For a flame, the above equation can be expressed as

$$\frac{\Delta I}{I_K} = \frac{f_2}{a_K n} \frac{n_0 - 1}{\rho_0} \int \frac{\partial \rho}{\partial y} dz \cong \frac{f_2}{a_K} \frac{n_0 - 1}{\rho_0} \int \frac{\partial \rho}{\partial y} dz \qquad (7.17)$$

On several occasions, we can consider $1/n$ to be equal to 1. In order to capture the density change in the flame along the x-direction, either the knife edge or the flame is to be turned 90°. By following this analysis, one can carry out qualitative analysis of temperature or density but it is not preferred for quantitative analysis. Let us understand this by considering that the minimum detectable intensity change (contrast) is around 15% but for a large contrast, a/f_2 must be as small as possible for a given density gradient. Thus for a given density gradient, the smallest deflection angle α (see Equation 7.15) that could be measured becomes equal to 0.15 (a_K/f_2). We know that for a particular Schlieren system during experiments, we cannot change the focal length of lens L_2. The knife edge aperture, a_K, cannot be reduced as the light intensity of the Schlieren image gets reduced with it. The change in the sign of deflection angle α affects the illumination of the Schlieren image, which can be detected easily as negative α accompanied by a decrease in illumination and vice versa. As well, this Schlieren system is a bit erroneous due to assumptions of no optical aberration but it is used routinely for the identification of a flame surface. Thus, both the burning velocity in the case of a premixed flame and the flame height in the case of a jet diffusion flame are being routinely determined from Schlieren images.

In this Schlieren system, a good-quality achromatic lens must be used to avoid the problem of chromatic aberration. Even in an expensive chromatic lens, one can face the problem of residual chromatic aberration due to partial spectrum cutoff by the knife edge. One way to overcome this problem is to use monochromatic light sources such as mercury lamps, sodium lamps, or laser. Another way of avoiding this problem is to use a narrowband color filter at the light source itself. However, for large flames like those in a furnace, it is quite cumbersome to have a larger diameter lens with a higher focal length, which is also quite expensive. Hence, instead of a lens, mirrors are used in the Schlieren system. A typical twin-mirror Schlieren system is discussed briefly in the next section.

Example 7.1

In a Schlieren system, lens 2 with a focal length of 100 mm is used for making the ray from a test section fall on the knife edge. The distance between lens 2 and the camera is 150 mm. If the minimum depth of light, a_K, not being cut off by the knife edge is equal to 6 mm, determine the deflection angle α_{max} subtended by the deflected ray between the test section and lens 2 for maximum contrast and distance p between the test section and lens 2.

Given: $f_2 = 100$ mm, $a_K = 6$ mm, $q = 150$ mm.

To Find: α_{max}, p.

Solution: By applying a lens expression for lens 2 at a focused condition, we have

$$\frac{1}{f_2} = \frac{1}{p} + \frac{1}{q}; \quad \Rightarrow p = \frac{f_2 q}{q - f_2} = \frac{100 \times 150}{150 - 100} = 300 \, \text{mm}$$

The sensitivity of this Schlieren system for determining the deflection angle can be obtained by diffentiating Equation 7.15 with respect to α, as given below:

$$\frac{d}{d\alpha}\left(\frac{\Delta I}{I_K}\right) = \frac{f_2}{a_K}$$

We know that for a particular optical system, the contrast can be maximized by minimizing a_k by the movement of the knife edge. For a minimum a_K value the maximum deflection angle would be

$$\alpha_{max} = \frac{a_K}{f_2} = \frac{6}{100} = 0.06 \text{ rad}$$

7.8.2 Twin-mirror Schlieren Arrangement

In order to take the Schlieren images of large flames, two spherical parabolic concave mirrors are generally used in place of the two-Schlieren lens (see Figure 7.15) because the lens can suffer from chromatic aberrations. Besides this, large-diameter mirrors with longer focal lengths are cheaper and easier to manufacture compared to lenses. Hence, a twin-mirror Schlieren system has been designed and developed that is used for scientific investigation of combustion systems. A typical twin-mirror Schlieren system is shown in Figure 7.15, which consists of a light source, condensing lens, slit, two Schlieren mirrors, knife edge, and camera. Figure 7.15 shows that the twin-mirror system is quite similar to that of the twin-lens system. For example, they have a similar light-source arrangement in which light is passed through a slit with a condensing lens to produce a well-defined light source with sharp boundaries at the focal plane of the first Schlieren mirrors. But the two opposite-paired parabolic mirrors M_1 and M_2 are tilted by the same angles in order to avoid distortion of the light source images, which is known as coma. As a result, the light source and its images are positioned off-axis from the axis between the centers of the two

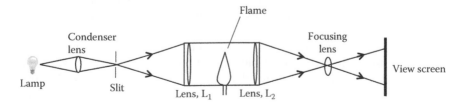

Figure 7.16

Lens shadow optical system.

mirrors. In other words, the angle between the axis (parallel beam) and the light source is the same as that between the parallel beam and the camera (see Figure 7.16). The angle must be as small as possible and hence its value is restricted up to 7° for getting better-quality images. The distance between the two mirrors is not very critical but greater than twice the focal length of the mirror is used routinely. It is customary to maintain a higher distance between mirror M_2 and the flame (test section) and that of the focal length of mirror M_2 because the parallel beam entering into the flame (test section) will no longer be parallel, as shown in Figure 7.15, due to variations in the refractive index. As discussed earlier, the camera, which can capture images of the flame using its focal lens in order to get suitable flame images, is placed just behind the knife edge. The sharpness of the images can be improved by adjusting the position and orientation of the knife edge. The knife edge is placed vertically so that its movement along the vertical direction can be adjusted easily. As a result, the position of the knife edge can be altered with respect to images and hence the sensitivity of the image can be enhanced easily. This type of twin-mirror Schlieren system, which consists of a diverging illuminating beam starting from a light source, a parallel light beam flanked by twin mirror pairs, and a converging light beam ending at the camera, resembles the English letter "Z" and therefore is often known as a Z-type twin-mirror Schlieren system.

7.9 Shadowgraph Methods

The shadowgraph method is one of the simplest forms of flame visualization methods used in industries and laboratories. You may have observed a shadow image from a plume rising from a hot body when either light from the sun or any other source is passed through it and projected into a screen because any change in temperature in the typical plume of a hot body causes a change in the refractive index. As a result the light rays passing through the plume get deflected, forming an image with sharp features. Similarly, when a light beam passes through the flame, there will be a deviation of the light beam's resulting redistribution of light intensity on the screen or camera due to a change in the density of hot gases in the flame. Of course, the deviation of the light beam cannot be the same if the light is being absorbed in the flame due to soot formation or certain gas molecules. The change in the light intensity on the screen is possible only if there will be changes in the refraction angle along the spatial domain. In other words, the shadow depth and its pattern are dependent on the change in the refraction angle rather than the angle itself as in the Schlieren image.

Shadowgraphs of the flame can be obtained by using several optical systems. We will be discussing two generic widely used optical systems: (i) the two-lens shadowgraph and (ii) the two-mirror shadowgraph, as shown in Figures 7.16 and 7.17, respectively. Both contain a light-beam arrangement that consists of a light source, condensing lens, and slit, as discussed earlier. In the case of a diverging direct shadowgraph optical system (see Figure 7.16), a light beam passing through the test section (flame) gets distorted due to the density gradient and falls on the screen/camera at a certain distance from it. Note that the flame (test section) is placed between the slit (aperture) of the light source and the screen at a certain distance from it. But in the case of a parallel direct shadowgraph optical system, light beams are made parallel that pass through the test section (flame) and fall on the screen/camera (Figure 7.16) at a certain distance from the flame after becoming distorted by the flame. The parallel beam shadowgraph system is preferred more due to its higher sensitivity because the illumination at the receptor plane (screen/camera) is independent of the distance between the flame and the camera. As discussed earlier, we know that the size of a collimated lens with a higher focal length must be large enough to capture a huge flame, and is quite difficult to manufacture and also quite expensive. Hence a collimated lens is substituted with a parabolic mirror, as shown in Figure 7.17. This parallel-light mirror shadowgraph imaging system is basically half of the Z-type twin-mirror Schlieren system, as shown in Figure 7.15.

We know that a light beam gets deflected by the density gradient in a test section. Hence, it is expected that the deflected light beam will affect the illumination on the shadow image because any photograph records the relative value of the light intensity that arrives on it. We need to examine the change in light intensity due to the shadow effect in a similar manner as discussed for the Schlieren image. For this purpose let us consider the schematic of a light beam deflected by the density gradient, as shown in Figure 7.18, in which we will have to consider the deflection of the light beam at the center of the test section. Note that the parallel light beam entering into the test section would not remain parallel due to deflection. Let α be the deflection angle at the center that varies along the y-direction. The width of two parallel beams Δy become Δy_{sc} at the screen placed at distance L_{sc} from the test section. The light intensity at the screen I_{sc} can be related to the initial light intensity at flame I_t as

$$I_{sc} = I_t \frac{\Delta y}{\Delta y_{sc}}. \tag{7.18}$$

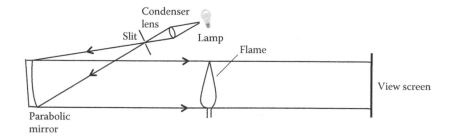

Figure 7.17

Mirror shadow optical system.

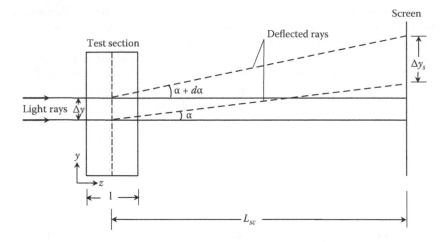

Figure 7.18

Deflection of a parallel light beam due to the shadow effect.

But the Δy_{sc} can be related to Δy and length L as

$$\Delta y_{sc} = \Delta y + L_{sc} d\alpha \tag{7.19}$$

Then the contrast of the shadow image can be expressed as

$$\frac{\Delta I}{I_t} = \frac{I_{sc} - I_t}{I_t} = \frac{\Delta y}{\Delta y_{sc}} - 1 \cong -L_{sc} \frac{\partial \alpha}{\partial y} \tag{7.20}$$

By using Equation 7.11, the above equation can be expressed in terms of the refractive index as

$$\frac{\Delta I}{I_t} = -\frac{L_{sc}}{n_a} \int \frac{\partial^2 n}{\partial y^2} dz \tag{7.21}$$

If there is variation of n along the x-direction, then the expression for contrast would be

$$\frac{\Delta I}{I_t} = -\frac{L_{sc}}{n_a} \int \left(\frac{\partial^2 n}{\partial x^2} + \frac{\partial^2 n}{\partial y^2} \right) dz \tag{7.22}$$

Note that the variation of the refractive index along both the x- and y-directions can be captured by the shadowgraph, unlike in the Schlieren photograph, as the variation normal to the knife edge is captured by the Schlieren image.

7.10 Interferometry

We have already discussed in the shadow and Schlieren methods how the deflection of a light beam due to the variation in density can be used for flame visualization. These two flame visualization methods are used for qualitative measurements. Now we will initiate a discussion about flame visualization using an interferometer, which is used for the quantitative determination of density and temperature field of a flame. Note that the refraction effect used in the shadow and Schlieren methods must be avoided in an interferometer. Rather, the change in the refractive index due to a change in the density of the flame is determined in this method for the quantitative measurement of density and temperature for which the reference density/temperature must be known. The interference of light waves is used for flow and flame visualization in this method. When two monochromatic light waves from the same source but traveling in different paths are reunited, dark and bright regions with formation of fringe patterns are formed on the screen. Based on these inferences of light waves, several types of interferometers have been designed and developed for flow and flame visualization. In this book, we will restrict our discussion to the Mach-Zehnder interferometer, as it can have a larger displacement of a reference beam from the flame (test section) itself compared to other interferometers. As well, it can have sharp images because the light beam can pass only once through the flame (disturbed) region. Therefore, the Mach-Zehnder interferometer is preferred over other interferometers in the studies of combustion systems for quantitative measurements of density and temperature on the flame surface itself.

Let us consider the optical system of a Mach-Zehnder interferometer, as shown in Figure 7.19, which consists of a monochromatic light source, a condenser lens, two beam splitters (BS_1 and BS_2), two mirrors (M_1 and M_2), and a screen/camera. We know that a monochromatic light source along with a condensing lens can produce a parallel beam of light. This parallel light beam strikes the beam splitter and gets divided into two parallel light beams covering two different optical paths, as shown in Figure 7.19. The light beam travels path 1 while passing through the flame (test section) after getting reflected through the mirror M_1 while the other parallel light beam passes through path 2 and gets reflected by 90°. The second beam splitter helps in combining both light beams that travel different paths to fall on the screen/camera and thus produces interference fringe patterns. Note that two beam splitters and two mirrors are placed on the corner of a rectangle for maintaining parallel light beams in this optical setup.

In order to understand the interference fringe pattern and how it can be used to measure properties like temperature and density quantitatively, we need to invoke the wave theory of light instead of the geometric theory of light as discussed in the Schlieren and Shadowgraph systems. For this purpose, we need to assume that the light beams in both path 1 and 2 are parallel and thus there would not be any variation in optical properties normal to any of these light beams. This can be ensured provided mirrors and beam splitters are placed perfectly parallel to each other. All the optical components must be of higher quality.

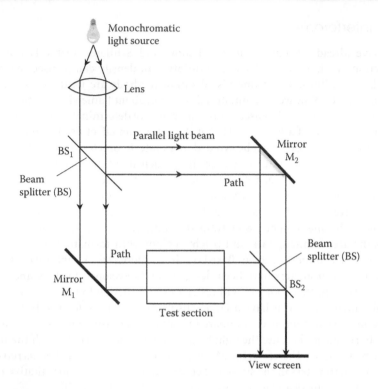

Monochromatic
light source

Lens

Parallel light beam

Mirror
M_2

BS_1

Beam
splitter (BS)

Path

Beam
splitter (BS)

Path

BS_2

Mirror
M_1

Test section

View screen

Figure 7.19

Schematic of a Mach-Zehnder interferometer.

Light beams arriving at a screen/camera must be in phase with each other for the basic configuration of the interferometer in which no combustion system (flame) is placed in the test section. Therefore, perfection is called for in setting up a requisite optical system of an interferometer. Two beams of light emerging from the beam splitter BS_2 must be parallel and fall on the screen/camera without an iota of distortion. From the wave theory of light, the amplitude of a plane light wave passing through a homogeneous medium can be modeled as

$$A = A_0 \frac{2\pi}{\lambda}(ct - z) \tag{7.23}$$

where A_0 is the maximum amplitude, λ is the wavelength of light, c is the speed of light, t is the time, and z is the distance. Let us consider the amplitude of light beam 1 at beam splitter BS_2 (see Figure 7.19) that can be represented as

$$A_1 = A_{01} \sin\left(\frac{2\pi ct}{\lambda}\right). \tag{7.24}$$

7. Optical Combustion Diagnostics

Similarly the amplitude of light beam 2 at the same beam splitter BS_2 can be represented as

$$A_2 = A_{02} \sin\left(\frac{2\pi ct}{\lambda} - \delta\right)$$ (7.25)

where δ is the phase difference due to a change in the path length for both beams and λ is the wavelength of light. Note that both light beams are coherent as they are emanated from the same source. Hence, they can interfere with each other only if the phase difference δ is not a function of time. These two waves can be added to form a new wave that is represented as

$$A_N = A_1 + A_2 = A_0 \left[\sin\left(\frac{2\pi ct}{\lambda}\right) + \sin\left(\frac{2\pi ct}{\lambda} - \delta\right) \right]$$ (7.26)

where A_N is the amplitude of the new wave, which is the same as that of both beams ($A_0 = A_{01} = A_{02}$). The above equation can be simplified and read as

$$A_N = 2A_0 \cos\left(\frac{\delta}{2}\right) \sin\left(\frac{2\pi ct}{\lambda} - \theta\right)$$ (7.27)

We will be more interested in the intensity of the combined light beam that is falling on the screen as it is being used to relate a physical quantity. It can be noted from the above equation that the intensity of the new combined light wave is proportional to the square of the peak amplitude as expressed as

$$I \propto 4A_0^2 \cos^2\left(\frac{\delta}{2}\right)$$ (7.28)

It can be noted from Equation 7.28 that the resultant intensity can be maximum when the phase difference δ is equal to $2n\pi$ ($\delta = 2n\pi$) and can be minimum ($\delta = (2n + 1)\pi$), where n is an integer. In the first case, the resultant beam becomes constructive interference (bright fringe) as intensity I is equal to four times the initial intensity. In contrast, in the second case, it results in destructive interference (dark fringe) of light beams as intensity I becomes zero. The optical path length PL along a light beam can be determined as

$$PL = \int n\,dz = \int \frac{c_0}{c}\,dz = \lambda_0 \int \frac{dz}{\lambda}$$ (7.29)

where c_0 is the speed of sound in a vacuum. It can be noted from Equation 7.29 that optical path length PL is equal to vacuum wavelength times actual path length with any other wavelength through an actual medium. Let us now see how optical path difference *PD* can be changed due to inhomogeneity created by placing a flame in the test section perpendicular to the parallel light beam (see Figure 7.19):

$$PD = PL_1 - PL_2 = \lambda_0 \left[\int \left(\frac{dz}{\lambda} \right)_1 - \int \left(\frac{dz}{\lambda} \right)_2 \right] \tag{7.30}$$

In the case of a Mach-Zehnder interferometer, when a flame is placed in the test section leading to inhomogeneity in density, there will be a path difference. By neglecting the diffraction on the cross section of the light beam, the change in optical path difference, *PD*, with respect to the vacuum wavelength λ_0 can be expressed by using Equation 7.30 in terms of refractive index change as

$$\varepsilon = \frac{PD}{\lambda_0} = \frac{1}{\lambda_0} \int (n - n_r) dz \tag{7.31}$$

where ε is the interferometer fringe shift, n is the refractive index, and n_r is the refractive index of the reference light beam 2. If the interferometer fringe shift, ε, becomes an integer, the fringe becomes bright, and when ε is half an integer, the fringe becomes dark. Hence when a flame is introduced in the test section, the initially uniform field becomes a series of bright dark fringes whose width will be dependent on the value of the interferometer fringe shift, ε. The fringe width can be measured from the photograph, which can be used to analyze the flame visualization both qualitatively and quantitatively. Let us consider now how temperature at any location of an interferogram can be evaluated. For this purpose, we need to invoke the Gladstone-Dale relation, as given below:

$$n - 1 = \rho C \tag{7.32}$$

where C is the Gladstone-Dale constant, which is dependent on the type of gas and varies slightly with wavelength. Then, combining Equations 7.31 and 7.32, we can have a relation for the interferometer fringe shift in terms of the density of gas as given below:

$$\varepsilon = \frac{C}{\lambda_0} \int (\rho - \rho_r) dz \tag{7.33}$$

For a two-dimensional field, the refractive index and thus density varies along the light beam (*z*-direction) and the above equation for the number of fringe shifts becomes

$$\varepsilon = \frac{C}{\lambda_0} (\rho - \rho_r) L; \quad \rho - \rho_r = \frac{\lambda_0 \varepsilon}{LC} = \frac{\lambda_0 \varepsilon}{L(n_0 - 1)} \rho_0 \tag{7.34}$$

7. Optical Combustion Diagnostics

For an ideal gas and isobaric process, Equation 7.33 can be expressed in terms of T as

$$\frac{P}{R}\left(\frac{1}{T}-\frac{1}{T_r}\right)=\frac{\lambda_0\varepsilon}{LC};\ T=\frac{PCLT_r}{PCL+\varepsilon\lambda_0RT_r} \tag{7.35}$$

where R is the specific gas constant, T_r is the reference temperature, P is the pressure, and C is the Gladstone-Dale constant. We can determine the temperature of gas by knowing other terms in the above expression. For example, we can determine the fringe shift in an interferogram created by a Mach-Zehnder interferometer of the light source of wavelength λ_0 of 632.8 nm (He-Ne CW laser) and a path length of 20 cm. At the reference temperature of air at 303 K, 0.1 MPa, and $C = 2.36 \times 10^{-4}$, the measured temperature would be 299.5 K. Note that a suitable interferogram can be achieved when the optical path difference is less than half of the wavelength of the light source. For flame application, it will be difficult to achieve this as the prevailing temperature is quite high. For quantitative measurements, it is important to know the properties of undisturbed regions. Besides this, it is not quite suitable for high-temperature flame regions because higher density differences do exist in the flame region. Therefore, it is used in a restricted sense for combustion systems.

Example 7.2

An interferometer with a light source ($\lambda = 532$ nm) is employed to visualize the plume during the combustion of a droplet under quiescent conditions at ambient pressure. The test section with a depth of 100 mm width is placed 300 mm above the flame. The hot gas temperature is 200°C while the ambient temperature is 35°C. Determine the number of fringe shifts by considering the Gladstone-Dale constant for plume gas to be $C = 2.31 \times 10^{-4}$ m³/kg.

Given: $\lambda = 532$ nm, $L = 100$ mm, $H = 300$ mm, $T = 200°C = 473$ K, $T_r = 35°C = 308$ K, $P = 101\ 325$ Pa.

To Find: Number of fringe shifts.

Solution: By using Equation 7.34, the number of fringe shifts can be determined as

$$\rho-\rho_r=\frac{\lambda_0\varepsilon}{LC_0};\ \Rightarrow\varepsilon=\frac{LC}{\lambda_0}(\rho-\rho_r)$$

But we need to determine the reference density by assuming the gas to obey the ideal gas law as

$$\rho=\frac{P}{RT}=\frac{101325}{287\times473}=0.746\ \text{kg/m}^3$$

Similarly we can determine the reference density by assuming the gas to obey the ideal gas law as

$$\rho_r = \frac{P}{RT} = \frac{101325}{287 \times 308} = 1.146 \text{ kg/m}^3$$

7.11 Laser-Based Optical Velocimetry

In the past few sections we learned to appreciate the efficacy of optical techniques within flow and flame visualization techniques that can be used for both qualitative and quantitative measurements without disturbing the flow and flame structure. Generally these methods are restricted to the measurement of temperature and density of flame. But the flow field is also equally important to understand the complex structure of flame and combustion systems. It is quite important to know the local velocity for a flame structure. We have also learned that the local velocity in a flow field can be measured using intrusive measurement systems like pitot-static and hot wire anemometers, which were covered in Chapter 6. Some of the flow and flame visualization techniques may be adopted to determine the velocity profile but they have to undergo major modifications for measurement of the local velocity in a combustion system. We know that the local velocity can be obtained from the continuity equation from the stream tube geometry through flow visualization provided the initial condition for mass conservation is known. Several laser-based optical techniques have been developed for the measurement of local velocity in both isothermal and reacting flow fields due to the following advantages over other intrusive measurement systems:

- Laser-based optical velocimetry can measure local fluid velocity without disturbing the flow even in a smaller measuring volume provided the flow medium is transparent with adequate concentration of seeding particles.
- The velocity measurement is not affected by the other physical parameters of temperature and pressure in combustion systems because the optical electromagnetic waves are quite linear and stable.
- The measuring direction of the fluid flow can be ascertained easily in optical methods.
- Higher spatial and temporal resolution can be achieved easily due to smaller measuring volume and faster signal-processing electronics. The time-resolved measurement of fluctuating velocity can be achieved easily.
- Three-dimensional velocity measurements can be possible, of course with added efforts.

As a result, several types of laser-based optical velocimetry methods have been designed and developed over the past four decades. The most common direct techniques for measurement of local flow velocity are based on the tracking of suspended particles in the flow. These particles are introduced deliberately and carefully so that they can move with almost the same velocity as that of the local fluid velocity. The path traced by these particles through the flame with time is visualized using laser light and a powerful camera and can be processed to determine the local fluid velocity. Two of these methods, (i) LDV and (ii) PIV, will be discussed briefly below.

7.12 Laser Doppler Velocimetry (LDV)

LDV is one of the most widely used methods of measuring local fluid velocity in both isothermal and reacting flow systems. The fundamental principle for the measurement of local velocity is based on the Doppler effects of light scattering from the moving particles in the flow field. The local velocity can be measured by sensing and measuring the Doppler frequency shift of laser light scattered from moving particles. It can provide information not only about steady velocity but also the fluctuating local velocity in a flow field. Hence it is preferred for characterizing a turbulent flow field over other laser-based methods of optical velocimetry. It has several advantages over other traditional methods such as pitot-static probes and hot wire anemometers, which are enumerated below:

- The local fluid velocity is measured directly without any inference either from pressure measurement in the case of a pitot-static probe or the heat transfer coefficient in the case of a hot wire anemometer.
- The flow would not be disturbed by this measurement as it is an optical technique. However, proper particles and its distribution and loading must be controlled so that it would not affect the flow field and combustion environment.
- A measuring volume as small as 0.2 mm³ can be easily possible.
- The frequency response in order of megahertz is easily achievable.

7.12.1 Doppler Effect

We know that LDA uses the principles of the Doppler frequency shift of laser light scattered from moving seeded particles. Let us first understand the concept of the Doppler effect by considering a small particle moving with a velocity, V_p. If this particle scatters or reflects light with velocity c and frequency f, which is detected by a stationary observer located at a point along the direction of velocity, then the distance traveled by the light waves toward the observer during time t is equal to $(c - V_p)t$. Note that the number of light waves emitted during time t would be ft. The light received by the observer has a different frequency, f_0, than that of the source due to the Doppler effect, which can be expressed as

$$f_0 = \frac{cft}{(c - V_p)ft} = \frac{cf}{(c - V_p)} \tag{7.36}$$

Let us consider now the other way around; that is, the light source is stationary and the observer (detector) is moving away from the light source. Then the frequency of light as detected by the observer f_P is given by

$$f_P = \frac{f_L(c - V_P)}{c} \tag{7.37}$$

where f_L is the frequency of light (laser) emitted by the light source, which is equal to the scattering light as elastic scattering is only considered. But in a practical situation, the laser light source cannot be placed along the velocity vector direction; rather, it can be placed making an angle β with the velocity vector, as shown in Figure 7.20. The frequency of light experienced by the particle would be

$$f_P = \frac{f_L(c - V_P \cos\beta)}{c} \tag{7.38}$$

The light scattered from the moving particle is detected by the observer (detector). The Doppler frequency of light received by the detector would then be expressed as

$$f_D = \frac{cf_P}{\left(c - V_P \cos(\beta + \theta)\right)} \tag{7.39}$$

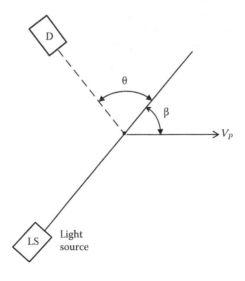

Figure 7.20

Schematic of light scattering from a particle from a single light beam.

7. Optical Combustion Diagnostics

By substituting the expression for f_P in the above equation, we have

$$f_D = \frac{f_L(c - V_P \cos\beta)}{\left(c - V_P \cos(\beta + \theta)\right)} \qquad (7.40)$$

Let us now consider the Doppler effect in a typical dual beam scatter system, as shown in Figure 7.21, which is used routinely in modern laser Doppler anemometry (LDA). Note that the detector (observer) can receive scattered light from both laser light beams. The frequency of light, f_{D_1}, received by the detector due to laser beam 1 can be expressed as

$$f_{D_1} = \frac{f_L(c - V_P \cos\beta_1)}{\left(c - V_P \cos(\phi)\right)} \qquad (7.41)$$

where $\phi = \beta_1 + \theta$. Similarly, the frequency of light, f_{D_2}, received by the detector due to laser beam 2 can be expressed as

$$f_{D_2} = \frac{f_L\left(c - V_P \cos(\beta_1 + \theta)\right)}{\left(c - V_P \cos(\phi)\right)} \qquad (7.42)$$

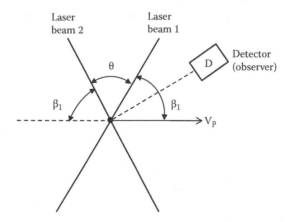

Figure 7.21

Schematic of light scattering from a particle from two light beams.

The frequency shift for this dual light beam scattering system can be determined as

$$\Delta f_D = f_{D_1} - f_{D_2} = \frac{f_L(c - V_P\cos\beta_1)}{\left(c - V_P\cos(\phi)\right)} - \frac{f_L\left(c - V_P\cos(\beta_1 + \theta)\right)}{\left(c - V_P\cos(\phi)\right)}$$
$$= \frac{f_L V_P\left(\cos(\beta_1 + \theta) - \cos\beta_1\right)}{\left(c - V_P\cos(\phi)\right)} \tag{7.43}$$

As the speed of light c is much larger than V_P, the above equation can then be simplified further as

$$\Delta f_D = f_{D_1} - f_{D_2} = \frac{f_L V_P}{c}\left(\cos(\beta_1 + \theta) - \cos\beta_1\right) \tag{7.44}$$

In a dual beam scatter system, the Doppler frequency does not depend on the detector angle unlike in a single beam scattering system. That is why a dual beam scatter system is preferred over LDV. We can simplify the above expression for the Doppler frequency shift by considering symmetry case, $\beta_1 = 90° - \dfrac{\theta}{2}$ as

$$\Delta f_D = \frac{f_L V_P}{c}\left(\cos\left(90° - \frac{\theta}{2} + \theta\right) - \cos\left(90° - \frac{\theta}{2}\right)\right) = -\frac{2V_P}{\lambda}\sin\left(\frac{\theta}{2}\right) \tag{7.45}$$

The Doppler frequency shift is dependent on the wavelength of the laser light beam, the angle between two laser beams, and velocity of a particle. Thus the velocity of a particle and the velocity in a fluid field can be determined by measuring the Doppler frequency shift using the following equation:

$$V_P = \frac{\lambda \Delta f_D}{2\sin\left(\dfrac{\theta}{2}\right)} \tag{7.46}$$

Note that the Doppler frequency is much smaller than the frequency of the light source that is being used in LDA. Hence, it will be quite difficult to sense this frequency and quantify it to measure the local velocity. Hence, the interference of two light beams leading to a fringe pattern are used to have the Doppler frequency shift, which can be measured easily to quantity local fluid velocity. We will be discussing the formation of a fringe pattern while discussing the dual beam differential LDA system.

Generally, the beam intersection volume is quite small through which scattering particles must pass for the measurement of velocity. Typical beam intersection

7. Optical Combustion Diagnostics

volumes are elliptical in shape and whose major axis dimension varies from 0.1 to 1 mm. As mentioned earlier, scattering particles passing through this beam volume are seeded into the flow as the properties of these scattering particles dictate the accuracy and quality of the velocity measurement. The type of particles to be used for seeding will be dependent on the flow and optical system employed for LDV. But for maintaining the continuum nature of flow and better quality of measurement, scattering particles must be seeded in the flow with the following properties:

i. Higher scattering capability
ii. Light enough to follow fluctuations of the flow
iii. Inexpensive, nontoxic, and noncorrosive
iv. Chemically inactive

It must be kept in mind that particles must be of lower density so that they can follow the fluctuations of the flow as accurately as possible. For this purpose, the diameter of particles must be quite small, in the range of the wavelength of visible light (μm), but must not be too small so as to reduce the scattering light intensity. Several types of particles, either solid or liquid droplets, as shown in Table 7.2, are used for LDV. For combustion applications, solid particles such as SiO_2, TiO_2, MgO, and Al_2O_3 are used while smoke and liquid droplets are used for seeding a nonreacting flow. Table 7.2 indicates the maximum size of seeding particle that can be used for measuring turbulent flow. Note that the typical frequencies of 1 and 10 kHz indicated in Table 7.2 are corresponding to the smallest size eddies in the turbulent flow.

7.12.2 Dual Beam LDV System

We know that several different configurations based on the principle of Doppler have been developed for the measurement of the local velocity of a flow field. The most common LDV is the dual beam differential Doppler system. The schematic of this LDA system is shown in Figure 7.22, which consists of a laser light source, beam splitters, lens, Bragg cell, and PMT (PMT is not shown in Figure 7.22). In the past, He-Ne (632-nm) lasers were used due to their low cost and better optical quality but in recent times, Ar-ion lasers (three colors: blue, 488 nm; green, 514.5 nm; and purple, 476.5 nm) are preferred as they can be used easily for 3D LDV due to their three colors, higher optical quality, and higher power level. Besides these two laser sources, Nd-YAG and diode lasers can also be used as

Table 7.2 Typical Seeding Particles for LDV

Particles	System	Density (ρ_P) (kg/m³)	Diameter (μm) for 1 kHz	Diameter (μm) for 10 kHz
Oil droplet (silicon)	Air	760	2.6	0.8
SiO_2	Air	2650	2.6	0.8
MgO	Flame	3580	2.6	0.8
TiO_2	Flame	4230	3.2	0.8
Al_2O_3	Flame	3950	2.6	0.8

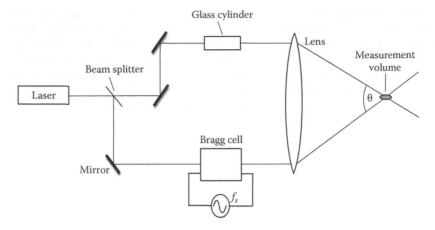

Figure 7.22

Schematic of a dual beam differential LDV.

light sources. But for LDV application the laser light must be a circular beam with a Gaussian intensity distribution and must be monochromatic in nature and linearly polarized. The monochromatic light beam from the laser source is passed through the beam splitter and gets divided into two equal-intensity parallel light beams. One light beam is passed through a glass cylinder and the other is passed through a Bragg cell. Note that the function of the Bragg cell is to increase or decrease the frequency of the incoming laser light beam that is required to detect the direction of velocity. It is basically an acoustic-optical modulator that needs a signal generator at 40 MHz. An achromatic lens is used to focus the shifted and unshifted beams of laser light to a point so that they can intersect each other, forming a requisite measuring volume at an angle θ, as shown in Figure 7.22. The seeding particles passing through the measuring volume scatter light from two laser beams that will undergo a Doppler shift due to particle velocity. The intensity of the scattered light can be sensed using a PMT. PMT can be placed either in forward- or backward-scattering points. In earlier days the forward-scattering method was preferred as a larger amount of light is scattered along the direction of the light transmitter (laser) compared to the backward position. But with the advent of bettor sensors, backward scattering is now used to detect Doppler frequency using PMT. Note that the intersecting angle θ of the two parallel beams can be varied using a lens of different focal length depending on the velocity and sensitivity range of the LDV. The frequency of each of the Doppler-shifted components may be high enough to be detected by the PMT. But the heterodyne (superposition) of two Doppler-shifted light beams result in a beat frequency, which is quite difficult to be detected by the PMT. Therefore, a fringe pattern is produced at the measuring volume when two coherent light beams following two different path lengths are allowed to intersect at the measurement volume, as shown in Figure 7.23. The intersection of the two

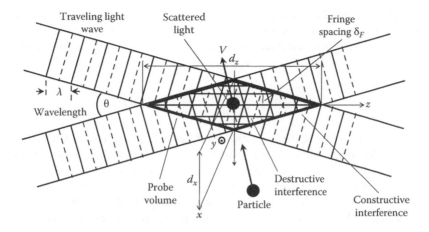

Figure 7.23

Interference fringe at the intersecting point of two coherent light beams.

beams takes place in their respective beam wastes, as shown in Figure 7.22, and parallel strips of bright- and dark-forming fringe patterns are formed on the light intersecting plane. In other words, these fringe patterns are oriented parallel to the optical axis (*y*-direction) and perpendicular to the plane defined by the two intersecting laser beams. By invoking the wave theory of light as discussed earlier in interferometers (see Section 7.10), it can be shown that the distance between two fringes, δ_F, is dependent on the angle between two interfering light beams and their included angle θ as

$$\delta_F = \frac{\lambda}{2\sin\left(\dfrac{\theta}{2}\right)} \tag{7.47}$$

As the fringe pattern is formed normal to the *x*-axis, the light scattered from the moving particle through the measuring volume will be proportional to the Doppler frequency due to particle velocity along the *x*-direction. Then the shift in Doppler frequency can be expressed as

$$\Delta f_D = \frac{V_x}{\delta_F} = \frac{2V_x}{\lambda}\sin\left(\frac{\theta}{2}\right) \tag{7.48}$$

Note that a similar expression can be obtained from Equation 7.46. By measuring the Doppler frequency shift from the fringe pattern in the measurement volume, the velocity in the *x*-direction can be determined easily as intersecting angle θ and the wavelength of the laser light are known. However, the measurement volume has a Gaussian intensity light distribution along the three dimensions, forming an ellipsoid consisting of dark and bright fringes. The size of this measurement volume can be determined easily from the beam waste diameter d_F of the focused laser beam and intersecting angle θ (see Figure 7.23) as given below

$$d_x = \frac{d_F}{\cos(\theta/2)}; \quad d_y = d_F; \quad d_z = \frac{d_F}{\sin(\theta/2)}; \tag{7.49}$$

where d_x is the height, d_y is the width, and d_z is the length of measuring volume. As a smaller intersection angle is usually used, d_x can be considered approximately equal to d_y. The dimensions of the measurement volume obtained from the above relations will be approximately one, which can be used as a guideline to judge whether enough particles are passing through it. The number of fringes, N_F, formed in the measurement volume can be calculated by knowing the fringe spacing, δ_F, and fringe height, d_x, as given below

$$N_F = \frac{d_x}{\delta_F} = \frac{d_F}{\cos(\theta/2)} \frac{2\sin(\theta/2)}{\lambda} = \frac{2d_F}{\lambda}\tan(\theta/2) \tag{7.50}$$

For example, in the case of a laser beam diameter of 100 μm with a wavelength of 488 nm and beam intersection angle of 30°, by using Equation 7.50 the number of fringes, N_F, can be found to be around 110. That means the seeding particles passing through the center of the measurement volume would cross 110 fringes. A typical number of fringes required for good results varies from 20 to 80 [11,12]. In recent times, with the use of the fast Fourier transformation algorithm, even a few fringes can provide an accurate measurement of frequency. Besides this, an adequate number of particles must pass through the center of this measuring volume to have a higher signal-to-noise level. If few particles pass through the periphery of the measurement volume, it will pass through few numbers of fringes and thus inaccuracies in determining the Doppler frequency will be enhanced.

7.12.3 Signal Processing and Data Analysis

Let us understand how scattered light is detected and related to the local fluid velocity. As discussed earlier, a PMT detector based on the principle of photoelectric effect is usually used to detect and quantify the light intensity. In this device light photons are converted into electrons that are directed toward an electron multiplier and thus the current produced by the incident light is amplified multifold. This current signal also contains Doppler frequency information related to local velocity along with other spurious signals. The photodetection shot noise is one of the sources of these spurious signals. The mean photocurrent is influenced by the interaction of the optical field and photosensitive materials of PMT known as the quantum process. Secondary electronic noise from the PMT and thermal noise from the preamplifier circuit can also contribute to the spurious signal in an LDV signal. Furthermore, the light scattered from outside the measurement volume or other reflecting surface can reach the face of the PMT and add a spurious current to the LDV signal.

The LDV is designed to be operated so that the signal level in the PMT due to scattered light from the measurement volume will be maximum. This condition can be achieved by correct selection of laser power, seeding particle size and its distribution, and optical system parameters. The performance of signal processors is dependent on the number of seeding particles present simultaneously in the measuring volume. It is preferred to have a single particle in the measuring volume that produces a burst-type Doppler signal, as shown in Figure 7.24a. Note that the Doppler burst signal results due to the superimposition of two Doppler-shifted components of laser beams in the measuring volume. Customarily the Doppler burst signal is transformed into the frequency plane before the identification of the Doppler frequency. In this domain, high-frequency noise and low-frequency pedestal noise from the signal are removed by using a bandpass filter (refer to Figure 7.24b). For meaningful and accurate measurement, a suitable limit for the bandpass filter must be chosen so that all ranges of velocity with its fluctuating components in a particular system can be determined easily. Once the DC portion of noise signal is removed, the Doppler frequency can be determined and thus local velocity can be measured. A typical Doppler signal after a DC component is removed is shown in Figure 7.24c with time. In real situations there will be occasions when multiple particles can pass through the measuring volume simultaneously and thus the PMT will be receiving the sum of all current bursts, resulting in multiple signals. This will definitely provide the weighted average velocity of particles. Note that these multiple signals will cause random phase fluctuations that are quite difficult to eliminate. Hence efforts must be made to have a single Doppler burst due to the presence of a single particle in the measuring volume at a time.

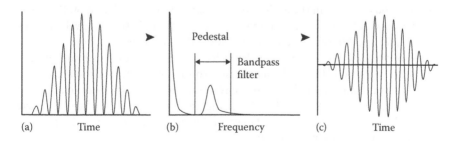

Figure 7.24

(a) Doppler burst signal, (b) frequency domain, and (c) filtered signal fringe at the intersecting point of two coherent light beams. (With kind permission from Springer Science+Business Media: *Optical Measurements: Techniques and Applications*, Second Edition, 2001, Mayinger, F. and Feldmann, O., eds.)

Example 7.3

A dual beam scattering LDV system using an argon-ion laser with a beam diameter of 0.2 mm is used to measure velocity in a flow has an intersection angle of 10°. Determine the number of fringes and the velocity of the flow if the Doppler shift frequency is 5.4 MHz.

Given: Ar-ion laser ($\lambda = 514.5$ nm), $d_F = 0.2$ mm, $\theta = 10°$, $\Delta f = 5.4 \times 10^6$ Hz.

To Find: Number of fringes and velocity of the flow.

Solution: The number of fringes, N_F, formed in the measurement volume can be calculated by using Equation 7.47 as given below:

$$N_F = \frac{2d_F}{\lambda}\tan(\theta/2) = \frac{2 \times 0.2 \times 10^{-3}}{514.5 \times 10^{-9}}\tan 5° = 68.018 \approx 68$$

Note that the number of fringes, N_F, is within the prescribed limit.

The flow velocity V can then be determined by using Equation 7.45 as

$$V = \frac{\Delta f_D \lambda}{2\sin(\theta/2)} = \frac{5.4 \times 10^6 \times 514.5 \times 10^{-9}}{2\sin 5°} = 15.94 \text{ m/s.}$$

7.13 Particle Image Velocimetry (PIV)

PIV is a nonintrusive type of optical technique for determining instantaneous velocity vectors in the entire flow field simultaneously. Unlike point measurement of velocity vectors using laser Doppler velocimetry, this optical method can provide all the information about velocity vectors over an entire two-dimensional plane simultaneously. Seeding particles like in LDA can also be used as tracers whose individual displacement along with fluid flow is measured with respect to time. The change in displacement of individual particles in a flow field captured by flow visualization images during a time interval can provide both local and instantaneous velocity in the entire flow field simultaneously. Based on this basic principle, several velocimetry techniques, such as particle tracking velocimetry (PTV), laser speckle velocimetry (LSV), and PIV, have been developed over the past four decades. But PIV is considered to be the most advanced and mature method of velocity field and hence is used routinely in various research areas such as aerodynamics, biofluids, and environment fluid dynamics apart from combustion systems. We will be discussing the fundamentals of this particle image velocimetry in this chapter. However, interested readers can refer to more advanced texts [5,6] for more details.

7.13.1 Basic Principles of a PIV System

Let us consider the basic principles of PIV, as shown in Figure 7.25, which consist of a laser light source, optical system, camera, imaging and digitization hardware, software, and a computer. In modern times, a double-pulsed Nd:YAG laser

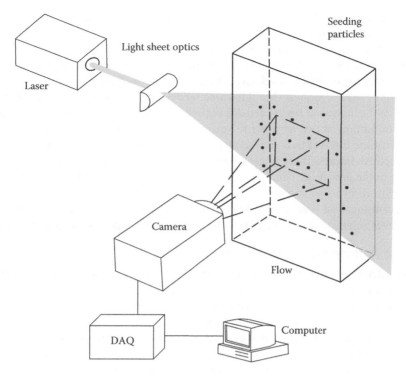

Figure 7.25

Schematic of a typical planar two-dimensional PIV system.

of wavelength $\lambda = 532$ nm, is used routinely. Of course, other lasers or even flash lamps or continuous light sources can be used to illuminate the seeding particles. It must be ensured that a sufficient amount of light intensity is provided to form visible images of micrometer-sized particles in the flow field. This can be made possible by using laser lights with energy between 50 to 500 mJ per pulse depending on the size of particle, types of particles, fluid velocity, and nature of the fluid. As discussed earlier, the flow must be seeded with particles small enough to follow fluid acceleration but large enough to scatter enough light to form bright images.

As discussed above, in a PIV system, a high-intensity light sheet must be used to illuminate seeding particles that can be generated from a collimating laser beam using cylindrical and spherical lenses, as shown in Figure 7.25. A cylindrical lens is employed to cause a light sheet to expand in only one direction while a spherical lens makes the light beam become expanded along the perpendicular direction at the focal plane of the beam. The electiveness intensity of a light sheet can be increased by sweeping a light beam to form a sheet, thereby concentrating the energy by a factor equal to the height of the light sheet divided by the height of the beam. The thickness of the light sheet must be small enough to minimize the error due to the out-of-plane velocity component. A sheet thickness of 1 mm is preferred in a PIV system. The light scattered from these seeded particles in the flow field illuminated by the laser sheet is captured using a camera. A charge-coupled device (CCD) camera with a proper zoom lens is used for acquisition of these images in a PIV system. Subsequently these images are processed for determination of the local velocity. The camera viewing direction must be normal to the laser light sheet through which seeded particles will be moving when it records the motion of these particles between certain time intervals. There must be a minimum of two or more recordings of these particle images during the known time interval that can be recorded either in a single frame or multiple consecutive frames. These images are processed to obtain the velocity vector at a plane in the flow field. Three-dimensional velocity vectors in a flow field can be determined by using several methods, such as stereoscopic PIV, dual-plane PIV, 3D scanning, and holographic PIV, which are beyond the scope of this book. However, interested readers can refer to other advanced books and papers [5,6].

7.13.2 Particle Image Locations

Let us determine how to evaluate velocity in a simple PIV system using a simplified mathematical model. In other words, we will be briefly discussing a simplified mathematical model of the recoding of images in a camera and subsequent statistical evaluation of velocity vectors in a flow field. For evaluation, the PIV images are subdivided into smaller subareas known as *interrogation spots* in the case of optical interrogation and *interrogation windows* for digital imaging. We will be discussing cross-correlation analysis for determining velocity from the displacement of particles in two images during certain known time intervals. These interrogation areas may not be located at the same positions in PIV images. The geometric back projection from the image is known as interrogation volume, as shown in Figure 7.26. In PIV, measurement volume is comprised of two interrogation volumes in two images, which are used for the statistical evaluation of velocity. Let us understand how to locate a particle $x_p(t)$ in the flowing fluid through the laser sheet that is being captured by the camera in a frame. First we need to get a relationship by which the location of this particle in the laser sheet can be related to the location of this particle $X_p(t)$ on an image acquired by the camera. Note that there is a lens between the laser sheet and the image plane of the camera. By using the basic principles of ray optics, the location of the particle in the laser light plane ($z = 0$) can be projected to the image plane of the camera, as shown in Figure 7.26, through the effective center of the lens. It can be observed in Figure 7.27 that the particle at point $(x,y,0)$ on the laser light sheet can be mapped to a point at (X, Y) in the camera as per the following relationship:

$$\begin{pmatrix} X \\ Y \end{pmatrix} = M_0 \begin{pmatrix} x \\ y \end{pmatrix} \quad \text{where, } M_0 = \frac{Z_0}{z_0} \tag{7.51}$$

M_0 is the paraxial lateral magnification factor, z_0 is the object distance from the laser sheet ($z = 0$) to the effective center of the lens, and Z_0 is the distance between

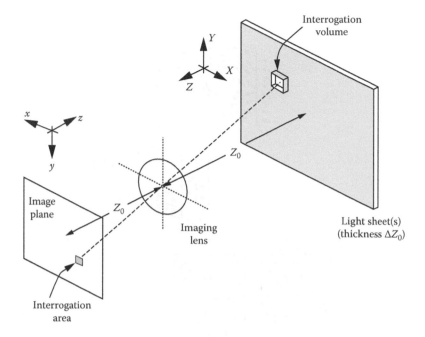

Figure 7.26

Schematic of geometric imaging. (With kind permission from Springer Science+Business Media: *Particle Image Velocimetry*, 1998, Raffel, M., Willert, C. and Kompenhans, J.)

the effective center of the lens and image plane in the camera. In order to have right mapping, the object point must be focused to the image plane during which not only the ray through the lens center but also other rays emanated from the object (particle) must be converged to the same point on the image plane. Both image distance Z_0 and object distance z_0 from the effective center of the lens can be related to the focal length of the lens by using the Gauss thin lens law, as given below:

$$\frac{1}{f} = \frac{1}{Z_0} + \frac{1}{z_0} \tag{7.52}$$

Note that when the laser light at time t is pulsed, light scatter from each particle in the laser light plane ($z = 0$) is captured by the image plane of the camera. At time $t + \Delta t$, a second set of images is captured by the camera using a second pulse of laser light. From these two images, the particle displacement within time interval Δt can be traced by considering the mapping property of the lens and camera system and the dynamics of the particles. The displacement between the

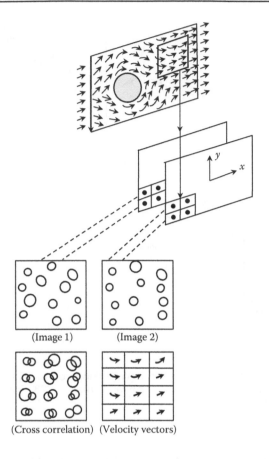

(Image 1)   (Image 2)

(Cross correlation) (Velocity vectors)

Figure 7.27

Schematic of interrogation windows for double frames and single-exposure recording.

images of a particle between a known interval time Δt is used for determination of velocity that can be expressed as

$$(V_x, V_y) \approx \frac{(\Delta x_p, \Delta y_p)}{\Delta t} = \frac{(\Delta X_p, \Delta Y_p)}{M_0 \Delta t} \tag{7.53}$$

This first-order finite difference assumed in the above expression can be justified only when the time interval between two laser light pulses is quite small. The magnification M_0 can be decided depending on the resolution required between the adjacent velocity vectors of the particles. It must be kept in mind that this ensuing discussion provides a very simplistic view of the entire gamut of PIV. Note that particles will be moving across the entire volume illuminated by the laser light sheet, not just on the $z = 0$ plane. Hence, all particles in this illuminated volume that can be captured by the images in the camera need not be in focus. The diameter of particles in the image can be affected by not only magnification M_0 of the lens but also by the diffraction effect and geometrical aberration of the lens. Besides these aspects, several types of other errors can be incurred during the determination of velocity using the PIV technique. For example, an image can be deformed between exposures, and noise can be introduced during the recoding of images. Proper determination of particle centers affects the correct determination of displacement during a time interval.

7.13.3 Image Processing in PIV

The images captured by the camera in a PIV system can be processed differently based on the basic principles used for forming images of particles in the field. Let us discuss the cross-correlation technique that is being used for determination of the displacement of particles on the images between two separate frames during known time interval, Δt. In this method, both frames are subdivided into interrogation windows as indicated in Figure 7.27. One way to determine the displacement from a local particle is to deconvolute these two images by dividing the respective Fourier transforms by each other. This may work only when noise in the signal is negligibly small but in a real situation, noise level is likely to be significant and thus incur errors during the determination of the displacement data. Signal peak is also too sharp to be used for determination of the true value of displacement in the subpixel level. In order to overcome these problems, the statistical method of the discrete cross-correlation function is adopted to find out the best match between two images for determination of displacement vectors for a particle. Let us consider two sequentially imposed images at t and $t + \Delta t$, as shown in Figure 7.27. Suppose we want to determine the displacement of a given particle at a given point $p(x,y)$. Let us consider integration area $(m \times n)$ around this point, consisting of $M \times N$ pixels. Let I_1 be the intensity of pixels in image 1 and I_2 be the intensity of pixels in image 2; the cross-correlation coefficient can be expressed mathematically as

$$R(m,n) = \sum_{i=0}^{M} \sum_{j0}^{N} I_1(i,j) I_2(i+m, j+n) \tag{7.54}$$

The cross-correlation coefficient of the maximum value is achieved only when sample particles will align with each other for a particular displacement. In other words, the cross-correlation coefficient indicates the statistical measure of degree of match of particles between two images for a particular displacement. Hence, the maximum value in the correlation plane can be employed to determine the displacement made by a particle during time interval Δt. Several types of methods have evolved for determining the exact location of the peak cross-correlation coefficient. Generally, three points are used to estimate the peak cross-correlation coefficient. In this category, three methods (i) peak centroid, (ii) parabolic fit, and (iii) Gaussian peak fit are commonly used. Among these, Gaussian peak fit is preferred over the others because properly focused particle images follow the Airy intensity function, which can be approximately similar to the Gaussian intensity distribution. The position of the cross-correlation coefficient can be evaluated to subpixel accuracy. The three-point estimators perform better for narrow correlation peaks in the 2–3 pixel diameter range. By identifying the location of the same particle in the two images and the time interval between the two exposures, the flow velocity can be determined as per Equation 7.53:

$$(V_x, V_y) = \frac{(\Delta X_p, \Delta Y_p)}{M_0 \Delta t} \tag{7.55}$$

where M_0 is the effective magnification factor of the lens. Note that the cross-correlation method can recover linear displacement only. It cannot handle the rotational motion or any nonlinear displacement of particles within interrogation windows. In order to overcome this problem, a smaller interrogation window must be chosen so that a second-order effect can be minimized. As a result, there will be a substantial increase in computational efforts. In order to overcome this problem, a frequency-domain-based correlation is used for determination of the displacement and velocity vectors in modern PIVs, which is beyond the scope of this book.

Errors in PIV

It might be construed that PIV, being nonintrusive, would not suffer from errors, but the velocity data obtained by PIV is in fact affected by several type of errors. Some of the sources of these errors are three-dimensional effects, distortion in the shape of the particle image pattern, particle size distribution, signal/noise ratio of recording, and systematical errors due to hardware setup and statistical methods. Note that random noise incurred during image recoding and image digitization can cause distortion in the particle image pattern between the first and second images. Random motion due to the presence of turbulence in the flow can impart random motions to the seeding particles, which can cause errors in the determination of velocity. Besides these issues, the velocity gradient may distort the particle image pattern, particularly in the second image, leading to errors in the PIV data. It is quite cumbersome to take account of all these issues during error estimation of the PIV data but researchers do attempt to do so. However, this subject is beyond the scope of this text.

7.13.4 Seeding Particles for PIV

The seeding particles are an inherently critical component of the PIV system. Depending on the fluid flow under investigation, the particles must be able to match the fluid properties reasonably well, as discussed in the case of LDV measurement (see Section 7.10 and Table 7.12). Otherwise, they will not follow the flow satisfactorily enough for the PIV analysis. The refractive index for the seeding particles should be different for different fluids so that the laser sheet incident on the fluid flow will reflect off the particles and be scattered toward the camera. The particles are typically of a diameter in the range of 0.5 to 10 µm. The size of the particles should not be too small, otherwise they will not scatter the light in an effective way and we will not get good-quality images. However, the size of the particles should not be too big, otherwise noise will be greater, multiple reflections will occur, and the particles will not follow the fluid flow. Thus the particle size needs to be balanced to scatter enough light to accurately visualize all particles within the laser sheet plane but small enough to accurately follow the flow. The seeding system also needs to be designed carefully as it is essential to seed the flow to a sufficient degree without disturbing the flow overtly.

Example 7.4

Two images of particle displacement during the time interval of 0.02 s are captured for determining velocity at a point using PIV. In this snapshot, a particle is displaced by 70 and 45 pixels along the x- and y-directions, respectively. A lens is used with a focal length of 120 mm. The image distance from the lens is 160 mm. Determine the velocity along the x- and y-directions, respectively, by considering 0.5 mm in the captured area to be 5 pixels on the camera.

Given: $\Delta t = 0.02$ s, $\Delta X_p = 70$ pixels, $\Delta Y_p = 45$ pixels, $Z_0 = 160$ mm, $f = 120$ mm.

To Find: V.

Solution: The particle displacement on the image would be

$$\Delta X_p = 14 \text{ mm}, \Delta Y_p = 9 \text{ mm}$$

By using the Gauss thin lens law as Equation 7.52, we can determine object distance z_0 as

$$\frac{1}{f} = \frac{1}{Z_0} + \frac{1}{z_0}; \Rightarrow z_0 = 1 \Big/ \left(\frac{1}{f} - \frac{1}{Z_0} \right) = 1 \Big/ \left(\frac{1}{120} - \frac{1}{160} \right) = 408 \text{ mm}$$

The magnification factor M_0 can be determined by using Equation 7.51:

$$M_0 = \frac{Z_0}{z_0} = \frac{120}{408} = 0.294$$

We can determine the velocity V_x by knowing the displacement between the images of a particle between known interval time Δt as per Equation 7.55:

$$V_x = \frac{\Delta X_p}{M_0 \Delta t} = \frac{14}{0.294 \times 0.02} = 2551 \text{ mm/s} = 2.55 \text{ m/s}$$

Similarly, we can determine the velocity V_y

$$V_y = \frac{\Delta X_p}{M_0 \Delta t} = \frac{9}{0.294 \times 0.02} = 1531 \, \text{mm/s} = 1.53 \, \text{m/s}$$

7.14 Planar-Laser-Induced Fluorescence Method

The complex processes occurring during the interactions of flow and chemistry in the flame of a typical combustion system can be unraveled by using laser-based diagnostic tools developed in recent years. Laser-induced fluorescence (LIF) based on the absorption-emission process of photons is considered to be one of the best-established diagnostic tools being used in modern combustion labs. In this nonintrusive method, laser light is tuned to a wavelength that matches with the specific absorption line of the species/molecules. In the process the energy level of the species becomes elevated to higher electronically excited states (Figure 7.7) by the resonate excitation with suitable laser radiation. The higher electronic states are metastable in nature with characteristic lifetimes that vary from 10^{-5} to 10^{-10} s and thus subsequently the excited states of the species return back to the original ground electronic state while undergoing spontaneous emission of light radiations (fluorescence) with different energies and frequencies (see Figure 7.7). We know that the energy of a photon is equal to the energy difference between two levels, which is directly proportional to the frequency of emitted light ($E = h\upsilon$), where h is Planck's constant and υ is the frequency of emitted light. As the species are excited in the resonant mode during this process, the scattering cross section during this fluorescence is much larger, in the order of six orders of magnitude, compared to that of a Rayleigh scattering cross section. Therefore the LIF method can offer higher sensitivity and thus a low concentration level (ppm level) of minor species in a typical flame can be measured easily. This is the reason that this method can find applications in measuring low concentrations of intermediate species (OH, CH, CN, H, N, C, NH, etc.) for characterization of the flame structure in a typical combustion system.

In order to perform the measurement of any species using LIF, one must know the emission spectrum of this species when excited with light with a certain wavelength. In other words, a species must have a known absorption wavelength by which it can be excited to emit a fluorescence signal with a known wavelength. Of course this absorption wavelength of light must be within a tunable laser source. The data for the main absorption wavelength of a few species used in combustion diagnostics is given in Table 7.3. In combustion systems, several other species apart from the measured species will be present and thus there will be loss in the fluorescence signal due to collision involving the excited state and other molecules, which is known as quenching. For quantitative measurement of species concentration, one has to make appropriate corrections for the fluorescence light intensity caused due to quenching, which is dependent on the partial pressure of

Table 7.3 Spectroscopic Data for Species to Be Measured in a
Combustion System

Species	Λ(nm)	A(sec⁻¹)	Q(sec⁻¹)
CH	$A^2\Delta - X^2\Pi$ (0,0) 431.5	1.8(6)	1.6(7) flame, 20 torr
			3(9) flame, 760 torr
OH	$A^2\Sigma - X^2\Pi$ (0,0) 306.4	1.4(6)	5.6(8) flame, 760 torr
			5(7) flame, 20 torr
CN	$B^2\Sigma - X^2\Sigma$ (0,0) 388.3	1.7(7)	4.7(9) NO–H_2O rate
C_2	$A^2\Pi g - X^3\Pi^u$ (0,0) 516.5	8.3(6)	1.2(12) flame, 760 torr
			2.4(10) flame, 760 torr
NH	$A^3\Pi - X^3\Sigma^-$ (0,0) 336.5	2(6)	19.9(9)
NO	$A^2\Sigma - X^2\Pi^{1/2}$ (0,0) 226.5	3.3(6)	4.7(9) NO – H_2O rate

Source: Eckbreth, A. C. et al., *Progress in Energy and Combustion Science*,
5(5):253–322, 1979. Eckbreth, A. C., *Laser Diagnostics for Combustion
Temperature and Species*, Abacus Press, Kent, United Kingdom, 1988.

all other species and deactivation rate of the excited state of the measured species
by other species. Note that this deactivation rate is also dependent on the temperature and therefore it is quite cumbersome to determine the quenching rate of
the excited rate of the measured species. As a result the fluorescence signals for
excited species are used for qualitative rather than quantitative evaluation of species concentration. Keep in mind that it would not be possible to measure several
species simultaneously using LIF because a specific wavelength is required for a
particular species to be excited and emit fluorescence signals.

7.14.1 Concentration Measurement

The concentration of species for which measurement is carried out using PLIF
will be dictating the fluorescence signal received by the detector. For measurement of the concentration of the ith species, we need to determine the number of
photons, N_{P_i}, reaching the detector for a laser pulse from the collected volume,
V_c, of the measured field, which can be related to various variables of the linear
fluorescence absorption-emission process [2,3], as given below:

$$N_{P_i} = \left(\eta_c \frac{\Omega R}{4\pi} \right) X_i N_{tot} V_c F_P B_{12} I_v S \tag{7.56}$$

where η_c is the optics collection efficiency, R is the ratio of active to total pixel
area, Ω is the collector solid angle, X_i is the mole fraction of species, N_{tot} is the
number of density of molecules, F_P is the absorbing state population factor, B_{12}
is the Einstein coefficient of spontaneous emission, I_v is the spectral intensity of
laser per unit frequency interval, and S is the Stern-Vollmer factor (fluorescence
yield). The term in the bracket in the right-hand side of Equation 7.56 is the
correction factor due to losses during the collection of photons. Note that part
of the fluorescence photons emitted with full solid angle 4π is collected by the

---

collection solid angle Ω. The collection efficiency η_c is dependent on the optical arrangement and various losses incurred on the optical path to the detector. The Stern-Vollmer factor can be expressed in terms of the quenching rate, Q_{21}, and the spontaneous emission rate, A_{21}, as given below:

$$S = \frac{A_{21}}{A_{21} + Q_{21}} \tag{7.57}$$

and the absorbing state population factor F_P can be expressed as

$$F_P = \left(\frac{g_1(2J''+1)}{Q_e Q_v Q_r} \right) \exp\left[-\frac{hc}{k_B} \frac{\left(Te(n) + G(v) + F(J'')\right)}{T} \right] \tag{7.58}$$

where Q_e is the electronic partition function, Q_v is the vibrational partition function, Q_r is the rotational partition function, $Te(n)$ is the electronic energy, $G(v)$ is the vibration term energy, $F(J'')$ is the rotational term energy, J'' is the ground state rotational quantum number, k_B is the Boltzmann constant, and T is the temperature. The total number density N_{tot} is dependent on temperature T. Hence, by using the perfect gas equation at a constant temperature, N_{tot} can be expressed in terms of temperature T as

$$N_{tot} \cong N_L \frac{273}{T} \tag{7.59}$$

where N_L is the Loschmidt number (constant) that is equal to $6 \times 10^{19}\,\text{cm}^{-3}$. By combining the above equations, we can get an expression for the number of photons, N_{Pi}, as

$$P_i = \left(\eta_c \frac{\Omega R}{4\pi} \right) X_i N_{tot} V_c \left[B_{12} I_v \frac{A_{21}}{A_{21} + Q_{21}} \right] \left(\frac{g_1(2J_1+1)}{Q_e Q_v Q_r} \right) \exp\left[-\frac{hc}{k_B} \frac{Te + G(v) + F(J)}{T} \right] \tag{7.60}$$

It can be noted from this equation that the fluorescence signal is dependent on the concentration (mole fraction) of the measured species and temperature apart from several other factors even though several simplified assumptions are made for deriving this expression [2,3,7]. In practice, we assume the fluorescence signal is proportional only to the mole fraction of the fluorescence species provided that the product of the Stern-Volmer factor S and population factor F_P would not vary significantly across the measured volume. However, the molecular transition can be chosen so that the temperature variation in the product $S \times F_P$ can be minimized. For example, a variation of the product $S \times F_P$ with a temperature for the molecule OH for various values of rotational quantum number,

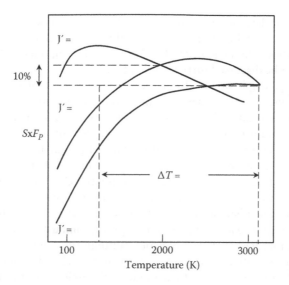

Figure 7.28

Variation of the product $S \times F_p$ with a temperature for the molecule OH for various values of rotational quantum number hc/k_B. (From Kryschakoff, G. et al., *Applied Optics*, 23(5):704–712, 1984.)

J'', as shown in Figure 7.28, indicates that $S \times F_p$ can be assumed to remain constant with a 10% variation between a temperature range of 1400 K and 3100 K [7]. Then the fluorescence signal can be considered to be dependent only on the mole fraction of the species under this restricted condition. Generally, two major steps are used to convert raw data into absolute values of the species number density. First, the raw data is placed in the correct relative scale by undertaking appropriate corrections for nonuniform pixel responsivity and laser illumination. For this purpose, the temperature dependence of $S \times F_p$ are chosen such that correction for temperature must be minimum, as described above. This relative data is mapped to an absolute value by undertaking the proper calibration or calculation of the unmeasured parameters in Equation 7.60. It is quite challenging and time consuming to get the absolute value of concentration of any species at any flow field. That is the reason why the quantitative analysis of a species is not being employed using this technique as it is quite cumbersome to calibrate the concentration measurement due to difficulties in determining various terms in Equation 7.60. Rather, quantitative interpretation of PLIF signals is used extensively, particularly for flame visualization, heat release rate, mixing analysis, and temperature. By using the above LIF method, several species in a combustion system can be detected but there are several other molecules, such as N_2, O_2, CO_2, H_2O, and most fuels (saturated hydrocarbons), known as dark compounds, which cannot be detected by LIF. However, tracer elements such as NO_2, NO, CO, acetone, and acetaldehyde that can fluorescence easily can be mixed homogenously with the dark compound to be detected and measured.

This method for detection measurement of dark species is known as tracer LIF. In this case, the signal intensity due to the fluorescence molecule as in normal LIF is proportional to the mole fraction of the dark compound to be measured, which will not be discussed further in this text.

The basic principle of LIF discussed above can be used for investigating combustion systems. For understanding of the reaction mechanism involving various stable and unstable species, a point-wise measurement of concentration and temperature in a laminar flame is carried out using LIF. However, it is quite convenient to use a two-dimensional planar laser sheet as in the PIV system for measurement of minor species distribution in a flame. PLIF is a widely used laser diagnostic technique for the characterization of minor species. It is mostly used along with measurement of the velocity field using a PIV system as it is inherent in PLIF. These two measurement systems are used widely for the investigation of several intriguing combustion concepts such as flame extinction, flashback and blowout, autoignition and cycle variation in IC engines, turbulent-chemistry interaction, and shock-laden combustion phenomena. Therefore, we will be focusing our discussion on PLIF in this book. Readers interested in other LIF systems such as saturation LIF, mulitphoton LIF, and predissociative LIF can refer to other books on combustion diagnostics [3,12,13].

7.14.2 Details of the PLIF System

In order to get a feel for how the measurement of species is carried out in a combustion system, let us consider a typical experimental setup for a PLIF system as shown in Figure 7.29. It can be observed from this figure that this setup, like the

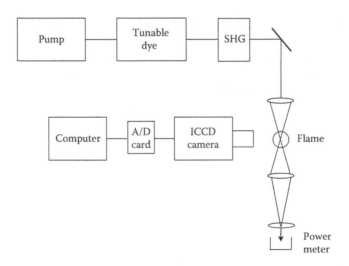

Figure 7.29

Schematic diagram of the PLIF system. SHG, Second Harmonic Generation Crystal.

PIV setup discussed in Section 7.13 consists of a laser light source, optical system, camera and imaging digitization hardware, software, and a computer. The laser light beam is converted into a thin light sheet using a long focal length spherical and a pair of cylindrical lens, as depicted in Figure 7.29. The size of the laser sheet depends on the requisite absorption cross section of the LIF process, which is eventually dependent on the specific species (OH, CH, NO, etc.) that undergoes the fluorescence process with the selected transition and selected fluorescence methods such as linear, saturated, and predissociated.

Besides these factors, the cross-section area on the test section (flame) depends on the angle of divergence of the light sheet, camera, and its associated zoom lens. The thickness of the laser sheet that dictates the spatial resolution along the perpendicular direction to the observing plane varies in the order of 100 μm. For measurement of minor species in combustion systems, intense and tunable pulsed laser light in a deep UV range is essential. The laser must be tunable to produce a requisite frequency corresponding to different species to ensure the resonant excitation process. The intensity of the laser light must be high enough to ensure high excitation efficiency. Short pulsed laser light is used to differentiate between laser-induced emission and radiation from the flame itself. As the spectrum of most combustion species is in the range of UV, laser light in a deep UV range (210–250 nm) is generally employed for the PLIF system. Traditionally, an Nd:YAG laser of wavelength $\lambda = 532$ nm is used to pump the tunable dye laser (see Figure 7.29), particularly in the visible range of the spectrum. Frequency mixing is employed to generate a requisite UV frequency for the PLIF system. Pulse duration in the range of 10 ns can be achieved in this laser system but the pulse energy is limited to a few hundred millijoules, which is sufficient to perform PLIF measurement. In recent times, a tunable excimer laser in the range of deep UV with a pulse power of a few hundred millijoules is being used in PLIF systems. The light intensity due to fluorescence emission is captured through a UV transparent lens, filters, and camera, as shown in Figure 7.29. Note that the camera should have a higher spatial resolution and collection efficiency. Both interference and cutoff filters are placed to allow only the fluorescence signal onto the detectors of the camera. In other words, other radiations from the flame and other sources would not be allowed to pass onto the camera. The CCD and complementary metal-oxide-semiconductor (CMOS) based cameras commonly used for PIV measurements would not be suitable for PLIF measurement due to non-UV sensitiveness and shorter gating times. Rather, image intensifier CCD cameras are being employed that can capture lower LIF intensities. Sometimes two-stage image intensifiers are employed, particularly at high laser pulse rates. A recently developed camera that combines a multichannel plate (MCP) with a booster that can capture images at a frame rate exceeding 20 kHz without any electron depletion can be employed for PLIF imaging. The digitized intensity data of planar fluorescence images are transferred from the camera to the computer by using an A/D card. Note that a control unit (not shown in Figure 7.30) plays an important role of timing the laser sequence, camera shutter opening, and data transfer. Generally, image data is stored in grayscale depending on the

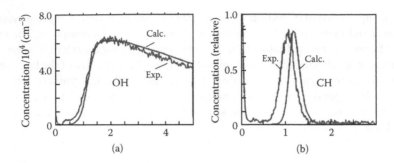

Figure 7.30

PLIF measured data of (a) OH and (b) CH concentrations for CH_4-air flame ($P =$ 40 mbar). (Reprinted from *Combustion and Flame*, 88, Heard, D. E., Jeffries, J. B., Smith, G. P. and Croslley, D. R., LIF measurement in methane/air flames of radicals important in prompt NO formation, 137–148, Copyright 1992, with permission from Elsevier.)

dynamic range of the camera and the intensity of the emitted light. This data of the fluorescence light intensity is used to determine the concentration and temperature fields of the flame. Generally, false colors are used to indicate the finer details of the concentration and temperature fields.

Nevertheless, in some cases, the quantitative measurement concentration for species OH and relative concentration of CH in a premixed laminar flat CH_4-air flame at low pressure can be seen in Figure 7.30 [14]. Note that the estimated OH concentration matches well with the data of OH obtained with PLIF. Important information about the variation and peak value of the CH species can also be obtained from the relative CH concentration profile that is itself obtained from the PLIF system. This qualitative information can be useful in developing an in-depth understanding of the complex processes that occur during combustion.

The PLIF system has several shortcomings that need to be overcome in the future by carrying out research in this field. Some of these limitations are

i. The flow field must contain molecular species with an optical resonance wavelength that can be accessed by a laser and a camera
ii. The signal-to-noise ratio is often limited by detector shot noise
iii. Fluorescence interferences from other species incurs error, especially from hydrocarbons in high-pressure reacting flows
iv. Attenuation of the laser sheet across the flow field or reabsorption of fluorescence before it reaches the detector can lead to systematic errors
v. Temperature measurements typically require two laser sources

7.15 Rayleigh Temperature Measurement

Recall that in Section 7.3 we discussed Rayleigh scattering, which comes under the category of elastic scattering, as there would not be any change in wavelength during the scattering process. This scattering from the gas molecules due to a change in density is used to measure temperature in combustion systems.

Let us consider first the Rayleigh scattering from the gases. As per Equation 7.2, the Rayleigh scattering intensity I can be expressed as

$$I = \beta I_0 NL\Omega_c \left(\frac{d\sigma}{d\Omega} \right)_{eff}$$

(7.61)

where $(d\sigma/d\Omega)_{eff}$ is the effective Rayleigh differential scattering cross section (cm²/sr) of a gas mixture. This is the summation of mole fraction X_i of the ith species and weighted cross section $(d\sigma/d\Omega)$, which can be expressed as [14,15]

$$\left(\frac{d\sigma}{d\Omega} \right)_{eff} = \sum_i X_i \left(\frac{d\sigma}{d\Omega} \right)$$

(7.62)

The Rayleigh scattering intensity is dependent on the location, effective differential cross section, and number density, N, at a spatial location in the flame with respect to time. Assuming the gas to behave as an ideal one, the number density, N, can be related to pressure P and temperature T as

$$N = \frac{PA_o}{R_u T}$$

(7.63)

where A_o is the Avogadro number and R_u is the universal gas constant. In a constant-pressure combustion system, pressure can be assumed to remain almost constant and hence number density, N, is inversely proportional to temperature T. By combining Equation 7.62 and 7.63, we have

$$I = \beta I_0 \left(\frac{PA_o}{R_u T} \right) L\Omega_c \left(\frac{d\sigma}{d\Omega} \right)_{eff} = \frac{C}{T} \left(\frac{d\sigma}{d\Omega} \right)_{eff}$$

(7.64)

where C is the constant that can be determined by calibration [16] and I_0 is assumed to be constant. Hence the Rayleigh scattering intensity is dependent on temperature and species concentration. As a result, in order to measure temperature unambiguously from the flame/combustion system, it is important to keep the Rayleigh cross section constant. In the case of a flame, both temperature and species vary compared to a single-gas case. The scattering cross section is dominant by one species, N_2 gas, particularly in the case of a lean fuel-air system. Therefore one can assume the Rayleigh scattering cross section to be almost constant. On the other hand, the Rayleigh scattering intensity varies mainly with temperature, as temperature across a flame varies by a factor of 7. The variation of an effective cross section from reactants to product varies by only 10% for a constant-pressure system and hence can be neglected compared to the variation in temperature. Of course by judicious selection of fuel composition, the variation of Rayleigh scattering intensity can be reduced to even 2%. Besides this simplification, the evolution of Rayleigh scattering intensity due to species from the predicted model can be incorporated in this method for determination of temperature.

A schematic of a simple and elegant experimental setup for a Rayleigh scattering based temperature measurement system is shown in Figure 7.31, which consists of a laser source, polarizer, filter, optical system, PMT, A/D card, and computer. The scattered light from the flame is collected with the help of a lens. Subsequently the scattered light is passed through slits, a polarizer, and a 1-nm bandwidth interference filter that helps in minimizing the background noise from flame luminescence and other sources. The Rayleigh scattering signal is sensed by a PMT whose analog signal is amplified and stored in a computer with the help of a suitable A/D card. Generally, the Rayleigh scattering intensity is determined from a fuel-air mixture without combustion at room temperature (i.e., $T_u = 300$ K) to determine the constant C in Equation 7.64. All other measurements with combustion can be carried out with reference to this value of constant for a particular set of fuel-air mixture. It is advisable to check the value of this constant at regular intervals. If it changes by more than 2%, then new constant C can be used for determination of temperature.

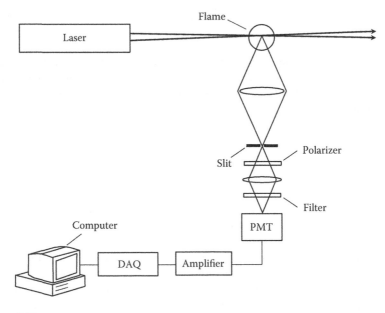

Figure 7.31

Illustration of a Rayleigh scattering setup.

As discussed above, the Rayleigh scattering technique for temperature and species measurements has several shortcomings:

   i. Interference of flame radiation with Rayleigh scattering
  ii. Scattering of soot particles will interfere with the Rayleigh scattering signal
 iii. Filtering of Mie scattering
  iv. Low signal-to-noise ratio at low temperatures
   v. Beam steering while transmitting through the flame region with a variable refractive index
  vi. Beam jumping around the turbulent flow
 vii. Reduction of signal-to-noise ratio due to the presence of a solid surface in the case of a combustor

Questions

1. What are the types of light sources used for flow visualization in combustion systems? Enumerate each of them with their respective advantages and limitations.

2. What is a laser? Why is it preferred over other light sources for both qualitative and quantitative measurements in combustion systems?

3. What are the types of lasers used for flow visualization in combustion systems? Enumerate each of them with their respective advantages and limitations along with specifications.

4. What are types of light scattering? Describe basic principles of light scattering.

5. How is Rayleigh scattering different from Mie scattering?

6. What is the difference between Stokes and anti-Stokes Raman scattering?

7. What is fluorescence? How is it different from Mie scattering?

8. What are the problems associated with direct flame visualization?

9. What are the differences between the Schlieren and shadowgraph methods?

10. What are the differences between twin-mirror and twin-lens Schlieren arrangements?

11. Derive an expression for the contrast in the case of a shadowgraph.

12. What is an interferometer? How is it different from a Schlieren system?

13. Derive an expression for the temperature to be measured for an isobaric ideal gas using a Mach-Zehnder interferometer.

14. What is an LDV? What is the basic principle of this measurement system?

---

15. Why is a dual beam scatter system preferred for LDV?

16. Derive an expression for velocity in terms of frequency shift in the case of a two-beam LDV system.

17. What are the advantages and disadvantages of an LDV system?

18. What is a PIV? What is the basic principle of this measurement system?

19. What are the advantages and disadvantages of a PIV system over an LDV system?

20. What is PLIF? What is the basic principle of this measurement system?

21. What are the advantages and disadvantages of a PLIF system?

22. Describe Rayleigh thermometry.

23. What are the advantages and disadvantages of Rayleigh thermometry?

Problems

1. An Ar-ion laser (λ = 514.5 nm) is to be used for designing a combustion diagnostic setup. Determine (i) the energy per photon and (ii) the energy per mol of this visible light.

2. In a Schlieren optical system, the distance between lens 2 and the camera is 100 mm. Lens 2, with a focal length of 75 mm, is used for making the ray from the test section fall on the knife edge. If the depth of light a_K not being cut off by the knife edge is equal to 9 mm and the subtending angle is 0.1 rad, determine distance p between the test section and lens 2 and the contrast.

3. In a Schlieren optical system, lens 2, with a focal length of 60 mm, is used for making the ray from the test section fall on the knife edge. The distance between lens 2 and the camera is 100 mm. If the depth of light a_K not being cut off by the knife edge is equal to 7 mm and the contrast is equal to 0.9, determine the deflection angle α subtended by the deflected ray between the test section and lens 2.

4. In a mirror shadow optical system the distance between the test section and the camera is 75 mm. If we consider that the vertical distance Δy of 1 mm at the test section is deflected and becomes 6 mm, determine the change in the deflection angle subtended by the deflected ray between the test section and the camera and the contrast obtained for this case.

5. An interferometer with a He-Ne laser (λ = 632.8 nm) is employed to visualize a two-dimensional plume during the combustion of a jet flame at ambient pressure and a temperature of 350°C. If it has a path length of 25 cm and the number of fringe shifts is –5.6, determine the temperature at a height of 20 cm by considering the Gladstone-Dale constant for plume gas to be $C = 2.31 \times 10^{-4}$ m³/kg.

6. An interferometer with a light source (λ = 514.5 nm) is employed to visualize the plume during the combustion of a droplet under a quiescent condition at ambient pressure. The test section with a depth of 10 mm width is placed 200 mm above the flame. The hot gas temperature is 500°C while the ambient temperature is 125°C. Determine the number of fringe shifts by considering the Gladstone-Dale constant for plume gas to be $C = 2.31 \times 10^{-4}$ m³/kg.

7. A dual beam LDV system that uses a He-Ne laser (λ = 632.8 nm) with a beam diameter of 0.3 mm to measure velocity at a point in a flow has an intersection angle of 12°. Determine the velocity of the flow if the Doppler shift frequency is 3.4 MHz.

8. A dual beam scattering LDV system with an argon-ion laser with a beam diameter of 0.2 mm is used to measure velocity at a point in a flow. A 150-mm lens having an angle of 10° is employed in this LDV. Determine the number of fringes and the velocity of the flow if the Doppler shift frequency is 2.5 MHz.

9. A dual beam LDV system that uses a He-Ne laser (λ = 632.8 nm) with a beam diameter of 0.3 mm to measure velocity at a point in a flow has an intersection angle of 12°. Determine the Doppler shift frequency for V = 1 m/s and 50 m/s.

10. In a 2D PIV system, two images of particle displacement are captured for a time interval of 0.01 s during which a particle is displaced by 80 and 55 pixels along the x- and y-directions, respectively. It uses a lens with focal length of 100 mm. The image distance from the lens is 150 mm. Determine the velocity along the x- and y-directions, respectively, by considering 1 mm in the captured area to be 5 pixels on the camera.

11. A Rayleigh thermometer system is employed to measure the temperature of a methane-air flame. The intensity ratio between the air and flame (I_{air}/I_{flame}) is measured to be 5.53. The atmospheric temperature is 300 K. The Rayleigh cross sections for CO_2, H_2O, N_2, and air are 1.50029×10^{-27}, 4.37685×10^{-28}, and 6.29651×10^{-28} cm²/st, respectively. Assuming complete combustion, determine the flame temperature at the postflame region.

References

1. Hanson, R. K., Combustion diagnostics: Planar flow field imaging, in *Twenty-First Symposium (International) on Combustion,* 21:1677–1691, 1987.
2. Ecberth, A. C., Bonczyk, P. A. and Verdieck, J. F., Combustion diagnostics by laser Raman and fluorescence techniques, *Progress in Energy and Combustion Science,* 5(5):253–322, 1979.
3. Eckbreth, A. C., *Laser Diagnostics for Combustion Temperature and Species,* Abacus Press, Kent, United Kingdom, 1988.

4. Mayinger, F. and Feldmann, O. (eds.), *Optical Measurements: Techniques and Applications,* Second Edition, Springer-Verlag, Berlin, 2001.
5. Adrian, R. J., Twenty years of particle image velocimetry, *Experiments in Fluids,* 39:159–169, 2005.
6. Raffel, M., Willert, C. and Kompenhans, J., *Particle Image Velocimetry,* Springer, Berlin, 1998.
7. Kryschakoff, G., Howe, R. D. and Hanson, R. K., Quantitative flow visualization technique for measurements in combustion gases, *Applied Optics,* 23(5):704–712, 1984.
8. Fristrom, R. M., *Flame Structure and Processes,* Oxford University Press, New York, 1995.
9. Gaydon, A. G., *Spectroscopy and Combustion Theory,* Chapman & Hall Ltd., London, 1948.
10. Settles, G. S., *Schlieren and Shadowgraph Techniques,* Springer, Berlin, 2001.
11. Goldstein, R. J. (ed.), *Fluid Mechanics Measurements,* Second Edition, Taylor & Francis, Philadelphia, 1996.
12. Kohse-Hoinghaus, K. and Jeffries, J. B. (eds.), *Applied Combustion Diagnostics,* Taylor & Francis, New York, 2002.
13. Chigier, N. (ed.), *Combustion Measurements,* Hemisphere Publishing Corporation, Washington, DC, 1991.
14. Heard, D. E., Jeffries, J. B., Smith, G. P. and Croslley, D. R., LIF measurement in methane/air flames of radicals important in prompt NO formation, *Combustion and Flame,* 88:137–148, 1992.
15. Zhao, F. and Hiroyuki, H., The applications of laser Rayleigh scattering to combustion diagnostics, *Progress in Energy and Combustion Science,* 19(6):447–485, 1993.
16. Sutton., G., Levick, A., Edwards, G. and Greenhalgh, D., A combustion temperature and species standard for the calibration of laser diagnostic techniques, *Combustion and Flame,* 147:39–48, 2006.

Appendix A: Physical Constants

Universal Gas Constant

$R_u = 8.314$ J/(g mol k) $= 1.987$ cal/(g mol) $= 82.05$ cm^3 atm/(g mol k)

Standard Acceleration due to Gravity

$$g = 1 \text{ kg m/(N s}^2) = 9.80665 \text{ m/s}^2$$

Stefan-Boltzmann Constant

$$\sigma = 5.6697 \times 10^{-8} \text{ W/(m}^2 \text{ K}^4)$$

Boltzmann Constant

$k_B = $ Boltzmann constant $= 1.38054 \times 10^{-16}$ erg/(molecule K)

Avogadro's Number

$$N_A = 6.02252 \times 10^{23} \text{ molecules/(g mol)}$$

Planck's Constant

$$h = 6.256 \times 10^{-27} \text{ erg s}$$

Speed of Light

$$c = 2.997925 \times 10^8 \text{ m/s (in vacuum)}$$

Standard Atmosphere Pressure

$$P_{atm} = 101\ 325 \text{ Pa} = 1 \text{ atm}$$

Appendix B: Conversion Factors

Energy: $1 \text{ J} = 9.47817.10^{-4} \text{ Btu}$
$= 2.3885.10^{-4} \text{ kcal}$
Energy rate: $1 \text{ W} = 3.41214 \text{ Btu/hr}$
Force: $1 \text{ N} = 0.224809$
Heat flux: $1 \text{ W/m}^2 = 0.3171 \text{ Btu/(hr.ft}^2)$
Kinematic viscosity and diffusivities: $1 \text{ m}^2/\text{s} = 3.875.10^4 \text{ ft}^2/\text{hr}$
Length: $1 \text{ m} = 39.370 \text{ in}$
$= 3.2808 \text{ ft}$
Mass: $1 \text{ kg} = 2.2046 \text{ lb}_m$
Mass density: $1 \text{ kg/m}^3\ 0.062428 \text{ lb}_m/\text{ft}^3$
Mass flow rate: $1 \text{ kg/s} = 7936.6 \text{ lb}_m/\text{hr}$
Pressure: $\text{Pa } 1 = 1 \text{ N/m}^2$
$= 0.20885\ 4 \text{ lbf/in}^2$
$= 1.4504.10^{-4} \text{ lbf/in}^2$
$= 4.015.10^{-3} \text{ in water}$
$1.10^5 \text{ N/m} = 1 \text{ bar}$
Specific heat: $1 \text{ J/kg K} = 2.3886.10^{-4} \text{ Btu/(lb}_m.{}^\circ\text{F)}$
Temperature: $\text{K} = (5/9)\ {}^\circ\text{R}$
$= (5/9)({}^\circ\text{F} + 459.67)$
$= {}^\circ\text{C} + 273.15$
Time: $3600_s = 1 \text{ hr}$

Appendix C

Table C.1 Selected Properties of Hydrocarbon Fuels

Fuel Symbol	Fuel Name	Mw (kg/kmol)	$\bar{h}_f^\circ$ (kJ/kmol)	HHV (MJ/kg)	LHV (MJ/kg)	Boiling Point (°C)	h_{fg} (kJ/kg)	ρ_{liq}^* (kg/m³)
CH_4	Methane	16.043	−74.831	55.528	50.016	−164	509	300
C_2H_2	Acetylene	26.038	226.748	49.923	48.225	−84	—	—
C_2H_4	Ethane	28.054	52.283	50.313	47.161	−103.7	—	—
C_2H_6	Ethane	30.069	−84.667	51.901	47.489	−88.6	488	370
C_3H_6	Propene	42.080	20.414	48.936	45.784	−47.4	437	514
C_3H_8	Propane	44.096	−103.847	50.368	46.357	−42.1	425	500
C_4H_8	1-Butene	56.107	1.172	48.471	45.319	−63	391	595
C_4H_{10}	n-Butane	58.123	−124.733	49.546	45.742	−0.5	386	579
C_5H_{10}	1-Pentene	70.134	−20.920	48.152	45.000	30	358	641
C_5H_{12}	n-Pentane	72.150	−146.440	49.032	45.355	36.1	358	626
C_6H_6	Benzene	78.113	82.927	42.277	40.579	80.1	393	879
C_6H_{12}	1-Hexene	84.161	−41.673	47.955	44.803	63.4	335	673
C_6H_{14}	n-Hexane	86.177	−167.193	48.696	45.05	69	335	659
C_7H_{14}	1-Heptene	98.188	−62.132	47.817	44.665	93.6	—	—
C_7H_{16}	n-Heptane	100.203	−187.820	48.456	44.926	98.4	316	684
C_8H_{16}	1-Octene	112.214	−82.97	47.712	44.560	121.3	—	—
C_8H_{18}	n-Octane	114.230	−208.447	48.275	44.791	125.7	300	703
C_9H_{18}	1-Nonene	126.241	−103.512	47.631	44.478	—	—	—
C_9H_{20}	n-Nonane	128.257	−229.032	48.134	44.686	150.8	295	718
$C_{10}H_{20}$	1-Decene	140.268	−124.139	47.565	44.413	170.6	—	—
$C_{10}H_{22}$	n-Decane	142.284	−249.659	48.020	44.602	174.1	277	730
$C_{11}H_{22}$	1-Undecene	154.295	−144.766	47.512	44.360	—	—	—
$C_{11}H_{24}$	n-Undecane	156.311	−270.286	47.926	44.532	195.9	265	740
$C_{12}H_{24}$	1-Dodecene	168.322	−165.352	47.468	44.316	213.4	—	—
$C_{12}H_{26}$	n-Dodecane	170.337	−292.162	47.841	44.467	216.3	256	749

Note: Heat of formation and higher and lower heating values all at 298.15 K and 1 atm; boiling points and latent heat of vaporization at 1 atm; liquid density. HHV = high heating value, LHV = low heating value.

Appendix D

Table D.1 Selected Properties of Air at 101.325 kPa

T (K)	ρ (kg/m³)	Cp (kJ/kg-k)	μ_{10}^{7} (N-s/m²)	$\upsilon \cdot 10^4$ (m²/s)	$k \cdot 10^3$ (w/m.K)	$\alpha \cdot 10^6$ (m²/s)	Pr
100	3.5562	1.032	71.1	2.00	9.34	2.54	0.786
150	2.3364	1.012	103.4	4.426	13.8	5.84	0.758
200	1.7458	1.007	132.5	7.590	18.1	10.3	0.737
250	1.3947	1.006	159.6	11.44	22.3	15.9	0.720
300	1.1614	1.007	184.6	15.89	26.3	22.5	0.707
350	0.9950	1.009	208.2	20.92	30.0	29.9	0.700
400	0.8711	1.014	230.1	26.41	33.8	38.3	0.690
450	0.7740	1.021	250.7	32.39	37.3	47.2	0.686
500	0.6964	1.030	270.1	38.79	40.7	56.7	0.684
550	0.6329	1.040	288.4	45.57	43.9	66.7	0.683
600	0.5804	1.051	305.8	52.69	46.9	76.9	0.685
650	0.5356	1.063	322.5	60.21	49.7	87.3	0.690
700	0.4975	1.075	338.8	68.10	52.4	98.0	0.695
750	0.4643	1.087	354.6	76.37	54.9	109	0.702
800	0.4354	1.099	369.8	84.93	57.3	120	0.709
850	0.4097	1.110	384.3	93.80	59.6	131	0.716
900	0.3868	1.121	398.1	102.9	62.0	143	0.720
950	0.3666	1.131	411.3	112.2	64.3	155	0.723
1000	0.3482	1.141	424.4	121.9	66.7	168	0.726
1100	0.3166	1.159	449.0	141.8	71.5	195	0.728
1200	0.2902	1.175	473.0	162.9	76.3	224	0.728
1300	0.2679	1.189	496.0	185.1	82	238	0.719
1400	0.2488	1.207	530	213	91	303	0.703
1500	0.2322	1.230	557	240	100	350	0.685
1600	0.2177	1.248	584	268	106	390	0.688

Source: Incorpera, F. P. and DeWitt, D. P., *Fundamentals of Heat and Mass Transfer*, Third Edition, John Wiley & Sons, New York, 1990.

Appendix E:
Revised Thermocouple Reference Tables

Table E.1 Type K (Chromel-Alumel)

°C	0	1	2	3	4	5	6	7	8	9	10
0	0.000	0.039	0.079	0.119	0.158	0.193	0.238	0.277	0.317	0.357	0.397
10	0.397	0.437	0.477	0.517	0.557	0.597	0.637	0.677	0.718	0.758	0.798
20	0.798	0.838	0.879	0.919	0.960	1.000	1.041	1.081	1.122	1.163	1.203
30	1.203	1.244	1.285	1.326	1.366	1.407	1.443	1.489	1.530	1.571	1.612
40	1.612	1.653	1.694	1.735	1.776	1.817	1.858	1.899	1.941	1.982	2.023
50	2.023	2.064	2.106	2.147	2.188	2.230	2.271	2.312	2.354	2.395	2.436
60	2.436	2.478	2.519	2.561	2.602	2.644	2.685	2.727	2.768	2.810	2.851
70	2.851	2.893	2.93-1	2.976	3.017	3.059	3.100	3.142	3.184	3.225	3.267
80	3.267	3.308	3.350	3.391	3.433	3.474	3.516	3.557	3.599	3.640	3.682
90	3.682	3.723	3.765	3.806	3.848	3.809	3.931	3.972	4.013	4.055	4.096
100	4.096	4.138	4.179	4.220	4.262	4.303	4.344	4.385	4.427	4.468	4.509
110	4.509	4.550	4.591	4.633	4.674	4.715	4.756	4.797	4.833	4.879	4.920
120	4.920	4.961	6.002	5.043	5.084	5.124	5.165	5.206	5.247	5.283	5.328
130	5.328	5.369	5.410	5.450	5.491	5.532	5.572	5.613	5.653	5.694	5.735
140	5.735	5.775	5.815	5.856	5.896	5.937	5.977	6.017	6.058	6.098	6.138
150	6.138	6.179	6.219	6.259	6.299	6.339	6.380	6.420	6.460	6.500	6.540
160	6.540	6.580	6.620	6.660	6.701	6.741	6.781	6.821	6.861	6.901	6.941
170	6.941	6.981	7.021	7.060	7.100	7.140	7.180	7.220	7.260	7.300	7.340
180	7.340	7.380	7.420	7.460	7.500	7.540	7.579	7.619	7.659	7.699	7.739
190	7.739	7.779	7.819	7.859	7.899	7.939	7.979	8.019	8.059	8.099	8.138
200	8.138	8.178	8.218	8.258	8.298	8.338	8.378	8.418	8.458	8.499	8.539
210	8.539	8.579	8.619	8.659	8.699	8.739	8.779	8.819	8.860	8.900	8.940
220	8.940	8.980	9.020	9.061	9.101	9.141	9.181	9.222	9.262	9.302	9.343
230	9.343	9.383	9.423	9.464	9.504	9.545	9.585	9.626	9.666	9.707	9.747
240	9.747	9.788	9.828	9.869	9.909	9.950	9.991	10.031	10.072	10.113	10.153

250	10.153	10.194	10.235	10.276	10.316	10.357	10.398	10.439	10.480	10.520	10.561
260	10.561	10.602	10.643	10.684	10.725	10.766	10.807	10.848	10.889	10.930	10.971
270	10.971	11.012	11.053	11.094	11.135	11.176	11.217	11.259	11.300	11.341	11.382
280	11.382	11.423	11.465	11.506	11.547	11.588	11.630	11.671	11.712	11.753	11.795
290	11.795	11.836	11.877	11.919	11.960	12.001	12.043	12.084	12.126	12.167	12.209
300	12.209	12.250	12.291	12.333	12.374	12.416	12.457	12.499	12.540	12.582	12.624
310	12.624	12.665	12.707	12.748	12.790	12.831	12.873	12.915	12.956	12.998	13.040
320	13.040	13.081	13.123	13.165	13.206	13.248	13.290	13.331	13.373	13.415	13.457
330	13.457	13.498	13.540	13.582	13.624	13.665	13.707	13.749	13.791	13.833	13.874
340	13.874	13.916	13.958	14.000	14.042	14.084	14.126	14.167	14.209	14.251	14.293
350	14.293	14.335	14.377	14.419	14.461	14.503	14.545	14.587	14.629	14.671	14.713
360	14.713	14.755	14.797	14.839	14.881	14.923	14.965	15.007	15.049	15.091	15.133
370	15.133	15.175	15.217	15.259	15.301	15.343	15.385	15.427	15.469	15.511	15.554
380	15.554	15.596	15.638	15.680	15.722	15.764	15.806	15.849	15.891	15.933	15.975
390	15.975	16.017	16.059	16.102	16.144	16.186	16.228	16.270	16.313	16.355	16.397
400	16.397	16.439	16.482	16.524	16.566	16.608	16.651	16.693	16.735	16.778	16.820
410	16.820	16.862	16.904	16.947	16.989	17.031	17.074	17.116	17.158	17.201	17.243
420	17.243	17.285	17.328	17.370	17.413	17.455	17.497	17.540	17.582	17.624	17.667
430	17.667	17.709	17.752	17.794	17.837	17.879	17.921	17.964	18.006	18.049	18.091
440	18.091	18.134	18.176	18.218	18.261	18.303	18.346	18.383	18.431	18.473	18.516
450	18.516	18.558	18.601	18.643	18.686	18.728	18.771	18.813	18.856	18.898	18.941
460	18.941	18.983	19.026	19.063	19.111	19.154	19.196	19.239	19.281	19.324	19.366
470	19.366	19.409	19.451	19.494	19.537	19.579	19.622	19.664	19.707	19.750	19.792
480	19.792	19.835	19.877	19.920	19.962	20.005	20.048	20.090	20.133	20.175	20.218
490	20.2018	20.261	20.303	20.346	20.389	20.431	20.474	20.516	20.559	20.602	20.644
500	20.644	20.687	20.730	20.772	20.815	20.857	20.900	20.943	20.985	21.028	21.071

(continued)

Table E.1 (Continued) Type K (Chromel-Alumel)

°C	0	1	2	3	4	5	6	7	8	9	10
510	21.071	21.113	21.156	21.199	21.241	21.284	21.326	21.369	21.412	21.454	21.497
520	21.497	21.540	21.582	21.625	21.668	21.710	21.753	21.796	21.838	21.881	21.924
530	21.924	21.966	22.009	22.052	22.094	22.137	22.179	22.222	22.265	22.307	22.350
540	22.350	22.393	22.435	22.478	22.521	22.563	22.606	22.649	22.691	22.734	22.776
550	22.776	22.819	22.862	22.904	22.947	22.990	23.032	23.075	23.117	23.160	23.203
560	23.203	23.245	23.288	23.331	23.373	23.416	23.458	23.501	23.544	23.586	23.629
570	23.629	23.671	23.714	23.757	23.799	23.842	23.884	23.927	23.970	24.012	24.055
580	24.055	24.097	24.140	24.182	24.225	24.267	24.310	24.353	24.395	24.438	24.480
590	24.480	24.523	24.565	24.603	24.650	24.693	24.735	24.778	24.820	24.863	24.905
600	24.905	24.948	24.990	25.033	25.075	25.118	25.160	25.203	25.245	25.288	25.330
610	25.330	25.373	25.415	25.458	25.500	25.543	25.585	25.627	25.670	25.712	25.755
620	25.755	25.797	25.840	25.882	25.924	25.967	26.009	26.052	26.094	26.136	26.179
630	26.179	26.221	26.263	26.306	26.348	26.390	26.433	26.475	26.517	26.560	26.602
640	26.602	26.644	26.687	26.729	26.771	26.814	26.856	26.898	26.940	26.983	27.025
650	27.025	27.067	27.109	27.152	27.194	27.236	27.278	27.320	27.363	27.405	27.447
660	27.447	27.489	27.531	27.574	27.616	27.658	27.700	27.742	27.784	27.826	27.869
670	27.860	27.911	27.953	27.995	28.037	28.079	28.121	28.163	28.205	28.247	28.289
680	28.289	28.332	28.374	28.416	28.458	28.500	28.542	28.584	28.626	23.668	28.710
690	28.710	28.752	28.794	28.835	28.877	28.919	28.961	29.003	29.045	29.087	29.129
700	29.129	29.171	29.213	29.255	29.297	29.338	29.380	29.422	29.464	29.506	29.548
710	29.543	29.589	29.631	29.673	29.715	29.757	29.798	29.340	29.332	29.924	29.965
720	29.965	30.007	30.049	30.090	30.132	30.174	30.216	30.257	30.299	30.341	30.382
730	30.332	30.424	30.466	30.507	30.549	30.590	30.632	30.674	30.715	30.757	30.793
740	30.798	30.840	30.881	30.923	30.964	31.006	31.047	31.039	31.130	31.172	31.213
750	31.213	31.255	31.296	31.338	31.379	31.421	31.462	31.504	31.545	31.586	31.628

760	31.628	31.669	31.710	31.752	31.793	31.834	31.876	31.917	31.958	32.000	32.041
770	32.041	32.082	32.124	32.165	32.206	32.247	32.289	32.330	32.371	32.412	32.453
780	32.453	32.495	32.536	32.577	32.618	32.659	32.700	32.742	32.783	32.824	32.865
790	32.865	32.906	32.947	32.988	33.029	33.070	33.111	33.152	33.193	33.234	33.275
800	33.275	33.316	33.357	33.398	33.439	33.480	33.521	33.562	33.603	33.644	33.685
810	33.685	33.726	33.767	33.808	33.848	33.889	33.930	33.971	34.012	34.053	34.093
820	34.093	34.134	34.175	34.216	34.257	34.297	34.338	34.379	34.420	34.460	34.501
830	34.501	34.542	34.582	34.623	34.664	34.704	34.745	34.786	34.826	34.867	34.903
840	34.908	34.948	34.989	35.029	35.070	35.110	35.151	35.192	35.232	35.273	35.313
850	35.313	35.354	35.394	35.435	35.475	35.516	35.556	35.596	35.637	35.677	35.718
860	35.718	35.758	35.798	35.839	35.879	35.920	35.960	36.000	36.041	36.081	36.121
870	36.121	36.162	36.202	36.242	36.282	36.323	36.363	36.403	36.443	36.484	36.524
880	36.524	36.564	36.604	36.644	36.685	36.725	36.765	36.805	36.845	36.885	36.925
890	36.925	36.965	37.006	37.046	37.086	37.126	37.166	37.206	37.246	37.286	37.326
900	37.326	37.366	37.406	37.446	37.486	37.526	37.566	37.606	37.646	37.686	37.725
910	37.725	37.765	37.805	37.845	37.885	37.925	37.965	38.005	38.044	38.084	38.124
920	38.124	38.164	38.204	38.243	38.283	38.323	38.363	38.402	38.442	38.482	38.522
930	38.522	38.561	38.601	38.641	38.680	38.720	38.760	38.799	38.839	38.878	38.918
940	38.918	38.953	38.997	39.037	39.076	39.116	39.155	39.195	39.235	39.274	39.314
950	39.314	39.353	39.393	39.432	39.471	39.511	39.550	39.590	39.629	39.669	39.708
960	39.708	39.747	39.787	39.826	39.866	39.905	39.944	39.984	40.023	40.062	40.101
970	40.101	40.141	40.180	40.219	40.259	40.298	40.337	40.376	40.415	40.455	40.494
980	40.494	40.533	40.572	40.611	40.651	40.690	40.729	40.768	40.807	40.846	40.885
990	40.885	40.924	40.963	41.002	41.042	41.081	41.120	41.159	41.198	41.237	41.276
1000	41.276	41.315	41.354	41.393	41.431	41.470	41.509	41.548	41.587	41.626	41.665
1010	41.665	41.704	41.743	41.781	41.820	41.859	41.898	41.937	41.976	42.014	42.053

(continued)

Table E.1 (Continued) Type K (Chromel-Alumel)

°C	0	1	2	3	4	5	6	7	8	9	10
1020	42.053	42.092	42.131	42.169	42.208	42.247	42.236	42.324	42.363	42.402	42.440
1030	42.440	42.479	42.518	42.556	42.595	42.633	42.672	42.711	42.749	42.788	42.826
1040	42.826	42.865	42.903	42.942	42.980	43.019	43.057	43.096	43.134	43.173	43.211
1050	43.211	43.250	43.288	43.327	43.365	43.403	43.442	43.480	43.518	43.557	43.595
1060	43.595	43.633	43.672	43.710	43.748	43.787	43.825	43.863	43.901	43.940	43.978
1070	43.978	44.016	44.054	44.092	44.130	44.169	44.207	44.245	44.283	44.321	44.359
1080	44.359	44.397	44.435	44.473	44.512	44.550	44.588	44.626	44.664	44.702	44.740
1090	44.740	44.778	44.816	44.853	44.891	44.929	44.967	45.005	45.043	45.081	45.119
1100	45.119	45.157	45.194	45.232	45.270	45.308	45.346	45.383	45.421	45.459	45.497
1110	45.497	45.534	45.572	45.610	45.647	45.685	45.723	45.760	45.798	45.836	45.873
1120	45.873	45.911	45.948	45.986	46.024	46.061	46.099	46.136	46.174	46.211	46.249
1130	46.249	46.286	46.324	46.361	46.398	46.436	46.473	46.511	46.548	46.585	46.623
1140	46.623	46.660	46.697	46.735	46.772	46.809	46.847	46.884	46.921	46.958	46.995
1150	46.995	47.033	47.070	47.107	47.144	47.181	47.218	47.256	47.293	47.330	47.367
1160	47.367	47.404	47.441	47.478	47.515	47.552	47.589	47.626	47.663	47.700	47.737
1170	47.737	47.774	47.811	47.848	47.884	47.921	47.958	47.995	48.032	48.069	48.105
1180	48.105	48.142	48.179	48.216	48.252	48.289	48.326	48.363	48.399	48.436	48.473
1190	48.473	48.509	48.546	48.582	48.619	48.656	48.692	48.729	48.765	438.802	48.838
1200	48.838	48.875	48.911	48.948	48.984	49.021	49.057	49.093	49.130	49.166	49.202

	0	1	2	3	4	5	6	7	8	9	10
1210	49.202	49.239	49.275	49.311	49.348	49.384	49.420	49.456	49.493	49.529	49.565
1220	49.565	49.601	49.637	49.674	49.710	49.746	49.782	49.818	49.854	49.890	49.926
1230	49.926	49.962	49.998	50.034	50.070	50.106	50.142	50.178	50.214	50.250	50.286
1240	50.286	50.322	50.358	50.393	50.429	50.465	50.501	50.537	50.572	50.608	50.644
1250	50.644	50.680	50.715	50.751	50.787	50.822	50.858	50.894	50.929	50.965	51.000
1260	51.000	51.036	51.071	51.107	51.142	51.178	51.213	51.249	51.284	51.320	51.355
1270	51.355	51.391	51.426	51.461	51.497	51.532	51.567	51.603	51.638	51.673	51.708
1280	51.703	51.744	51.779	51.814	51.849	51.885	51.920	51.955	51.990	52.025	52.060
1290	52.060	52.095	52.130	52.165	52.200	52.235	52.270	52.305	52.340	52.375	52.410
1300	52.410	52.445	52.480	52.515	52.550	52.585	52.620	52.654	52.689	52.724	52.759
1310	52.759	52.794	52.828	52.863	52.898	52.932	52.967	53.002	53.037	53.071	53.106
1320	53.106	53.140	53.175	53.210	53.244	53.279	53.313	53.348	53.382	53.417	53.451
1330	53.451	53.486	53.520	53.555	53.589	53.623	53.658	53.692	53.727	53.761	53.795
1340	53.795	53.830	53.864	53.898	53.932	53.967	54.001	54.035	54.069	54.101	54.138
1350	54.138	54.172	54.206	54.240	54.274	54.308	54.343	54.377	54.411	54.445	54.479
1360	54.479	54.513	54.547	54.581	54.615	54.649	54.683	54.717	54.751	54.785	54.819
1370	54.819	54.852	54.886								

Source: N.I.S.T. Monograph 175 Revised to ITS-90.

Table E.2 Type R (Platinum, 13% Rhodium vs. Platinum)

°C	0	1	2	3	4	5	6	7	8	9	10
0	0.000	0.005	0.011	0.016	0.021	0.027	0.032	0.038	0.043	0.049	0.054
10	0.054	0.060	0.065	0.071	0.077	0.082	0.088	0.094	0.100	0.105	0.111
20	0.111	0.117	0.123	0.129	0.135	0.141	0.147	0.153	0.159	0.165	0.171
30	0.171	0.177	0.183	0.189	0.195	0.201	0.207	0.214	0.220	0.226	0.232
40	0.232	0.239	0.245	0.251	0.258	0.264	0.271	0.277	0.284	0.290	0.296
50	0.296	0.303	0.310	0.316	0.323	0.329	0.336	0.343	0.349	0.356	0.363
60	0.363	0.369	0.376	0.383	0.390	0.397	0.403	0.410	0.417	0.424	0.431
70	0.431	0.438	0.445	0.452	0.459	0.466	0.473	0.480	0.487	0.494	0.501
80	0.501	0.508	0.516	0.523	0.530	0.537	0.544	0.552	0.559	0.566	0.573
90	0.573	0.581	0.588	0.595	0.603	0.610	0.618	0.625	0.632	0.640	0.647
100	0.647	0.655	0.662	0.670	0.677	0.685	0.693	0.700	0.708	0.715	0.723
110	0.723	0.731	0.738	0.746	0.754	0.761	0.769	0.777	0.785	0.792	0.800
120	0.800	0.808	0.816	0.824	0.832	0.839	0.847	0.855	0.863	0.871	0.879
130	0.879	0.887	0.895	0.903	0.911	0.919	0.927	0.935	0.943	0.951	0.959
140	0.969	0.976	0.984	0.992	1.000	1.008	1.016	1.025	1.033	1.041	1.041
150	1.041	1.049	1.058	1.066	1.074	1.082	1.091	1.099	1.107	1.116	1.124
160	1.124	1.132	1.141	1.146	1.158	1.166	1.175	1.183	1.191	1.200	1.208
170	1.208	1.217	1.225	1.234	1.242	1.251	1.260	1.268	1.277	1.285	1.294
180	1.294	1.303	1.311	1.320	1.329	1.337	1.349	1.355	1.363	1.372	1.381
190	1.381	1.389	1.398	1.407	1.416	1.425	1.433	1.442	1.451	1.460	1.469
200	1.469	1.477	1.486	1.495	1.504	1.513	1.522	1.531	1.540	1.549	1.558
210	1.558	1.567	1.575	1.584	1.593	1.602	1.611	1.620	1.629	1.639	1.648
220	1.648	1.657	1.666	1.675	1.684	1.693	1.702	1.711	1.720	1.729	1.739
230	1.739	1.748	1.757	1.766	1.775	1.784	1.794	1.803	1.812	1.821	1.831
240	1.831	1.840	1.849	1.858	1.868	1.877	1.886	1.895	1.905	1.914	1.923

250	1.923	1.933	1.942	1.951	1.961	1.970	1.980	1.989	1.998	2.008	2.017
260	2.017	2.027	2.036	2.046	2.055	2.064	2.074	2.083	2.093	2.102	2.112
270	2.112	2.121	2.131	2.140	2.150	2.159	2.169	2.179	2.188	2.198	2.207
280	2.207	2.217	2.226	2.236	2.246	2.255	2.265	2.275	2.284	2.294	2.304
290	2.304	2.313	2.323	2.333	2.342	2.352	2.362	2.371	2.381	2.391	2.401
300	2.401	2.410	2.420	2.430	2.440	2.449	2.459	2.469	2.479	2.488	2.498
310	2.498	2.508	2.518	2.528	2.538	2.547	2.557	2.567	2.577	2.587	2.597
320	2.597	2.607	2.617	2.626	2.636	2.646	2.656	2.666	2.676	2.686	2.696
330	2.696	2.706	2.716	2.726	2.736	2.746	2.756	2.766	2.776	2.786	2.796
340	2.796	2.806	2.816	2.826	2.836	2.846	2.856	2.866	2.876	2.886	2.896
350	2.896	2.906	2.916	2.926	2.937	2.947	2.957	2.967	2.977	2.987	2.997
360	2.997	3.007	3.018	3.028	3.038	3.048	3.058	3.068	3.079	3.089	3.099
370	3.099	3.109	3.119	3.130	3.140	3.150	3.160	3.171	3.181	3.191	3.201
380	3.201	3.212	3.222	3.232	3.242	3.253	3.263	3.273	3.284	3.294	3.304
390	3.304	3.315	3.325	3.335	3.346	3.356	3.366	3.377	3.387	3.397	3.408
400	3.408	3.418	3.428	3.439	3.449	3.460	3.470	3.480	3.491	3.501	3.512
410	3.512	3.522	3.533	3.543	3.553	3.564	3.574	3.585	3.595	3.606	3.616
420	3.616	3.627	3.637	3.648	3.658	3.669	3.679	3.690	3.700	3.711	3.721
430	3.721	3.732	3.742	3.753	3.764	3.774	3.785	3.795	3.806	3.816	3.827
440	3.827	3.838	3.848	3.859	3.869	3.880	3.891	3.901	3.912	3.922	3.933
450	3.933	3.944	3.954	3.965	3.976	3.986	3.997	4.008	4.018	4.029	4.040
460	4.040	4.050	4.061	4.072	4.083	4.093	4.104	4.115	4.125	4.136	4.147
470	4.147	4.158	4.168	4.179	4.190	4.201	4.211	4.222	4.233	4.244	4.255
480	4.255	4.265	4.276	4.287	4.298	4.309	4.319	4.330	4.341	4.352	4.363
490	4.363	4.373	4.384	4.395	4.406	4.417	4.428	4.439	4.449	4.460	4.471
500	4.471	4.482	4.493	4.504	4.515	4.526	4.537	4.548	4.558	4.569	4.580

(continued)

Table E.2 (Continued) Type R (Platinum, 13% Rhodium vs. Platinum)

°C	0	1	2	3	4	5	6	7	8	9	10
510	4.580	4.591	4.602	4.613	4.624	4.635	4.646	4.657	4.668	4.679	4.690
520	4.690	4.701	4.712	4.723	4.734	4.745	4.756	4.767	4.778	4.789	4.800
530	4.800	4.811	4.822	4.833	4.844	4.855	4.866	4.877	4.888	4.899	4.910
540	4.910	4.922	4.933	4.944	4.955	4.966	4.977	4.988	4.999	5.010	5.021
550	5.021	5.033	5.044	5.055	5.066	5.077	5.088	5.099	5.111	5.122	5.133
560	5.133	5.144	5.155	5.166	5.178	5.189	5.200	5.211	5.222	5.234	5.245
570	5.245	5.256	5.267	5.279	5.290	5.301	5.312	5.323	5.335	5.346	5.357
580	5.357	5.369	5.380	5.391	5.402	5.414	5.425	5.436	5.448	5.459	5.470
590	5.470	5.481	5.493	5.504	5.515	5.527	5.538	5.549	5.561	5.572	5.583
600	5.583	5.595	5.606	5.618	5.629	5.640	5.652	5.663	5.674	5.686	5.697
610	5.697	5.709	5.720	5.731	5.743	5.754	5.766	5.777	5.789	5.800	5.812
620	5.812	5.823	5.834	5.846	5.857	5.869	5.880	5.892	5.903	5.915	5.926
630	5.926	5.938	5.949	5.961	5.972	5.984	5.995	6.007	6.018	6.030	6.041
640	6.041	6.053	6.065	6.076	6.088	6.099	6.111	6.122	6.134	6.146	6.157
650	6.157	6.169	6.180	6.192	6.204	6.215	6.227	6.238	6.250	6.262	6.273
660	6.273	6.285	6.297	6.308	6.320	6.332	6.343	6.355	6.367	6.378	6.390
670	6.390	6.402	6.413	6.425	6.437	6.448	6.460	6.472	6.484	6.495	6.507
680	6.507	6.519	6.531	6.542	6.554	6.566	6.578	6.589	6.601	6.613	6.625
690	6.625	6.636	6.648	6.660	6.672	6.684	6.695	6.707	6.719	6.731	6.743
700	6.743	6.755	6.766	6.778	6.790	6.802	6.814	6.826	6.838	6.849	6.861
710	6.861	6.873	6.885	6.897	6.909	6.921	6.933	6.945	6.956	6.968	6.980
720	6.980	6.992	7.004	7.016	7.028	7.040	7.052	7.064	7.076	7.088	7.100
730	7.100	7.112	7.124	7.136	7.148	7.160	7.172	7.184	7.196	7.208	7.220
740	7.220	7.232	7.244	7.256	7.268	7.280	7.292	7.304	7.316	7.328	7.340
750	7.340	7.352	7.364	7.376	7.389	7.401	7.413	7.425	7.437	7.449	7.461

760	7.461	7.473	7.485	7.496	7.510	7.522	7.534	7.546	7.558	7.570	7.583
770	7.583	7.595	7.607	7.619	7.631	7.644	7.656	7.668	7.680	7.692	7.705
780	7.705	7.717	7.729	7.741	7.753	7.766	7.778	7.790	7.802	7.815	7.827
790	7.827	7.839	7.851	7.864	7.876	7.888	7.901	7.913	7.925	7.938	7.950
800	7.950	7.962	7.974	7.987	7.999	8.011	8.024	8.036	8.048	8.061	8.073
810	8.073	8.086	8.098	8.110	8.123	8.135	8.147	8.160	8.172	8.185	8.197
820	8.197	8.209	8.222	8.234	8.247	8.259	8.272	8.284	8.296	8.309	8.321
830	8.321	8.334	8.346	8.359	8.371	8.384	8.396	8.409	8.421	8.434	8.446
840	8.446	8.459	8.471	8.484	8.496	8.509	8.521	8.534	8.546	8.559	8.571
850	8.571	8.584	8.597	8.609	8.622	8.634	8.647	8.659	8.672	8.685	8.697
860	8.697	8.710	8.722	8.735	8.748	8.760	8.773	8.785	8.798	8.811	8.823
870	8.823	8.836	8.849	8.861	8.874	8.887	8.899	8.912	8.925	8.937	8.950
880	8.950	8.963	8.975	8.988	9.001	9.014	9.026	9.039	9.052	9.065	9.077
890	9.077	9.090	9.103	9.115	9.128	9.141	9.154	9.167	9.179	9.192	9.205
900	9.205	9.218	9.230	9.243	9.256	9.269	9.282	9.294	9.307	9.320	9.333
910	9.333	9.346	9.359	9.371	9.384	9.397	9.410	9.423	9.436	9.449	9.461
920	9.461	9.474	9.487	9.500	9.513	9.526	9.539	9.552	9.565	9.578	9.590
930	9.590	9.603	9.616	9.629	9.642	9.655	9.668	9.681	9.694	9.707	9.720
940	9.720	9.733	9.746	9.759	9.772	9.785	9.798	9.811	9.824	9.837	9.850
950	9.850	9.863	9.876	9.889	9.902	9.915	9.928	9.941	9.954	9.967	9.980
960	9.980	9.993	10.006	10.019	10.032	10.046	10.059	10.072	10.085	10.098	10.111
970	10.111	10.124	10.137	10.150	10.163	10.177	10.190	10.203	10.216	10.229	10.242
980	10.242	10.255	10.268	10.282	10.295	10.308	10.321	10.334	10.347	10.361	10.374
990	10.374	10.387	10.400	10.413	10.427	10.440	10.453	10.466	10.480	10.493	10.506
1000	10.506	10.519	10.532	10.546	10.559	10.572	10.585	10.599	10.612	10.625	10.638
1010	10.638	10.652	10.665	10.678	10.692	10.705	10.718	10.731	10.745	10.758	10.771

(continued)

Table E.2 (Continued) Type R (Platinum, 13% Rhodium vs. Platinum)

°C	0	1	2	3	4	5	6	7	8	9	10
1020	10.771	10.785	10.798	10.811	10.825	10.838	10.851	10.865	10.878	10.891	10.905
1030	10.905	10.918	10.932	10.945	10.958	10.972	10.985	10.998	11.012	11.025	11.039
1040	11.039	11.052	11.065	11.079	11.092	11.106	11.119	11.132	11.146	11.159	11.173
1050	11.173	11.186	11.200	11.213	11.227	11.240	11.253	11.267	11.280	11.294	11.307
1060	11.307	11.321	11.334	11.348	11.361	11.375	11.388	11.402	11.415	11.429	11.442
1070	11.442	11.456	11.469	11.483	11.496	11.510	11.524	11.537	11.551	11.564	11.578
1080	11.573	11.591	11.605	11.618	11.632	11.646	11.659	11.673	11.686	11.700	11.714
1090	11.714	11.727	11.741	11.754	11.768	11.782	11.795	11.809	11.822	11.836	11.850
1100	11.850	11.863	11.877	11.891	11.904	11.918	11.931	11.945	11.959	11.972	11.986
1110	11.986	12.000	12.013	12.027	12.041	12.054	12.068	12.082	12.096	12.109	12.123
1120	12.123	12.137	12.150	12.164	12.178	12.191	12.205	12.219	12.233	12.246	12.260
1130	12.260	12.274	12.288	12.301	12.315	12.329	12.342	12.356	12.370	12.384	12.397
1140	12.397	12.411	12.425	12.439	12.453	12.466	12.480	12.494	12.508	12.521	12.535
1150	12.535	12.549	12.563	12.577	12.590	12.604	12.618	12.632	12.646	12.659	12.673
1160	12.673	12.687	12.701	12.715	12.729	12.742	12.756	12.770	12.784	12.798	12.812
1170	12.812	12.825	12.839	12.853	12.867	12.881	12.895	12.909	12.922	12.936	12.950
1180	12.950	12.964	12.978	12.992	13.006	13.019	13.033	13.047	13.061	13.075	13.089
1190	13.089	13.103	13.117	13.131	13.145	13.158	13.172	13.186	13.200	13.214	13.228
1200	13.228	13.242	13.256	13.270	13.284	13.298	13.311	13.325	13.339	13.353	13.367
1210	13.367	13.381	13.395	13.409	13.423	13.437	13.451	13.465	13.479	13.493	13.507
1220	13.507	13.521	13.535	13.549	13.563	13.577	13.590	13.604	13.618	13.632	13.646
1230	13.646	13.660	13.674	13.688	13.702	13.716	13.730	13.744	13.758	13.772	13.786
1240	13.786	13.800	13.814	13.828	13.842	13.856	13.870	13.884	13.898	13.912	13.926
1250	13.926	13.940	13.954	13.968	13.982	13.996	14.010	14.024	14.038	14.052	14.066
1260	14.066	14.081	14.095	14.109	14.123	14.137	14.151	14.165	14.173	14.193	14.207

1270	14.207	14.221	14.235	14.249	14.263	14.277	14.291	14.305	14.319	14.333	14.347
1280	14.347	14.361	14.375	14.390	14.404	14.418	14.432	14.446	14.460	14.474	14.488
1290	14.488	14.502	14.516	14.530	14.544	14.558	14.572	14.586	14.601	14.615	14.629
1300	14.629	14.643	14.657	14.671	14.685	14.699	14.713	14.727	14.741	14.755	14.770
1310	14.770	14.784	14.798	14.812	14.826	14.840	14.854	14.868	14.882	14.896	14.911
1320	14.911	14.925	14.939	14.953	14.967	14.981	14.995	15.009	15.023	15.037	15.052
1330	15.052	15.066	15.080	15.094	15.108	15.122	15.136	15.150	15.164	15.179	15.193
1340	15.193	15.207	15.221	15.235	15.249	15.263	15.277	15.291	15.306	15.320	15.334
1350	15.334	15.348	15.362	15.376	15.390	15.404	15.419	15.433	15.447	15.461	15.475
1360	15.475	15.489	15.503	15.517	15.531	15.546	15.560	15.574	15.588	15.602	15.616
1370	15.616	15.630	15.645	15.659	15.673	15.687	15.701	15.715	15.729	15.743	15.758
1380	15.758	15.772	15.786	15.800	15.814	15.828	15.842	15.856	15.871	15.885	15.899
1390	15.899	15.913	15.927	15.941	15.955	15.969	15.984	15.998	16.012	16.026	16.040
1400	16.040	16.054	16.068	16.082	16.097	16.111	16.125	16.139	16.153	16.167	16.181
1410	16.181	16.196	16.210	16.224	16.238	16.252	16.266	16.280	16.294	16.309	16.323
1420	16.323	16.337	16.351	16.365	16.379	16.393	16.407	16.422	16.436	16.450	16.464
1430	16.464	16.478	16.492	16.506	16.520	16.534	16.549	16.563	16.577	16.591	16.605
1440	16.605	16.619	16.633	16.647	16.662	16.676	16.690	16.704	16.718	16.732	16.746
1450	16.746	16.760	16.774	16.789	16.803	16.817	16.831	16.845	16.859	16.873	16.887
1460	16.887	16.901	16.915	16.930	16.944	16.958	16.972	16.986	17.000	17.014	17.028
1470	17.028	17.042	17.056	17.071	17.085	17.099	17.113	17.127	17.141	17.155	17.169
1480	17.169	17.183	17.197	17.211	17.225	17.240	17.254	17.268	17.282	17.296	17.310
1490	17.310	17.324	17.338	17.352	17.366	17.380	17.394	17.408	17.423	17.437	17.451
1500	17.451	17.465	17.479	17.493	17.507	17.521	17.535	17.549	17.563	17.577	17.591
1510	17.591	17.605	17.619	17.633	17.647	17.661	17.676	17.690	17.704	17.718	17.732
1520	17.732	17.746	17.760	17.774	17.788	17.802	17.816	17.830	17.844	17.858	17.872

(continued)

Table E.2 (Continued) Type R (Platinum, 13% Rhodium vs. Platinum)

°C	0	1	2	3	4	5	6	7	8	9	10
1530	17.872	17.886	17.900	17.914	17.928	17.942	17.956	17.970	17.984	17.998	18.012
1540	18.012	18.026	18.040	18.054	18.068	18.082	18.096	18.110	18.124	18.138	18.152
1550	18.152	18.166	18.180	18.194	18.208	18.222	18.236	18.250	18.264	18.278	18.292
1560	18.292	18.306	18.320	18.334	18.348	18.362	18.376	18.390	18.404	18.417	18.431
1570	18.431	18.445	18.459	18.473	18.487	18.501	18.515	18.529	18.543	18.557	18.571
1580	18.571	18.585	18.599	18.613	18.627	18.640	18.654	18.668	18.682	18.696	18.710
1590	18.710	18.724	18.738	18.752	18.766	18.779	18.793	18.807	18.821	18.835	18.849
1600	18.849	18.863	18.877	18.891	18.904	18.918	18.932	18.946	18.960	18.974	18.988
1610	18.988	19.002	19.015	19.029	19.043	19.057	19.071	19.085	19.098	19.112	19.126
1620	19.126	19.140	19.154	19.168	19.181	19.195	19.209	19.223	19.237	19.250	19.264
1630	19.264	19.278	19.292	19.306	19.319	19.333	19.347	19.361	19.375	19.388	19.402
1640	19.402	19.416	19.430	19.444	19.457	19.471	19.485	19.499	19.512	19.526	19.540
1650	19.540	19.554	19.567	19.581	19.595	19.609	19.622	19.636	19.650	19.663	19.677
1660	19.677	19.691	19.705	19.718	19.732	19.746	19.759	19.773	19.787	19.800	19.814
1670	19.814	19.828	19.841	19.855	19.869	19.883	19.896	19.910	19.923	19.937	19.951
1680	19.951	19.964	19.978	19.992	20.005	20.019	20.032	20.046	20.060	20.073	20.087
1690	20.087	20.100	20.114	20.127	20.141	20.154	20.168	20.181	20.195	20.208	20.222
1700	20.222	20.235	20.249	20.262	20.275	20.289	20.302	20.316	20.329	20.342	20.356
1710	20.356	20.369	20.382	20.396	20.409	20.422	20.436	20.449	20.462	20.475	20.488
1720	20.488	20.502	20.515	20.528	20.541	20.554	20.567	20.581	20.594	20.607	20.620
1730	20.620	20.633	20.646	20.659	20.672	20.685	20.698	20.711	20.724	20.736	20.749
1740	20.749	20.762	20.775	20.788	20.801	20.813	20.826	20.839	20.852	20.864	20.877
1750	20.877	20.890	20.902	20.915	20.928	20.940	20.953	20.965	20.978	20.990	21.003
1760	21.003	21.015	21.027	21.040	21.052	21.065	21.077	21.089	21.101		

Source: N.I.S.T. Monograph 175 Revised to ITS-90.

Table E.3 Type S (Platinum, 10% Rhodium vs. Platinum)

°C	0	1	2	3	4	5	6	7	8	9	10
0	0.000	0.005	0.011	0.016	0.022	0.027	0.033	0.038	0.044	0.050	0.055
10	0.055	0.061	0.067	0.072	0.078	0.084	0.090	0.095	0.101	0.107	0.113
20	0.113	0.119	0.125	0.131	0.137	0.143	0.149	0.155	0.161	0.167	0.173
30	0.173	0.179	0.185	0.191	0.197	0.204	0.210	0.216	0.222	0.229	0.235
40	0.235	0.241	0.248	0.254	0.260	0.267	0.273	0.280	0.286	0.292	0.299
50	0.299	0.305	0.312	0.319	0.325	0.332	0.338	0.345	0.352	0.358	0.365
60	0.365	0.372	0.378	0.385	0.392	0.399	0.405	0.412	0.419	0.426	0.433
70	0.433	0.440	0.446	0.453	0.460	0.467	0.474	0.481	0.488	0.495	0.502
80	0.502	0.509	0.516	0.523	0.530	0.538	0.545	0.552	0.559	0.566	0.573
90	0.573	0.580	0.588	0.595	0.602	0.609	0.617	0.624	0.631	0.639	0.646
100	0.646	0.653	0.661	0.668	0.675	0.683	0.690	0.698	0.705	0.713	0.720
110	0.720	0.727	0.735	0.743	0.750	0.758	0.765	0.773	0.780	0.788	0.795
120	0.795	0.803	0.811	0.818	0.826	0.834	0.841	0.849	0.857	0.865	0.872
130	0.872	0.880	0.888	0.896	0.903	0.911	0.919	0.927	0.935	0.942	0.950
140	0.950	0.958	0.966	0.974	0.982	0.990	0.998	1.006	1.013	1.021	1.029
150	0.646	0.653	0.661	0.668	0.675	0.683	0.690	0.698	0.705	0.713	0.720
160	0.720	0.727	0.735	0.743	0.750	0.758	0.765	0.773	0.780	0.788	0.795
170	0.795	0.803	0.811	0.818	0.826	0.834	0.841	0.849	0.857	0.865	0.872
180	0.872	0.880	0.888	0.896	0.903	0.911	0.919	0.927	0.935	0.942	0.950
190	0.950	0.958	0.966	0.974	0.982	0.990	0.998	1.006	1.013	1.021	1.029
150	1.029	1.037	1.045	1.053	1.061	1.069	1.077	1.085	1.094	1.102	1.110
160	1.110	1.118	1.126	1.134	1.142	1.150	1.158	1.167	1.175	1.183	1.191
170	1.191	1.199	1.207	1.216	1.224	1.232	1.240	1.249	1.257	1.265	1.273
180	1.273	1.282	1.290	1.298	1.307	1.315	1.323	1.332	1.340	1.348	1.357

(*continued*)

Table E.3 (Continued) Type S (Platinum, 10% Rhodium vs. Platinum)

°C	0	1	2	3	4	5	6	7	8	9	10
190	1.357	1.365	1.373	1.382	1.390	1.399	1.407	1.415	1.424	1.432	1.441
200	1.441	1.449	1.458	1.466	1.475	1.483	1.492	1.500	1.509	1.517	1.526
210	1.526	1.534	1.543	1.551	1.560	1.569	1.577	1.586	1.594	1.603	1.612
220	1.612	1.620	1.629	1.638	1.646	1.655	1.663	1.672	1.681	1.690	1.698
230	1.698	1.707	1.716	1.724	1.733	1.742	1.751	1.759	1.768	1.777	1.786
240	1.786	1.794	1.803	1.812	1.821	1.829	1.838	1.847	1.856	1.865	1.874
250	1.874	1.882	1.891	1.900	1.909	1.918	1.927	1.936	1.944	1.953	1.962
260	1.962	1.971	1.980	1.989	1.998	2.007	2.016	2.025	2.034	2.043	2.052
270	2.052	2.061	2.070	2.078	2.087	2.096	2.105	2.114	2.123	2.132	2.141
280	2.141	2.151	2.160	2.169	2.178	2.187	2.196	2.205	2.214	2.223	2.232
290	2.232	2.241	2.250	2.259	2.268	2.277	2.287	2.296	2.305	2.314	2.323
300	2.323	2.332	2.341	2.350	2.360	2.369	2.378	2.387	2.396	2.405	2.415
310	2.415	2.424	2.433	2.442	2.451	2.461	2.470	2.479	2.488	2.497	2.507
320	2.507	2.516	2.525	2.534	2.544	2.553	2.562	2.571	2.581	2.590	2.599
330	2.599	2.609	2.618	2.627	2.636	2.646	2.655	2.664	2.674	2.683	2.692
340	2.692	2.702	2.711	2.720	2.730	2.739	2.748	2.758	2.767	2.776	2.786
350	2.786	2.795	2.805	2.814	2.823	2.833	2.842	2.851	2.861	2.870	2.880
360	2.880	2.889	2.899	2.908	2.917	2.927	2.936	2.946	2.955	2.965	2.974
370	2.974	2.983	2.993	3.002	3.012	3.021	3.031	3.040	3.050	3.059	3.069
380	3.069	3.078	3.088	3.097	3.107	3.116	3.126	3.135	3.145	3.151	3.164
390	3.164	3.173	3.183	3.192	3.202	3.212	3.221	3.231	3.240	3.250	3.259
400	3.259	3.269	3.279	3.288	3.298	3.307	3.317	3.326	3.336	3.346	3.355
410	3.355	3.365	3.374	3.384	3.394	3.403	3.413	3.423	3.432	3.442	3.451
420	3.451	3.461	3.471	3.480	3.490	3.500	3.509	3.519	3.529	3.538	3.548
430	3.548	3.558	3.567	3.577	3.587	3.596	3.606	3.616	3.626	3.635	3.645

440	3.645	3.655	3.664	3.674	3.684	3.694	3.703	3.713	3.723	3.732	3.742
450	3.742	3.752	3.762	3.771	3.781	3.791	3.801	3.810	3.820	3.830	3.840
460	3.840	3.850	3.859	3.869	3.879	3.889	3.898	3.908	3.918	3.928	3.938
470	3.938	3.947	3.957	3.967	3.977	3.987	3.997	4.006	4.016	4.026	4.036
480	4.036	4.046	4.056	4.065	4.075	4.085	4.095	4.105	4.116	4.125	4.134
490	4.134	4.144	4.154	4.164	4.174	4.184	4.194	4.204	4.213	4.223	4.233
500	4.233	4.243	4.253	4.263	4.273	4.283	4.293	4.303	4.313	4.323	4.332
510	4.332	4.342	4.352	4.362	4.372	4.382	4.392	4.402	4.412	4.422	4.432
520	4.432	4.442	4.452	4.462	4.472	4.482	4.492	4.502	4.512	4.522	4.532
530	4.532	4.542	4.552	4.562	4.572	4.582	4.592	4.602	4.612	4.622	4.632
540	4.632	4.642	4.652	4.662	4.672	4.682	4.692	4.702	4.712	4.722	4.732
550	4.732	4.742	4.752	4.762	4.772	4.782	4.793	4.803	4.813	4.823	4.833
560	4.833	4.843	4.853	4.863	4.873	4.883	4.893	4.904	4.914	4.924	4.934
570	4.934	4.944	4.954	4.964	4.974	4.984	4.995	5.005	5.015	5.025	5.035
580	5.035	5.045	5.055	5.066	5.076	5.086	5.096	5.106	5.116	5.127	5.137
590	5.137	5.147	5.157	5.167	5.178	5.188	5.198	5.208	5.218	5.228	5.239
600	5.239	5.249	5.259	5.269	5.280	5.290	5.300	5.310	5.320	5.331	5.341
610	5.341	5.351	5.361	5.372	5.382	5.392	5.402	5.413	5.423	5.433	5.443
620	5.443	5.454	5.464	5.474	5.485	5.495	5.505	5.515	5.526	5.536	5.546
630	5.546	5.557	5.567	5.577	5.588	5.598	5.608	5.618	5.629	5.639	5.649
640	5.649	5.660	5.670	5.680	5.691	5.701	5.712	5.722	5.732	5.743	5.753
650	5.753	5.763	5.774	5.784	5.794	5.805	5.815	5.826	5.836	5.846	5.857
660	5.857	5.867	5.878	5.888	5.898	5.909	5.919	5.930	5.940	5.950	5.961
670	5.961	5.971	5.982	5.992	6.003	6.013	6.024	6.034	6.044	6.055	6.065
680	6.065	6.076	6.086	6.097	6.107	6.118	6.128	6.139	6.149	6.160	6.170
690	6.170	6.181	6.191	6.202	6.212	6.223	6.233	6.244	6.254	6.265	6.275

(continued)

Table E.3 (Continued) Type S (Platinum, 10% Rhodium vs. Platinum)

°C	0	1	2	3	4	5	6	7	8	9	10
700	6.275	6.286	6.296	6.307	6.317	6.328	6.338	6.349	6.360	6.370	6.381
710	6.381	6.391	6.402	6.412	6.423	6.434	6.444	6.455	6.465	6.476	6.486
720	6.466	6.497	6.508	6.518	6.529	6.539	6.550	6.561	6.571	6.582	6.593
730	6.593	6.603	6.614	6.624	6.635	6.646	6.656	6.667	6.678	6.688	6.699
740	6.699	6.710	6.720	6.731	6.742	6.752	6.763	6.774	6.784	6.795	6.806
750	6.806	6.817	6.827	6.838	6.849	6.859	6.870	6.881	6.892	6.902	6.913
760	6.913	6.924	6.934	6.945	6.956	6.967	6.977	6.988	6.999	7.010	7.020
770	7.020	7.031	7.042	7.053	7.064	7.074	7.085	7.096	7.107	7.117	7.128
780	7.128	7.139	7.150	7.161	7.172	7.182	7.193	7.204	7.215	7.226	7.236
790	7.236	7.247	7.258	7.269	7.280	7.291	7.302	7.312	7.323	7.334	7.345
800	7.345	7.356	7.367	7.378	7.398	7.399	7.410	7.421	7.432	7.443	7.454
810	7.454	7.465	7.476	7.487	7.497	7.508	7.519	7.530	7.541	7.552	7.563
820	7.563	7.574	7.585	7.596	7.607	7.618	7.629	7.640	7.651	7.662	7.673
830	7.673	7.684	7.695	7.706	7.717	7.728	7.739	7.750	7.761	7.772	7.783
840	7.783	7.794	7.805	7.816	7.827	7.838	7.849	7.860	7.871	7.882	7.893
850	7.893	7.904	7.915	7.926	7.937	7.948	7.959	7.970	7.981	7.992	8.003
860	8.003	8.014	8.026	8.037	8.048	8.059	8.070	8.081	8.092	8.103	8.114
870	8.114	8.125	8.137	8.148	8.159	8.170	8.181	8.192	8.203	8.214	8.226
880	8.226	8.237	8.248	8.259	8.270	8.281	8.293	8.304	8.315	8.326	8.337
890	8.337	8.348	8.360	8.371	8.382	8.393	8.404	8.416	8.427	8.438	8.449
900	8.449	8.460	8.472	8.483	8.494	8.505	8.517	8.528	8.539	8.550	8.562
910	8.562	8.573	6.584	8.595	8.607	8.618	8.629	8.640	8.652	8.663	8.674
920	8.674	8.685	8.697	8.708	8.719	8.731	8.742	8.753	8.765	8.776	8.787
930	8.787	8.798	8.810	8.821	8.832	8.844	8.855	8.866	8.878	8.889	8.900
940	8.900	8.912	8.923	8.935	8.946	8.957	8.969	8.980	8.991	9.003	9.014

950	9.014	9.025	9.037	9.048	9.060	9.071	9.082	9.094	9.105	9.117	9.128
960	9.128	9.139	9.151	9.162	9.174	9.185	9.197	9.208	9.219	9.231	9.242
970	9.242	9.254	9.265	9.277	9.288	9.300	9.311	9.323	9.334	9.345	9.357
960	9.357	9.368	9.380	9.391	9.403	9.414	9.426	9.437	9.449	9.460	9.472
990	9.472	9.483	9.495	9.506	9.518	9.529	9.541	9.552	9.564	9.576	9.587
1000	9.587	9.599	9.610	9.622	9.633	9.645	9.656	9.668	9.680	9.691	9.703
1010	9.703	9.714	9.726	9.737	9.749	9.761	9.772	9.784	9.795	9.807	9.819
1020	9.819	9.830	9.842	9.853	9.865	9.877	9.888	9.900	9.911	9.923	9.935
1030	9.935	9.946	9.958	9.970	9.981	9.993	10.005	10.016	10.028	10.040	10.051
1040	10.051	10.063	10.075	10.086	10.098	10.110	10.121	10.133	10.145	10.156	10.168
1050	10.168	10.180	10.191	10.203	10.215	10.227	10.238	10.250	10.262	10.273	10.285
1060	10.285	10.297	10.309	10.320	10.332	10.344	10.356	10.367	10.379	10.391	10.403
1070	10.403	10.414	10.426	10.438	10.450	10.461	10.473	10.485	10.497	10.509	10.520
1080	10.520	10.532	10.544	10.556	10.567	10.579	10.591	10.603	10.615	10.626	10.638
1090	10.638	10.650	10.662	10.674	10.686	10.697	10.709	10.721	10.733	10.745	10.757
1100	10.757	10.768	10.780	10.792	10.804	10.816	10.828	10.839	10.851	10.863	10.875
1110	10.875	10.887	10.899	10.911	10.922	10.934	10.946	10.958	10.970	10.982	10.994
1120	10.994	11.006	11.017	11.029	11.041	11.053	11.065	11.077	11.089	11.101	11.113
1130	11.113	11.125	11.136	11.148	11.160	11.172	11.184	11.196	11.208	11.220	11.232
1140	11.232	11.244	11.256	11.268	11.280	11.291	11.303	11.315	11.327	11.339	11.351
1150	11.351	11.363	11.375	11.387	11.399	11.411	11.423	11.435	11.447	11.459	11.471
1160	11.471	11.483	11.495	11.507	11.519	11.531	11.542	11.554	11.566	11.578	11.590
1170	11.590	11.602	11.614	11.626	11.638	11.650	11.662	11.674	11.686	11.698	11.710
1180	11.710	11.722	11.734	11.746	11.758	11.770	11.782	11.794	11.806	11.818	11.830
1190	11.830	11.842	11.854	11.866	11.878	11.890	11.902	11.914	11.926	11.939	11.951
1200	11.951	11.963	11.975	11.987	11.999	12.011	12.023	12.035	12.047	12.059	12.071

(continued)

Table E.3 (Continued) Type S (Platinum, 10% Rhodium vs. Platinum)

°C	0	1	2	3	4	5	6	7	8	9	10
1210	12.071	12.083	12.095	12.107	12.119	12.131	12.143	12.155	12.167	12.179	12.191
1220	12.191	12.203	12.216	12.228	12.240	12.252	12.264	12.276	12.288	12.300	12.312
1230	12.312	12.324	12.336	12.348	12.360	12.372	12.384	12.397	12.409	12.421	12.433
1240	12.433	12.445	12.457	12.469	12.481	12.493	12.505	12.517	12.529	12.542	12.554
1250	12.554	12.566	12.578	12.590	12.602	12.614	12.626	12.638	12.650	12.662	12.675
1260	12.675	12.687	12.699	12.711	12.723	12.735	12.747	12.759	12.771	12.783	12.796
1270	12.796	12.808	12.820	12.832	12.844	12.856	12.868	12.880	12.892	12.905	12.917
1280	12.917	12.929	12.941	12.953	12.965	12.977	12.989	13.001	13.014	13.026	13.038
1290	13.038	13.050	13.062	13.074	13.086	13.098	13.111	13.123	13.135	13.147	13.159
1300	13.159	13.171	13.183	13.195	13.208	13.220	13.232	13.244	13.256	13.268	13.280
1310	13.280	13.292	13.305	13.317	13.329	13.341	13.353	13.365	13.377	13.390	13.402
1320	13.402	13.414	13.426	13.438	13.450	13.462	13.474	13.487	13.499	13.511	13.523
1330	13.523	13.535	13.547	13.559	13.572	13.584	13.596	13.608	13.620	13.632	13.644
1340	13.644	13.657	13.669	13.681	13.693	13.705	13.717	13.729	13.742	13.754	13.766
1350	13.766	13.778	13.790	13.802	13.814	13.826	13.839	13.851	13.863	13.875	13.887
1360	13.887	13.899	13.911	13.924	13.936	13.948	13.960	13.972	13.984	13.996	14.009
1370	14.009	14.021	14.033	14.045	14.057	14.069	14.081	14.094	14.106	14.118	14.130
1380	14.130	14.142	14.154	14.166	14.178	14.191	14.203	14.215	14.227	14.239	14.251
1390	14.251	14.263	14.276	14.288	14.300	14.312	14.324	14.336	14.348	14.360	14.373
1400	14.373	14.385	14.397	14.409	14.421	14.433	14.445	14.457	14.470	14.482	14.494
1410	14.494	14.506	14.518	14.530	14.542	14.554	14.567	14.579	14.591	14.603	14.615
1420	14.615	14.627	14.639	14.651	14.664	14.676	14.688	14.700	14.712	14.724	14.736
1430	14.736	14.748	14.760	14.773	11.785	14.797	14.809	14.821	14.833	14.845	14.857
1440	14.857	14.869	14.881	14.894	14.906	14.918	14.930	14.942	14.954	14.966	14.978
1450	14.978	14.990	15.002	15.015	15.027	15.039	15.051	15.063	15.075	15.087	15.099

1460	15.099	15.111	15.123	15.135	15.148	15.160	15.172	15.184	15.196	15.208	15.220
1470	15.220	15.232	15.244	15.256	15.268	15.280	15.292	15.304	15.317	15.329	15.341
1480	15.341	15.353	15.365	15.377	15.389	15.401	15.413	15.425	15.437	15.449	15.461
1490	15.461	15.473	15.485	15.497	15.509	15.521	15.534	15.546	15.558	15.570	15.582
1500	15.582	15.594	15.606	15.618	15.630	15.642	15.654	15.666	15.678	15.690	15.702
1510	15.702	15.714	15.726	15.738	15.750	15.762	15.774	15.786	15.798	15.810	15.822
1520	15.822	15.834	15.846	15.858	15.870	15.882	15.894	15.906	15.918	15.930	15.942
1530	15.942	15.954	15.966	15.978	15.990	16.002	16.014	16.026	16.038	16.050	16.062
1540	16.062	16.074	16.086	16.098	16.110	16.122	16.134	16.146	16.158	16.170	16.182
1550	16.182	16.194	16.205	16.217	16.229	16.241	16.253	16.265	16.277	16.289	16.301
1560	16.301	16.313	16.325	16.337	16.349	16.361	16.373	16.385	16.396	16.408	16.420
1570	16.420	16.432	16.444	16.456	16.468	16.480	16.492	16.504	16.516	16.527	16.539
1580	16.539	16.551	16.563	16.575	16.587	16.599	16.611	16.623	16.634	16.646	16.658
1590	16.658	16.670	16.682	16.694	16.706	16.718	16.729	16.741	16.753	16.765	16.777
1600	16.777	16.789	16.801	16.812	16.824	16.836	16.848	16.860	16.872	16.883	16.895
1610	16.895	16.907	16.919	16.931	16.943	16.954	16.966	16.978	16.990	17.002	17.013
1620	17.013	17.025	17.037	17.049	17.061	17.072	17.084	17.096	17.108	17.120	17.131
1630	17.131	17.143	17.155	17.167	17.178	17.190	17.202	17.214	17.225	17.237	17.249
1640	17.249	17.261	17.272	17.284	17.296	17.308	17.319	17.331	17.343	17.355	17.366
1650	17.366	17.378	17.390	17.401	17.413	17.425	17.437	17.448	17.460	17.472	17.483
1660	17.483	17.495	17.507	17.518	17.530	17.542	17.553	17.565	17.577	17.588	17.600
1670	17.600	17.612	17.623	17.635	17.647	17.658	17.670	17.682	17.693	17.705	17.717
1680	17.717	17.728	17.740	17.751	17.763	17.775	17.786	17.798	17.809	17.821	17.832
1690	17.832	17.844	17.855	17.867	17.878	17.890	17.901	17.913	17.924	17.936	17.947
1700	17.947	17.959	17.970	17.982	17.993	18.004	18.016	18.027	18.039	18.050	18.061
1710	18.061	18.073	18.084	18.095	18.107	18.118	18.129	18.140	18.152	18.163	18.174

(continued)

Table E.3 (Continued) Type S (Platinum, 10% Rhodium vs. Platinum)

°C	0	1	2	3	4	5	6	7	8	9	10
1720	18.174	18.185	18.196	18.208	18.219	18.230	18.241	18.252	18.263	18.274	18.285
1730	18.285	18.297	18.308	18.319	18.330	18.341	18.352	18.362	18.373	18.384	18.395
1740	18.395	18.406	10.417	18.428	18.439	18.449	18.460	18.471	18.482	18.493	18.503
1750	18.503	18.514	18.525	18.535	18.546	18.557	18.567	18.578	16.588	18.599	18.609
1760	18.609	18.620	18.630	18.641	18.651	18.661	18.672	18.682	18.693		

Source: N.I.S.T. Monograph 175 Revised to ITS-90.

Index

Page numbers followed by f and t indicate figures and tables, respectively.

A

Voltages, 44
    across capacitance, 164–165
    bias, 165
    common mode, 89
    differential mode, 89
    input, 81, 82, 83, 84, 85, 86, 92
    input signal, 110
    in instruments, 81–82
    inverting, 85
    measuring, 147, 179–180
    noninverting, 85
    offset, 86
    output, 81, 82, 83, 84, 85, 86, 92
    producing, 159
    of sensor, 84
Volumetric flow meter. *See* Mass flow
       meters
Volumetric flow rate, 24, 28, 199, 202, 203,
       204, 206
Volumetric unit of species, 216

W

Water cooled porous plug burner, 25
Wheatstone bridge, 179–180, 229–230
Word, defined, 104

X

Xenon lamps, 248f, 249

Y

Yaw angle, 173, 174, 176
Yaw probes. *See* Three-hole (yaw) probes
Yaw sphere, 176, 176f

Z

Zero drift, 53, 53f
Zero offset, 119
Zeroth-order instrument, 60
Z-type twin-mirror Schlieren system,
      271f, 274